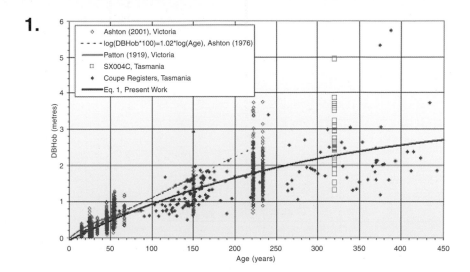

Plate 1. Observed DBH_{ob} versus age for individual trees, plus various models. (Data from Patton, 1919 and Ashton, 2001.)

Plate 2. (a) Images before and after rectification. The focal length was 28 mm, taken 44 m from the tree, DBH_{ob} = 3.85 m, height = 57.5 m. (b) Stages in the process of digitizing the trunk to obtain taper.

Plate 3. Examples of decay in mature (400 ± 50 years) trees (location in Brown, 2001). (a) The 'Chapel Tree', DBH_{ob} = 5.92 m. (b) The 'Big Stump', DBH_{ob} = 5.87 m. (c) (appears overleaf) The 'Maydena Buttlog', DBH_{ob} = 4.08 m.

3c.

4.

5.

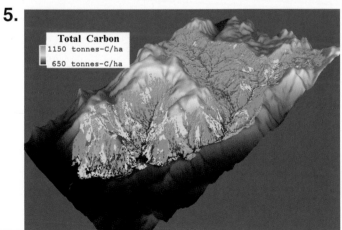

Plate 4. Area of *Eucalyptus regnans* clearfelled, burnt and reforested (low vegetation) below an area subject to a stand replacement fire (tall white stags, probably resulting from an escaped 'regen' burn), Styx Valley, Tasmania.
Plate 5. Frame from animation of carbon sequestration for the O'Shannessy catchment in the Central Highlands of Victoria at starting year (2002). The animation of the total carbon accumulation was calculated within about 20 seconds for a period of 200 years. The advantage of animations will be mainly in visualizing different land-use scenarios.

MODELLING FOREST SYSTEMS

MODELLING FOREST SYSTEMS

Edited by

A. Amaro

Department of Mathematics
Instituto Superior de Gestão
Lisbon, Portugal

D. Reed

School of Forest Resources and Environmental Science
Michigan Technological University
USA

and

P. Soares

Department of Forestry
Instituto Superior de Agronomia
Lisbon, Portugal

CABI Publishing

CABI Publishing is a division of CAB International

CABI Publishing
CAB International
Wallingford
Oxon OX10 8DE
UK

CABI Publishing
44 Brattle Street
4th Floor
Cambridge, MA 02138
USA

Tel: +44 (0)1491 832111
Fax: +44 (0)1491 833508
E-mail: cabi@cabi.org
Website: www.cabi-publishing.org

Tel: +1 617 395 4056
Fax: +1 617 354 6875
E-mail: cabi-nao@cabi.org

A catalogue record for this book is available from the British Library, London, UK.

Library of Congress Cataloging-in-Publication Data

Modelling forest systems / edited by A. Amaro, D. Reed and P. Soares.
        p. cm.
Papers from a workshop held in Sesimbra, Portugal, June 2–5, 2002.
Includes bibliographical references.
    ISBN 0-85199-693-0
    1. Forests and forestry--Mathematical models--Congresses.
    2. Forest ecology--Mathematical models--Congresses. 3. Forest manage-
ment--Mathematical models--Congresses. I. Amaro, A. (Ana), 1960– II.
Reed, D. (David), 1956– III. Soares, P. (Paula), 1967–

    SD387.M33 M63 2003
    634.9′01′5118--dc21

                                                                2002154262

ISBN 0 85199 693 0

Typeset in Palatino by Columns Design Ltd, Reading
Printed and bound in the UK by Biddles Ltd, Guildford and King's Lynn

Contents

Principal Contributors

V. Alenius, *Finnish Forest Research Institute, Rovaniemi Research Station, PO Box 16, FIN – 96300 Rovaniemi, Finland*

A.C. Almeida, *Aracruz Celulose, S.A., Rodovia Aracruz/B. Riacho, PO Box 331011, 291-9500 Aracruz – ES, Brazil*

A. Amaro, *Instituto Superior de Gestão, Departamento de Matematica, Rua Vitorino Nemesio, N° 5, 1750-306 Lisbon, Portugal*

R. Amateis, *Virginia Tech, Department of Forestry, College of Natural Resources, Blacksburg, VA 24061-0324, USA*

H. Burkhart, *Department of Forestry, College of Natural Resources, Blacksburg, VA, 24061-0324, USA*

H. Carino, *Auburn University, School of Forestry & Wildlife Sciences, Auburn, AL 36849, USA*

G. Chirici, *Università di Firenze, DISTAF – Dipart. Scienze Tecnologie Ambientali Forestali, Via San Bonaventura 13, 50145 Florence, Italy*

J. Colbert, *USDA – Forest Service, Disturbance Ecology & Management of Oak-Dominated Forests, 180 Canfield Street, Morgantown, WV 26505-3101, USA*

F. de Coligny, *INRA, UMR botAnique et bioinforMatique de l'Architecture des Plantes (AMAP), TA 40/PS2 Boulevard de la Lironde, 34398 Montpellier Cedex 5, France*

C. Dean, *CRC for Greenhouse Accounting, GPO Box 475, Canberra ACT 2601, Australia*

G. Deckmyn, *University of Antwerp UIA, Research Group Plant and Vegetation Ecology, Universiteitsplein 1, 2620 Wilrijk, Belgium*

T. Eid, *Agricultural University of Norway, Department of Forest Sciences, PO Box 5044, NO-1432 Ås, Norway*

R. Fleming, *Canadian Forest Service, Great Lakes Forest Research Centre, 1219 Queen St East, Sault Ste Marie, Ontario P6A 2E5, Canada*

J.H. Gove, *USDA Forest Service, Northeastern Research Station, PO Box 640, 271 Mast Road, Durham, NH 03824, USA*

M. Hauhs, *University of Bayreuth, Bitoek, Dr – Hans – Frisch – Str. 1–3, D-95448 Bayreuth, Germany*

S. Huang, *Forest Management Branch, Land and Forest Service, Alberta Sustainable Resource Development, 8th Floor – 9920 – 108 Street Edmonton, Alberta T5K 2M4, Canada*

A. Kiviste, *Estonian Agricultural University, Kreutzwaldi 5, 51014 Tartu, Estonia*

F.-J. Knauft, *University of Bayreuth, Bayreuth Institute for Terrestrial Ecosystem Research, D-95440 Bayreuth, Germany*

H. Kotze, *Komatiland Forests, Research Division, PO Box 574, 1260 Sabie, South Africa*

S. Magnussen, *Canadian Forest Services, Natural Resources of Canada, 506 West Burnside Road, Victoria, BC V8Z 1M5, Canada*

C. Meredieu, *INRA, Unité de Recherches Forestières, 69 Route d'Arcachon, 33612 Cestas Cedex, France*

R.A. Monserud, *USDA Forest Service, PNW Research Station, 620 SW Main St, Suite 400, Portland, OR 97205, USA*

N. Nanos, *Unidad de Anatomia, Fisiologia y Genética Forestal, ETS de Ingenieros de Montes, Ciudad Universitária, 28080 Madrid, Spain*

N. Picard, *CIRAD – Forêt, Bamako, BP 1813, Mali*

F. Rodriguez, *Universitat de Lleida, ETSEA-Udl, Dept. Producción Vegetal y Ciencia Forestal, Av. Rovira Roure 177, 25198 Lleida, Spain*

P. Radonja, *SE Serbiaforest – Institute of Forestry, Kneza Viseslava 3, 11030 Belgrade, Yugoslavia*

D. Reed, *Michigan Technological University, School of Forest Resources and Environmental Science, 1400 Townsend Drive, Houghton, MI 49931-1295, USA*

K. Rennolls, *University of Greenwich, School of Computing and Mathematical Sciences, 30 Park Row, London SE10 9LS, UK*

K. Reynolds, *USDA Forest Service, Pacific Northwest Research Station, 3200 SW Jefferson Way, Corvallis, OR 97331, USA*

N.A. Ribeiro, *Universidade de Évora, Departamento de Fitotecnia, Apartado 94, 7002-554 Evora, Portugal*

P. Soares, *Instituto Superior de Agronomia, Departamento de Engenharia Florestal, Tapada de Ajuda, 1349 017 Lisbon, Portugal*

N. Tchebakova, *Institute of Forest Siberian Branch, Russian Academy of Sciences, Akademgorodok, 660036 Krasnoyarsk, Russia*

P. van Gardingen, *University of Edinburgh, CECS John Muir Building, Mayfield Road, Edinburgh EH9 3JK, UK*

Preface

The real world – the forest ecosystem – can be analysed in multiple dimensions (e.g. from the cellular to the landscape level). The 'thinking', in order to understand and simplify the forest's complex reality so that it can be modelled, can begin from the simple univariate scenario and proceed to multivariate and complex scenarios. Different mathematical and philosophical approaches can thus be adopted in order to design and obtain a model. The mathematical tools selected to quantify the rates, coefficients and constants that are used in the models can range from the simple use of the sample mean, to complex numerical algorithms that may violate every possible statistical assumption. Once the model has been built, the process of validating it and understanding its limits or ranges of applicability must be defined, particularly if it will be used for management decision support.

A workshop on the interface between reality, modelling and the parameter estimation processes, jointly sponsored by IUFRO (4.01 and 4.11), Instituto Superior de Gestão and Instituto Superior de Agronomia (Portugal), was held in Sesimbra, Portugal (2–5 June 2002). The contents of this book are based on this workshop and both reflect the state of the art and identify the key issues that should be considered as priorities for further research in this area.

We would like to acknowledge and thank all the involved institutions,[1] all the scientists involved as authors and the following referees:

> Alexander Fedorec, Bob Monserud, Chris Dean, Fréderic Raulier, Glen W. Armstrong, Harold Burkhart, Isabel Cañellas, Jeff Gove, Jim Colbert, J.P. Skovsgaard, Keith Rennolls, Keith Reynolds, Kevin O'Hara, Lianjun Zhang, Margarida Tomé, Marion Reynolds, Mark J. Ducey, Michael Rauscher, Mike R. Strub, Mingliang Wang, Nadarajah Ramesh, Paul Doruska, Paul van Gardingen, Quang V. Cao, Quao Cao, Ralph Amateis, Rod Keenan, Shongming

[1] Fundação para a Ciência e Tecnologia (Apoio do Programa Operacional Ciência, Tecnologia e Inovação do Quadro Comunitário de Apoio III); Fundação Calouste Gulbenkian; British Council; Fundação Luso-Americana; Embaixada do Canadá; Ministério da Agricultura, do Desenvolvimento Rural e das Pescas; Raiz, Instituto de Investigação da Floresta e do Papel; Aliança Florestal; Soporcel, SA; Grupo Amorim; Costa Azul; Câmara Municipal de Sesimbra.

Huang, Steen Magnussen, Stephen Roxburgh, Susan S. Hummel, Thomas J. Seablom, Timo Pukkala, Tony Ackroyd, Valerie Le May, Victor Lieffers, William B. Leak, Yue Wang,

and Luisa Santos for her invaluable editorial assistance. All together they made the development of this volume possible.

Modelling forest systems is a powerful discipline that promotes knowledge and understanding of forest functioning. This is by no means a simple or small task. The purpose of this book is to promote and advance the systematic understanding of these complex, but critical, systems.

Ana Amaro, David Reed, Paula Soares
The Editors

Part 1

Forest Reality and Modelling Strategies

Modelling objectives is one of the most important considerations when determining modelling strategies. The importance of user interaction and the definition of the modelling objectives allow the identification of the three main types of users with whom the modellers interact: scientists, forestry practitioners and managers, and 'Administration' and society. It is essential to expand the scope of modelling to address the needs of these various users.

Until now, modelling has mainly focused on wood production. New approaches are needed: for instance, modelling natural forests, modelling the evolution of the sustainability of the systems, modelling deadwood, modelling wildlife habitats and modelling unusual events (e.g. catastrophes, diseases).

It is also important to interact with other professions and disciplines, to get in touch with specialists in other areas and to communicate with them to a greater extent than has happened in the past. The enhancement of forest structure modelling, e.g. connecting with habitat modelling and deadwood and taking into consideration boundary effects, will be critical as demands for information increase.

It is essential to communicate the theory and reporting of the modelling process. Modellers work with models – when these models fail, as would be expected on occasion, these failures may promote understanding of the faulty model functioning. However, this phase is almost never reported. An absence of published 'negative results' is a consequence of the reporting system; no-one is prepared to publish reports on modelling exercises that did not go well. However, it is important to discuss the reasons why a model did not evolve well, as this is a source of understanding for the entire modelling community.

It is important to think ahead about possible uses and to collect data that will be useful and flexible in application. Data at the extremes – very low and very high stand densities and site qualities, and young and old ages – are essential for broad-scale inferences and precise parameter estimates.

It is very difficult to decide which criteria should be followed in order to plan data acquisition. The balance between data quality and quantity is always difficult to define: there is always a need for quantity in order to have representative samples but, at each sampling point, quality is needed. From a practical standpoint, quantity and quality push in opposite directions, creating the difficult task of defining the balance between both objectives.

There are contrasts between the data used for modelling and the data available for using the model, especially when the models are included in decision-support systems. It is very important to define the data characteristics. Often the models are good, as are the decision tools, but the decisions may still not be as good as they should be due to a lack of specific quality in the data.

When publishing the results, documentation about the data is often reduced to simple tables, making it difficult to evaluate the importance of the results, due to the lack of information on the data sets that were used to build the models. It is important to define minimum standards for data description, which should be presented along with the report of the research. In order to complement this and to make sure that data characterization is self-explanatory, a Web-based tool could be built in order to allow data publication or at least further details that could help the users to understand the possibilities of the application.

For some applications, there are links that modellers should or could have with national forest inventories (NFIs), both on using data from national inventories to calibrate models, as well as using data from the NFI for model evaluation and scenario development. However, this can result in extensive assumptions and serious extrapolations being made, the repercussions of which must be considered.

On the other hand, when planning a forest inventory, it is important to be aware that there are a lot of data that can be collected, much of which are not directly usable by the growth models. Such data may be very interesting for other modelling objectives. It is therefore important to plan to acquire data that may be useful for many objectives.

It is also important to work on different scales, e.g. time scale, spatial resolution, diversity, but especially on the linkage between the different scales (i.e. how the links between the different scales are going to be established).

The following chapters address these and other issues concerning modelling strategies and relationships between models and actual system behaviour. The editors wish to acknowledge the efforts of Harold Burkhart (USA) and Margarida Tomé (Portugal) for coordinating this section, and thank them for their contributions to this volume.

1 Suggestions for Choosing an Appropriate Level for Modelling Forest Stands

Harold E. Burkhart[1]

Abstract

Models are abstractions of reality. In order to be useful, models must include essential elements of the real world system that are to be mimicked to meet some specified modelling objective. The pattern in a data set can often be described with a relatively simple model. Models of forests have been constructed for numerous management and research objectives. To determine an appropriate modelling unit (e.g. cell, organ, tree, stand, landscape), one must define the modelling objective and the forecasting time frame. Often the level of modelling detail possible is dictated by the data available. However, there are guiding principles that can aid in selecting an appropriate level for modelling. These principles include: (i) developing as parsimonious a model as possible, and (ii) adjusting the number of state variables for the forecasting period involved. The application of these principles is discussed within the framework of forest growth and yield models.

Introduction

Scientific inquiry often consists of describing reality through models. No natural system is so simple that it can be grasped and controlled without abstraction. Abstraction involves replacing the natural system under consideration by a model of similar but simpler structure. In order to be useful, models must include essential elements of the real world system that are to be mimicked in order to meet some specific modelling objective.

Prediction, Parsimony and Noise

To have high utility, models must be accurate. Model accuracy can be improved by increasing the size of a data set, improving the quality of the data obtained, or by applying more sophisticated modelling techniques to existing data. Collecting data

[1] Department of Forestry, Virginia Polytechnic Institute and State University, USA
Correspondence to: burkhart@vt.edu

is expensive relative to performing analyses, so it behoves us to make the best possible use of the data available. When planning data collection efforts, it is important that a wide range of site and stand conditions be included in the sample. Vanclay *et al.* (1995) recommend that permanent plot systems for developing growth models include experimental observations that are manipulated to provide data on a wide range of stand densities and thinning treatments, and data from stands which have been allowed to develop beyond normal rotation ages.

It is commonly believed that a model can be no more accurate than the data on which it is based. However, because models can amplify patterns and discard unwanted noise, they can be more accurate than the data used to build them. Gauch (1993) emphasized that models being more accurate than the data available is dependent on: (i) the precise question being asked of the model, (ii) the design of the experiment, and (iii) the quantity and accuracy of the available data.

Typical modelling efforts attempt to enhance prediction by amplifying a pattern and discarding noise. Ordinarily, most of the pattern in a data set is recovered quickly with relatively simple models. Patterns usually depend on a few main causal factors that can be summarized readily. Noise, on the other hand, is recovered slowly as a model's complexity increases. The accuracy of prediction, therefore, increases quickly as parameters are added to relatively simple models. Predictive ability tends to peak rather quickly (this point is sometimes referred to as 'Ockham's Hill') and then decrease with increasingly complex models (Gauch, 1993). Figure 1.1 illustrates how model accuracy (in terms of predictive ability) is affected by the recovery of pattern and noise through increasing model complexity. Modelling thus offers its greatest benefits when a parsimonious, simple model captures the essence of the data's pattern in a large, noisy data set.

An Example

As an illustration of the relationship between model accuracy and complexity, data from a loblolly pine spacing trial (Amateis *et al.*, 1988) were used to predict stand volume. This spacing trial consists of three replicates at four locations. Each replicate contains 16 plots of varying spacing with number of trees per hectare ranging from 747 to 6725. Two locations were randomly selected for model fitting and the

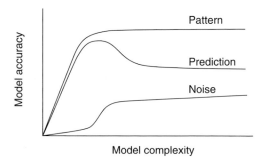

Fig. 1.1. Pattern is generally determined by a few main factors, meaning that relatively simple models recover much of it in a data set; noise is more idiosyncratic and complex and is thus recovered more slowly as model complexity increases; predictive ability arises from pattern recovery minus noise recovery so it is expected to increase quickly in simple models but to peak on 'Ockham's Hill' and then decrease in increasingly complex models. (Adapted from Gauch, 1993.)

observations from the other two locations were used to evaluate model accuracy. For this example, accuracy was defined as the reciprocal of the square root of mean squared error (MSE), a combination of bias and precision. As shown in Fig. 1.2, the pattern of increasing mean volume per hectare over time (stand age) is clearly discernible, but there is large variability at each age; this variability is assumed to stem from differences in numbers of trees and site index levels of the individual plots. Stand volume was, accordingly, to be predicted over time using the variables age, trees per hectare and site index. The models fitted, in sequence, were:

$$\ln(Y) = \beta_0 + \beta_1(1/A) \tag{1}$$

$$\ln(Y) = \beta_0 + \beta_1(1/A) + \beta_2(N) \tag{2}$$

$$\ln(Y) = \beta_0 + \beta_1(1/A) + \beta_2(N) + \beta_3(S) \tag{3}$$

where Y is stand volume (m³/ha), A is stand age (years), N is trees/ha and S is site index (m at base age 25 years). Data from the individual spacing plot measurements from ages 6 to 16 years were utilized in fitting Equations 1–3 and the fitted equations were applied to the plot observations from the two locations not used in fitting to compute the MSE for the differences between observed and predicted volume per hectare. The resulting values at stand age 16 years (trends in results were the same at other ages) were:

Model	R^2	√MSE	Accuracy
1	0.84	15.89	0.0629
2	0.90	6.44	0.1553
3	0.93	30.02	0.0333

Although the R^2 value increased for models 1–3 from 0.84 to 0.90 to 0.93, respectively, and all parameter estimates were significantly different from zero ($P < 0.001$) using the t-test, the accuracy of the models, as measured by the reciprocal of the square root of MSE, reached a peak (Ockham's Hill) with two predictor variables, age and trees per hectare, but declined when site index was added. Following the dictum of parsimony and relying on the results from the MSE computations, the

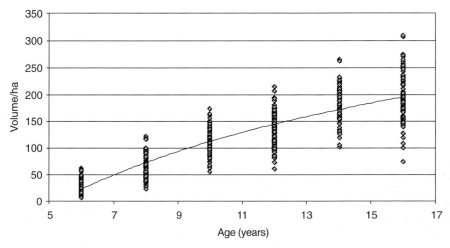

Fig. 1.2. Plotting of data from a loblolly pine spacing trial shows a readily identifiable pattern of increasing mean volume per hectare with increasing age, but with fairly large variability at each age.

two predictor variable model is best. In these spacing trial data, the variation in site index is small; furthermore, site index was not observed but rather it was estimated, with error, from the height measurements at age 16 years (the oldest observation available). Inclusion of a rather 'noisy' (imprecise) variable deteriorated model performance for predicting volume per hectare in the locations not used in fitting.

Model complexity can also be increased by adding more terms involving the basic variables chosen. The model utilizing age and trees per hectare as predictors (model 2) was expanded to include the first-order interaction term for the regressors; that is

$$\ln(Y) = \beta_0 + \beta_1(1/A) + \beta_2(N) + \beta_3(1/A)(N) \tag{4}$$

The square root of the MSE values for models 1, 2 and 4 were computed as before, and the reciprocal taken to compute the defined measure of accuracy, with the following results:

Model	R^2	$\sqrt{MSE}$	Accuracy
1	0.84	15.89	0.0629
2	0.90	6.44	0.1553
4	0.91	8.81	0.1135

The R^2 value for Equation 4 was 0.91 and all coefficients were significant ($P < 0.07$), but the simple two variable model (model 2) again performed best when predicting with independent data.

This simple illustration shows that, for these data, the pattern of stand volume over time at varying densities can be described with a very parsimonious model. Adding more predictor variables or more terms with the age and density predictors caused a loss in model accuracy. Although results from this example are specific to the yield functions chosen and data used for fitting, they do demonstrate that accuracy of prediction can be expected to increase with increasing model complexity, but only up to a point (Ockham's Hill).

Forest Models

Models of forests have been constructed for a host of management and research objectives. Forest management decision making is predicated on accurate forecasting of growth and yield. While growth and yield forecasts enter into virtually all decisions, the primary uses of models can be categorized as: inventory updating, management planning, evaluation of silvicultural alternatives and harvest scheduling.

Growth and yield information is used for a variety of purposes; no single database or modelling approach can be optimal for all applications. As an example, if one were primarily interested in evaluation of silvicultural alternatives, designed-experiment type data with the relevant silvicultural treatments included would probably be used. The model structure would probably be quite detailed in terms of the component equations and the types of output produced in order that the full range of treatments could be evaluated under varying assumptions. If, on the other hand, one were primarily interested in inventory updating, the data should be obtained from a representative sample of the types of stands to which the model is going to be applied. The input to the model would necessarily need to be consistent and compatible with the inventory data definitions and quantities available. The component equations should be as simple and straightforward as possible for producing the output (updated stand statistics) that is needed for and consistent with the inventory database. What is 'best' depends primarily on the objective(s) for developing the model; obviously, the objectives should be specified clearly before any data collection or analyses are initiated.

Growth and yield models produced to date can be broadly grouped as follows: (1) whole-stand models: (a) aggregated values only, (b) size class information; (2) individual tree models: (a) distance-independent, (b) distance-dependent.

In the whole-stand approach, quantities such as volume, basal area and/or number of trees per unit area are forecast. The basic input or predictor variables for these models for even-aged stands are generally age, site index and stand density (number of trees planted per unit area for plantations; some initial basal area for natural stands). Often only aggregated volume growth and/or yield is predicted for the total stand. As a variation on this approach, several researchers have applied probability density functions to estimate the number of trees by diameter at breast height (DBH) class, given that an estimate of the total number per unit area is available. This approach, commonly termed the 'diameter distribution approach', still relies on overall stand values as the basic modelling unit.

Approaches to predicting stand growth and yield that use individual trees as the modelling unit are referred to as 'individual-tree models'. The components of tree growth (e.g. diameter increment, height increment) in these models are commonly linked through a computer program that simulates the growth of each tree and then aggregates these to provide estimates of stand growth and yield. Individual-tree models are divided into two classes, distance-independent and distance-dependent, depending on whether or not individual tree locations are used. Distance-independent models project tree growth either individually or by size classes, usually as a function of present size and stand level variables (e.g. age, site index, number of trees per unit area). In distance-dependent models, initial stand conditions are input or generated and each tree is assigned a coordinate location. Typically, the growth of each tree is predicted as a function of its attributes, the site quality and a measure of competition from neighbours. Additional information about the growth and yield modelling approaches in common use can be found in the books by Vanclay (1994), Gadow and Hui (1999), Avery and Burkhart (2002, Chapter 17), Clutter *et al.* (1983, Chapter 4) and Davis *et al.* (2001, Chapter 5).

Level for Modelling Forest Stands

The level at which forest stands can be modelled is often dictated by the data available. If, for instance, individual trees are not numbered and identified, individual-tree-based approaches are not possible. Permanent plots established in the past have sometimes had limited usefulness because of inadequacies in the measurements taken. In permanent plots established for growth estimation purposes, the minimum data measurements should include DBH, height, crown measures, stem quality assessment and tree spatial locations, to allow for flexibility in modelling approaches.

The typical approach taken in past growth and yield studies was to define a population of interest, obtain a sample from the defined population (the sample could consist of temporary plots, permanent plots, or both), and estimate coefficients (usually with least squares) in specified equation forms. This approach produces satisfactory prediction tools for many purposes, but it may not be adequate in circumstances where forest management practices and objectives are changing rapidly. Given that growth and yield models are used to project the present forest resource and to evaluate alternative treatment effects, data both of the inventory type (which describe operational stands of interest) and of the experimental or research type (which describe response to treatment) are needed. The amount of effort that should be devoted to each type of data collection is not immediately obvious. Nor is it at all clear whether the data should be combined and a single model produced or the data kept separate and different models produced for different uses or objectives.

There are some guiding principles that can aid in selecting an appropriate level for modelling. One concept that has been advanced is to model at one level of detail below the level desired for prediction. Following that principle for forest stand modelling, one would be led to individual-tree models. The use of individual-based models in ecology has gained considerable acceptance in recent years (DeAngelis and Gross, 1992), and individual-tree models have been developed for a number of forest types (examples for North America are Mitchell (1975) for Douglas-fir and Burkhart *et al.* (1987) for loblolly pine). While individual-tree models offer a great deal of flexibility in describing stand structure and simulating silvicultural operations such as thinnings, they may not estimate overall stand values (volume and basal area per hectare) as accurately as whole-stand equations.

Whether one should model at the tree level and aggregate for stand estimates or model at an aggregated level depends on the specific objectives for modelling. The use for which a growth model is intended, it is generally argued, should determine the resolution level at which one should operate. However, as Leary (1979) discussed, another consideration is the relationship between dimensionality of the model (or resolution level) and the time horizon over which projections are to be made. The following relationship from Kahne (1976) has been found to be useful in a number of large-scale modelling efforts:

$$d = \left(\frac{k_a}{h} \right) \qquad\qquad (5)$$

where h is the time horizon over which projections are to be made, d is the dimension of the model state vector and k_a is a constant for a given accuracy or precision level. This relationship indicates that the model dimension should be reduced for long-term projections and increased for short-term projections, to give the same level of accuracy.

The relationship between model dimensionality and projection length has been fairly well accepted by scientists working in certain areas with large-scale models, but it has not received much attention by researchers involved in forest projection. Forest projections are often made for any time horizon of interest without regard to the dimensionality of the model.

Insight into the influence of dimensionality of growth and yield models and performance over increasing projection length can be gained from the study of Shortt and Burkhart (1996). They evaluated a whole-stand (Sullivan and Clutter, 1972) and an individual-tree (Amateis *et al.*, 1989) growth and yield model for loblolly pine for the purpose of updating forest inventory data. Both of the growth and yield models were evaluated at varying projection periods by using permanent plots measured at 0, 3, 6 and 9 years after initial plot establishment. Evaluations were based solely on the capability of each model to predict merchantable volume. The individual-tree model produced the best results until the 6-year period, at which time it was approximately equal to the whole-stand model. After 6 years, the whole-stand model produced the most reliable results. Only the whole-stand model appeared to be unbiased. Both models displayed rapidly increasing error of prediction with increasing projection length. As expected from the general relationship of model dimension to projection length, for short-term projections the more detailed individual-tree model performed best, but for long-term projections the simple whole-stand model performed best.

Conclusions

From experience and information reported in the scientific literature, the following conclusions can be drawn.

- There is no easy answer to the question of what is the appropriate modelling level but, within cost constraints, data should be collected to allow for modelling at the highest level of resolution that might be needed.
- Relatively simple, parsimonious models are generally adequate.
- In determining the level of resolution for modelling, one should also consider the time horizon for projection; as a rule, when the projection period increases the model dimensionality should decrease.

Although the science of modelling has advanced greatly, and continues to advance, there is still a great deal of art involved. It is impossible to overemphasize the necessity of having clearly stated objectives for modelling, because no approach can satisfy all purposes. Regardless of the modelling objective, one should strive to: (i) fit as parsimonious a model as possible to describe the population trends of interest, and (ii) adjust the dimensionality of the model to suit the projection length.

Acknowledgements

The support of the Loblolly Pine Growth and Yield Research Cooperative at Virginia Tech and the assistance of Dr Mahadev Sharma with analyses of the Cooperative's spacing trial data reported here are gratefully acknowledged.

References

Amateis, R.L., Burkhart, H.E. and Zedaker, S.M. (1988) Experimental design and early analyses for a set of loblolly pine spacing trials. In: *Forest Growth Modelling and Prediction*. North Central Forest Experiment Station General Technical Report NC-120, pp. 1058–1065.

Amateis, R.L., Burkhart, H.E. and Walsh, T.A. (1989) Diameter increment and survival equations for loblolly pine trees growing in thinned and unthinned plantations on cutover, site-prepared lands. *Southern Journal of Applied Forestry* 13, 170–174.

Avery, T.E. and Burkhart, H.E. (2002) *Forest Measurements*, 5th edn. McGraw-Hill, New York, 456 pp.

Burkhart, H.E., Farrar, K.D., Amateis, R.L. and Daniels, R.F. (1987) *Simulation of Individual Tree Growth and Stand Development in Loblolly Pine Plantations on Cutover, Site-prepared Areas*. Virginia Polytechnic Institute and State University, Publication FWS-1-87, 47 pp.

Clutter, J.L., Fortson, J.C., Pienaar, L.V., Brister, G.H. and Bailey, R.L. (1983) *Timber Management: a Quantitative Approach*. John Wiley & Sons, New York, 333 pp.

Davis, L.S., Johnson, K.N., Bettinger, P.S. and Howard, T.E. (2001) *Forest Management*, 4th edn. McGraw-Hill, New York, 804 pp.

DeAngelis, D.L. and Gross, L.J. (eds) (1992) *Individual-based Models and Approaches in Ecology*. Chapman and Hall, New York, 525 pp.

Gadow, K. von and Hui, G. (1999) *Modelling Forest Development*. Kluwer Academic Publishers, Dordrecht, The Netherlands, 213 pp.

Gauch, H.G. Jr (1993) Prediction, parsimony and noise. *American Scientist* 81, 468–478.

Kahne, S. (1976) Model credibility for large-scale systems. *IEEE Transactions on Systems, Man and Cybernetics* 6(8), 53–57.

Leary, R.A. (1979) Design. In: *A Generalized Forest Growth Projection System Applied to the Lake States Region*. US Forest Service, North Central Forest Experiment Station General Technical Report NC-49, pp. 5–15.

Mitchell, K.J. (1975) Dynamics and simulated yield of Douglas-fir. *Forest Science Monograph 17*, 39 pp.

Shortt, J.S. and Burkhart, H.E. (1996) A comparison of loblolly pine plantation growth and yield models for inventory updating. *Southern Journal of Applied Forestry* 20, 15–22.

Sullivan, A.D. and Clutter, J.L. (1972) A simultaneous growth and yield model for loblolly pine. *Forest Science* 18, 76–86.

Vanclay, J.K. (1994) *Modelling Forest Growth and Yield: Applications to Mixed Tropical Forests*. CAB International, Wallingford, UK, 312 pp.

Vanclay, J.K., Skovsgaard, J.P. and Hansen, C.P. (1995) Assessing the quality of permanent sample plot databases for growth modelling in forest plantations. *Forest Ecology and Management* 71, 177–186.

2 Mapping Lodgepole Pine Site Index in Alberta

Robert A. Monserud[1] and Shongming Huang[2]

Abstract

We demonstrate methods and results for broad-scale mapping of forest site productivity for the Canadian province of Alberta. Data are observed site index (SI) for lodgepole pine (*Pinus contorta* var. *latifolia* Engelm.) based on stem analysis (observed height at an index breast height age of 50 years). A total of 2624 trees at nearly 1000 site locations were available for the analysis. Mapping methods are based on ANUSPLIN, Hutchinson's thin-plate smoothing spline in four dimensions (latitude, longitude, elevation, site index). Although this approach is most often used for modelling climatic surfaces, the high density of the site productivity network in Alberta made this an appropriate application of the method. Maps are presented for lodgepole pine, the major forest species of Alberta. Although map patterns are highly complex, predicted SI decreases regularly and continuously as elevation increases from the parklands, through the foothills, to the mountains, conforming to field observations and a shorter growing season. In the high mountains, SI predictions are the lowest (<10 m), again conforming to field observations. Thus, the map correctly represents the inverse relationship between SI and elevation exhibited in the data. Analysis of residuals revealed no bias in the predictions. Furthermore, residuals were homogeneous and had no apparent pattern. The standard deviation of the observed site index values was 3.23 m, and the root mean squared error of the spline surface predictions was 1.16 m.

Introduction

Lodgepole pine is Alberta's provincial tree. It is the most common tree species in the Rocky Mountains and Foothills regions (Alberta Environmental Protection, 1994), occurring on the eastern slopes of the Canadian Rocky Mountains (Fig. 2.1). It also occurs in a large zone to the north in boreal regions where it hybridizes with jack pine (*Pinus banksiana* Lamb.). Although lodgepole pine comprises about 20% of the

[1] USDA Forest Service, Pacific Northwest and Rocky Mountain Research Stations, USA
Correspondence to: rmonserud@fs.fed.us
[2] Forest Management Branch, Ministry of Sustainable Resource Development, Government of Alberta, Canada

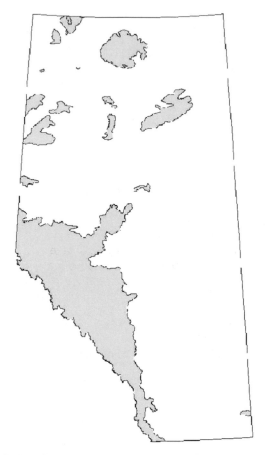

Fig. 2.1. The distribution of lodgepole pine in Alberta.

mature standing timber in Alberta, it accounts for approximately 40% of the annual harvest in the province (Huang *et al.*, 2001).

Lodgepole pine grows well on a wide range of both edaphic and climatic site conditions, a quality that Wheeler and Critchfield (1985) describe as remarkable. In Alberta, it has optimal growth on sites similar to those of aspen (*Populus tremuloides* Michx.) and white spruce (*Picea glauca* (Moench) Voss): moist, rich, well-aerated sites with friable soils and long, warm growing seasons. Unlike aspen and white spruce, however, lodgepole pine can tolerate sites that have short and irregular growing seasons, poor to very poor nutrient regimes and a distinct moisture deficit (Huang *et al.*, 2001).

Lodgepole pine is a major forest component in four of the 20 subregions of Alberta (Alberta Environmental Protection, 1994): the Upper and Lower Foothills subregions, the Montane subregion and the Subalpine subregion. It is a minor component in the Boreal Forest region. A brief description of these regions follows:

● The Upper Foothills subregion occurs on strongly rolling topography along the eastern edge of the Rocky Mountains (Alberta Environmental Protection, 1994). Bedrock outcrops of marine shales and non-marine sandstones are frequent. This subregion has the greatest amount of summer precipitation in Alberta at about 340 mm and a mean annual precipitation of about 540 mm. Winter is colder than in the Lower Foothills subregion. Upland forests are nearly all coniferous and are dominated by lodgepole pine, white spruce, black spruce (*Picea mariana* (Mill.) B.S.P.) and subalpine fir (*Abies lasiocarpa* (Hook.) Nutt.).

- The Lower Foothills subregion is a rolling landscape created by deformed sandstone and shale along the edge of the Rocky Mountains, as well as erosional remnants with flat-lying bedrock (Alberta Environmental Protection, 1994). Although this subregion is somewhat cooler in summer than the adjacent, lower-elevation Boreal Forest subregions, it is warmer in winter because it is influenced less often by cold Arctic air masses and more frequently moderated by chinook winds. Mixed forests reflect the transitional nature of this subregion. Lodgepole pine forests occupy extensive portions of the uplands, especially following fire.

- The Montane subregion occurs in the lower elevations of the Rocky Mountain region in southwestern Alberta, above the Foothills region. This subregion has two components. The first is warm and dry, a structural montane landscape of grassland communities interspersed with dry forest communities (Alberta Environmental Protection, 1994). Lodgepole pine is present but is not abundant; aspen and interior Douglas-fir (*Pseudotsuga menziesii* var. *glauca* (Biessn.) Franco) are more frequent, with minor admixtures of limber pine (*Pinus flexilis* James). The second montane component is also mild but is moister, and consists of closed-canopy forests. They are composed of a large proportion of lodgepole pine and white spruce, with lesser amounts of aspen, limber pine and occasionally subalpine fir. This closed-canopy montane forest is often quite productive.

- The Subalpine subregion occupies a band between the Montane and Alpine subregions in the south and between the Upper Foothills and Alpine subregions in the north (Alberta Environmental Protection, 1994). Freezing temperatures occur in all months and the frost-free period probably lasts for less than 30 days. Winter precipitation is greater in this subregion than in any other part of Alberta, often with more than 200 cm of snowfall. Soils are highly variable because of complex topography and parent materials. Closed forests of lodgepole pine, Engelmann spruce (*Picae engalmannii* Parry ex Engelm.) and subalpine fir are characteristic of lower elevations within this subregion, and open forests are typical of higher elevations near the treeline.

- Lodgepole pine is also a minor component of the Boreal Forest Natural Region, which consists of broad lowland plains and discontinuous hill systems (Alberta Environmental Protection, 1994). Extensive wetlands are characteristic of this natural region. Climatic conditions reflect a strong boreal influence. Typically, the winters are long and cold, the summers are short and cool, and the majority of precipitation falls in the summer. The vegetation is typically dominated by aspen, with mixed or coniferous forests at higher elevations or in wetlands. Jack pine outcompetes lodgepole on dry, nutrient-poor sites. Lodgepole pine also occurs in association with black spruce in fringes around bog margins on wetter sites before disappearing in the boreal lowlands.

The increased values and pressures placed on the timber resources in Alberta call for accurate information relevant to management of the forest resource. A key first step is accurate estimation of site productivity for a given location. To choose the best management strategy, it is necessary to have a quantitative means for estimating site productivity for a range of management objectives.

Over the last few decades, both the government and the industry in Alberta have allocated a sizeable amount of time and resources to the collection and analysis of data on productivity, growth and yield of lodgepole pine stands. Thousands of felled-tree stem analyses have been conducted. One product has been the development of site index (SI)/height growth curves useful for indexing site productivity (Huang, 1997; Huang *et al.*, 1997, 2001). We concentrate on the observed SIs from these stem analyses. In particular, we set out first to map them, and then to develop isoclines of SI within the range of lodgepole pine in Alberta.

We are faced with two practical constraints in this mapping project. The first is the natural range of lodgepole pine (Fig. 2.1). This excludes two large agricultural areas: the prairie and parklands in the southeast quarter of Alberta (too dry for forests), and the agricultural lands in the Peace River area (west central Alberta). Although lodgepole pine could certainly survive in favourable locations outside its natural range, our intention for now is to stay inside this range. A climate-change scenario analysis in a subsequent study will remove this constraint. Secondly, destructive stem analysis sampling is not allowed in National Parks (e.g. Jasper, Banff), many of which include large numbers of lodgepole pine (Fig. 2.2). Both constraints exclude the enormous Wood Buffalo National Park in northeast Alberta (Fig. 2.2).

Data

Site quality describes the carrying capacity of the environment and is determined by the biogeoclimatic conditions of a site. We used a standard measure in forestry, SI, defined as dominant tree height at 50 years breast height age.

In the 1980s, Alberta began a vigorous programme of measuring stand growth and site productivity. Thousands of trees were felled and the stems were sectioned (stem analysis). Guidelines called for destructively sampling approximately three dominant/codominant trees per location. Most of the felled trees were sectioned at stump height (0.3 m), breast height (1.3 m), 1.5 m above breast height, and at equal lengths of 2.5 m thereafter to the top of the tree (Huang *et al.*, 2000). A small proportion of the felled trees was sectioned more intensively at 0.3, 0.8, 1.3, 2.0 and 2.8 m above ground and then at equal lengths of 1.0 m thereafter to the top of the tree. Ring count and height were the variables recorded for each section that we used in this study. By 1988, over 11,000 trees had been sectioned throughout the inventoried areas of the Province (Alberta Forest Service, 1988). A large number of these sectioned trees were sampled in the buffer zone around permanent sample plots (see Alberta Forest Service, 1988). A second set was obtained on temporary sample plots. Recently, the Foothills Growth and Yield Cooperative has conducted additional stem analyses on the Foothills Model Forest (Forestry Corp., 2002).

We combined the lodgepole pine in these three data sets for a common analysis, after determining the longitude, latitude and elevation of each sample stand. The breakdown by data set is $n_1 = 1225$ from permanent sample plot buffers, $n_2 = 1129$ from temporary sample plots and $n_3 = 270$ from the Foothills Model Forest. This left us with $n = 2624$ observations of lodgepole pine SI in Alberta (Fig. 2.2). A summary of the data by natural subregions of Alberta (Alberta Environmental Protection, 1994) is provided in Table 2.1. The variation in number of sample trees by subregions is an indication of the relative abundance and importance of lodgepole pine in that subregion.

Site index determination

Much of this stem analysis data was used by Huang *et al.* (1997) to develop species-specific height growth and SI curves for Alberta. Recall that the stem analysis data consist of multiple pairs of observations of height and age for each tree. We began by estimating SI for each pair using the appropriate SI curve from Huang *et al.* (1997,

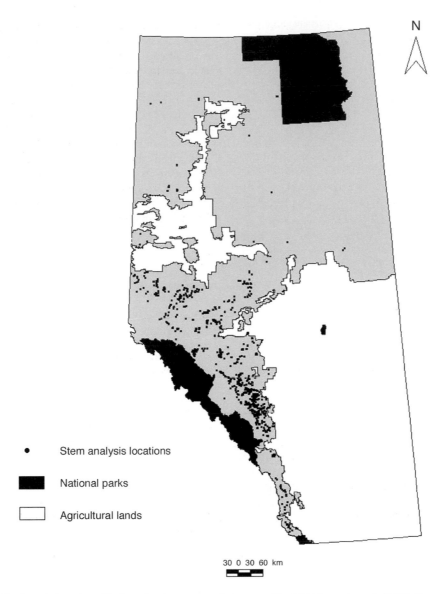

Fig. 2.2. Sample location of lodgepole pine stem analysis trees. $N = 2624$ trees at nearly 1000 site locations. National parks: Banff and Jasper in the Rocky Mountains in the southwest, Wood Buffalo in the northeast, and Elk Island in central Alberta near Edmonton. Agricultural lands: prairie grasslands in the southeast and Peace River area in the west-northwest.

2001). SI for each of the 2624 stem analysis trees was determined as the mean of the individual SI estimates from the sectional height–age observations. We averaged these estimates from each of the three site trees at each location to obtain plot SI for 995 locations. Because Huang *et al.* (1997) found that SI estimates below age 20 years had much higher variability (Fig. 2.3), we confined our estimates to observations that were at least 20 years old. This method for estimating SI had the advantage of replicating the field application of the site curves of Huang *et al.* (1997), and of giving equal importance to all stem sections beyond age 20.

Table 2.1. Summary statistics for the lodgepole pine stem analysis data.

Subregion[a]	No. of trees	HT (m)[b]				SI (m)[c]			
		Mean	Min.	Max.	SD	Mean	Min.	Max.	SD
Central Mixedwood	20	20.42	8.92	27.71	5.36	17.09	13.71	22.77	3.10
Dry Mixedwood	17	17.18	5.50	27.15	8.92	18.02	14.66	21.03	1.98
Wetland Mixedwood	14	15.33	11.28	20.78	2.74	11.49	8.77	16.21	2.42
Peace River Lowlands	3	21.53	19.90	23.08	1.59	14.64	12.72	16.75	2.02
Boreal Highlands	3	16.55	14.21	18.76	2.28	14.50	12.05	16.50	2.26
Alpine	2	17.75	17.70	17.80	0.07	7.74	7.30	8.18	0.63
Subalpine	305	15.65	8.45	25.00	3.29	11.38	4.30	18.29	3.05
Montane	47	15.01	9.90	18.41	1.71	13.74	8.14	19.12	2.90
Lower Foothills	1327	18.74	5.99	31.78	4.41	14.07	5.41	24.41	3.43
Upper Foothills	883	21.46	6.10	36.80	4.47	15.90	7.90	26.46	3.13
Central Parkland	3	15.93	14.87	17.02	1.08	12.25	10.56	13.47	1.51
Grand total	2624	19.21	5.50	36.80	4.73	14.39	4.30	26.46	3.56

[a]Subregions of Alberta are defined in Alberta Environmental Protection (1994).
[b]HT, total tree height (m).
[c]SI, site index, which refers to tree height at 50 years breast height age.

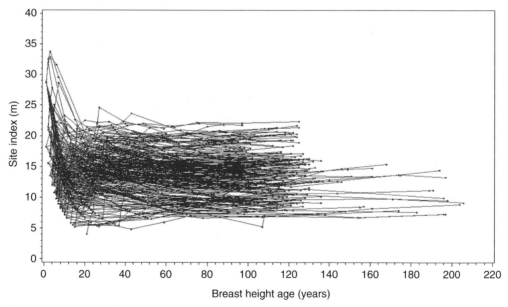

Fig. 2.3. Consistency of site index estimates vs. age. Each line represents repeated site index estimates for one stem analysis tree (this graph was based on 10% of the stem analysis trees).

Methods

Our goal was to produce a map of SI isoclines within and bordering the natural range of lodgepole pine in Alberta. We began by assembling a database of longitude, latitude, elevation and SI for each of the stem analysis stands. Elevation was obtained from a digital elevation model (DEM) using a 100 m grid for the province. SI at each sample location ($n = 995$) was estimated directly using felled-tree stem analysis data (see previous paragraph). We then used a thin-plate smoothing spline program (ANUSPLIN; see Hutchinson, 2002) to fit this SI surface as a function of the three topographic variables: longitude, latitude and elevation. Finally, we combined this fitted surface with a province-wide DEM to predict isoclines of SI for the entire range of lodgepole pine in Alberta. Maps were produced using the ArcView geographical information system.

ANUSPLIN

The aim of the ANUSPLIN package (acronym: Australian National University SPLINe) is to provide a facility for transparent analysis and interpolation of noisy multivariate data using thin-plate smoothing splines (Hutchinson, 2002). The surface-fitting procedure was primarily developed for fitting climate surfaces such as temperature and precipitation. Thus, there are normally at least two independent spline variables, longitude and latitude (in decimal degrees). A third independent variable, elevation above sea level, is normally included as a third independent spline variable (in kilometre units). Our problem is directly akin to the standard problem of fitting a temperature surface using a network of weather stations, except that our weather stations are actually stem analysis locations and temperature is replaced by SI.

The original thin-plate (formerly Laplacian) smoothing spline surface-fitting technique was described by Wahba (1979), with modifications for larger data sets due to Hutchinson and de Hoog (1985). Thin-plate smoothing splines can be viewed as a generalization of standard multivariate linear regression, in which the parametric model is replaced by a suitably smooth non-parametric function (Hutchinson, 2002). The degree of smoothness (or complexity) of the fitted function is determined by minimizing a measure of predictive error of the fitted surface given by the generalized cross-validation (Craven and Wahba, 1979; Wahba, 1983).

Wahba (1990) provides a comprehensive introduction to the technique of thin-plate smoothing splines. A brief overview of the basic theory and applications to spatial interpolation of monthly mean climate is given in Hutchinson (1991). A comprehensive discussion of the algorithms and associated statistical analyses, and comparisons with kriging, are given in Hutchinson (1993) and Hutchinson and Gessler (1994).

Results

The network of SI observations (995 locations) had a standard deviation of 3.32 m. This collection of observations was then fitted with a thin-plate smoothing spline using ANUSPLIN. Residuals between the resulting surface and the observations had a root mean squared error of 1.16 m. Analysis of residuals revealed no bias in the predictions. Furthermore, residuals were homogeneous and had no apparent pattern. Using this ANUSPLIN surface of lodgepole pine SI, predictions of SI were made on a 1-km grid across Alberta. This grid of predictions was then mapped using ArcView.

The first problem we encountered was somewhat expected: the negative correlation between elevation and SI (Fig. 2.4) resulted in some negative SI predictions in the high Rockies. This is an instance of using an interpolation tool for unintended extrapolation. The Canadian Rockies contain a seemingly endless chain of dramatic 3000+ m peaks, all well above the timberline and above the elevational limits of lodgepole pine. The elevation limit for lodgepole in our data set is 2200 m, with an

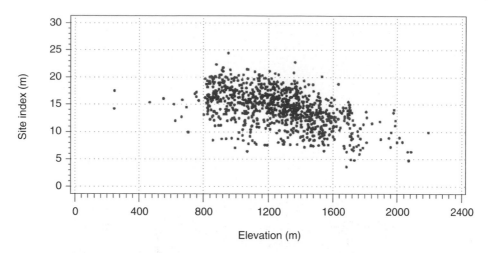

Fig. 2.4. The relationship between elevation and plot site index for lodgepole pine.

observed minimum SI near 4 m. Our solution was simply to impose a linear constraint setting SI = 0 if elevation exceeds 2400 m. The revised map appeared reasonable for the mountains (Figs 2.5 and 2.6).

Next, we used ArcView to overlay a block-out map of the agricultural lands and Wood Buffalo National Park (Fig. 2.2) on top of the ANUSPLIN predictions of SI (Fig. 2.6). We intentionally left Jasper and Banff National Parks in the map to illustrate the predictions in the high Rocky Mountains (Fig. 2.6), even though we have no data there. Finally, we overlaid the ANUSPLIN predictions of SI with the natural distribution of lodgepole pine (Fig. 2.7).

Discussion

We have a very large sample of SIs from the most important regions for lodgepole pine (Table 2.1): 2210 observations from the Foothills Natural Region (Upper Foothills and Lower Foothills subregions) and 352 observations from the Rocky Mountain Natural Region (Montane and Subalpine subregions). These subregions are all in southwestern and western Alberta. Sample size is relatively small (57 observations) in the largely roadless Boreal Forest Natural Region (mostly the Central Mixedwood, Dry Mixedwood and Wetland Mixedwood subregions), which includes much of northern Alberta.

Although map patterns are complex, predicted SI decreases regularly and continuously as elevation increases from the parklands, through the foothills, to the mountains, conforming to field observations and a shorter growing season. In the high mountains, SI predictions are the lowest (<10 m), again conforming to field observations. Thus, the map correctly represents the inverse relationship between SI and elevation exhibited in the data (Fig. 2.4).

Predictions on agricultural land warrant closer examination. Because we have no observations in agricultural land, these predictions are clearly extrapolations and should be viewed with caution. Generally, they fall into two categories. First, the Grassland Natural Region (the prairies of southeastern Alberta) is clearly too dry to support the establishment and growth of forest trees. Moisture stress would also limit lodgepole pine establishment and growth in sections of the Montane subregion

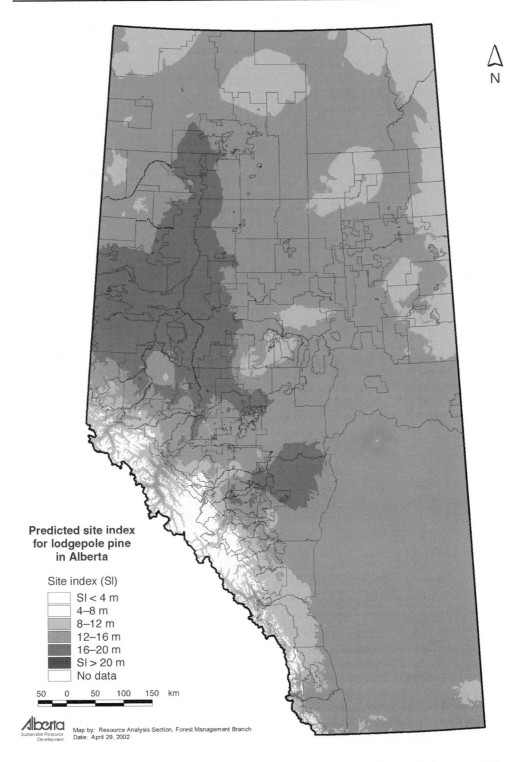

Fig. 2.5. Unblocked site index predictions from ANUSPLIN. These maps are mostly extrapolations beyond the range of the data (compare Fig. 2.2), and are intended only to examine the limitations of the predictions.

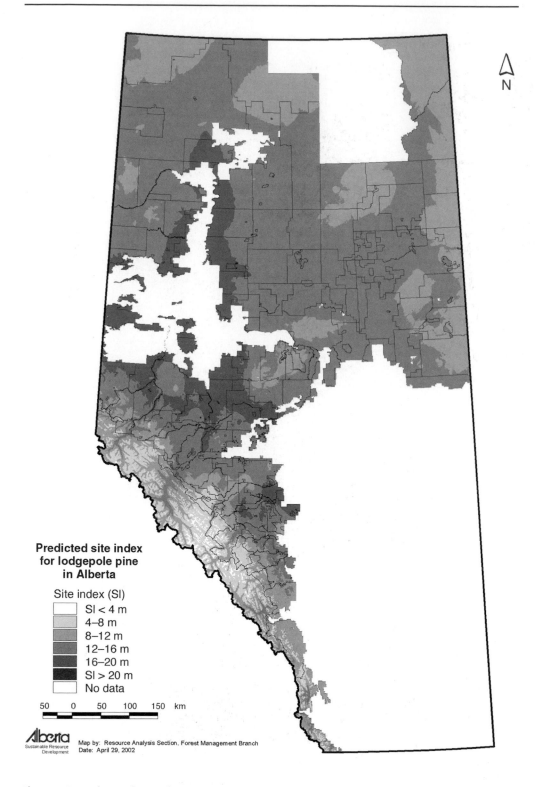

Fig. 2.6. Site index predictions from ANUSPLIN, excluding agricultural lands and Wood Buffalo National Park.

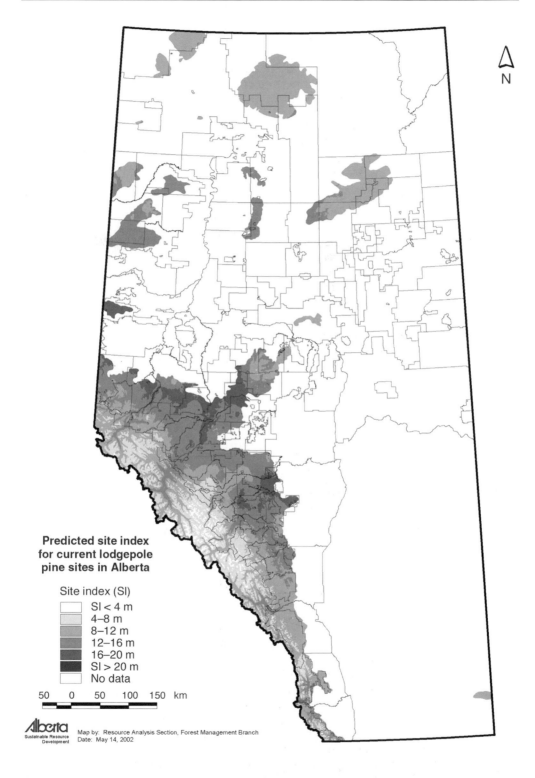

**Predicted site index
for current lodgepole
pine sites in Alberta**

Site index (SI)

SI < 4 m
4–8 m
8–12 m
12–16 m
16–20 m
SI > 20 m
No data

50   0   50   100   150   km

Alberta
Sustainable Resource
Development

Map by: Resource Analysis Section, Forest Management Branch
Date: May 14, 2002

Fig. 2.7. Site index predictions within the natural distribution of lodgepole pine (see Fig. 2.1).

and Parkland Natural Region (e.g. parkland areas between Grande Prairie and Peace River). These parklands are both too dry and probably have soils with a hard-pan, both of which would limit the establishment of trees. Our solution is simply to refrain from making predictions in the agricultural land (Figs 2.6 and 2.7).

The second type of agricultural land is not too dry for forests, however. Both the Peace River Valley and the agricultural area west of Edmonton belong to the Dry Mixedwood or Central Mixedwood subregions. Because these cultivated areas also support pure deciduous and mixed white spruce and aspen forests, it is likely that interspecific competition (and cultivation) rather than moisture stress is the prime factor limiting lodgepole pine. ANUSPLIN predicts high SI levels (16–20 m) in the Peace River Valley and the agricultural area west of Edmonton (Fig. 2.5), an extrapolation that cannot be confirmed by the data.

When we applied the ANUSPLIN predictions to the entire 1-km grid of Alberta (Fig. 2.5), we were forced to use vast extrapolations (compare the lodgepole distribution map in Fig. 2.1). This was intentional, so as to examine the geographical limits of the predictions. We succeeded in finding two limits: all elevations >2400 m, and most predictions into agricultural land. Note that the model extrapolates quite well into the mountains of Banff and Jasper National Parks (Figs 2.6 and 2.7), where we have no data. The elevation effect is correctly captured by ANUSPLIN, provided that SI predictions are set to zero above 2400 m.

Northern Alberta (55–60° N, 110–120° W) presents another interesting case. It is sparsely represented in the data, so predictions there should be viewed with caution. We do have 26 plots in the area, but the vastness of the mostly unroaded landscape makes this a relatively small sample for inference. Because lodgepole pine occurs in isolated hilltops to the north and east of its present continuous range, it probably had a broader distribution in the immediate period following the retreat of the Laurentide ice mass, and later retreated, leaving scattered outliers on favourable sites (Rudolph and Yeatman, 1982). L. Barnhardt (Edmonton, 2002, personal communication) points out that the persistent pockets of lodgepole on the hilltops in the north (as in the Cypress Hills in the southeast) are the only evidence of its suitability in this region, but that he expects that it would grow better lower down the slopes. It probably does not occur there because it is out-competed in early succession by aspen. Interspecific (density-dependent) competition becomes the limiting factor, rather than climate (Rehfeldt et al., 1999). The predictive world of ANUSPLIN is one-dimensional (lodgepole pine SI, with no competing species to complicate things. ANUSPLIN predicts intermediate lodgepole SI for most of northern Alberta (Fig. 2.6), even though most of this boreal forest/wetland is outside the species' distribution (Fig. 2.1; Little, 1971). Clearly, this is an extrapolation, except for the isolated hill systems inside the species' range (Fig. 2.7).

Northeastern Alberta is actually in the natural range of jack pine, a closely related serotinous species. Lodgepole and jack pine hybridize extensively in north-central Alberta, where the two species' ranges overlap (Little, 1971; Wheeler, 1981; Rudolph and Yeatman, 1982). Although our predictions of lodgepole pine SI could possibly be viewed as jack pine predictions in the northeast, we do not have data from jack pine in this analysis, so such predictions thus remain extrapolations.

Within the observed range of lodgepole pine, the ANUSPLIN predictions conform generally to expectations (L. Barnhardt, Edmonton, 2002, personal communication) and to field data (Table 2.1). Lodgepole does not grow well at higher elevations, where it is the most prevalent. At the upper elevation limits of its distribution, climatic factors and poor soil determine its range. At the lower elevational extremes, density-dependent competition (primarily from aspen) rather than climatic factors is limiting. The largest lodgepole trees are often in mixed stands (or adjacent to them) in the foothills. In early succession, aspen frequently outcompetes lodgepole pine on these low-elevation, mesic sites in the foothills and lowlands (L. Barnhardt, Edmonton, 2002, personal communication). Stand history (e.g. frequency and intensity of burning) sometimes tips the balance toward persistence of lodgepole, although it cannot outcompete white spruce in later successional stages. The model predicts lodgepole pine SI to be high in the Foothills subregions.

Finally, Barnhardt expects that the best sites for lodgepole may very well be bordering or outside its present range and be occupied by aspen in early succession and white spruce as succession proceeds. This is supported by genetic results for British Columbia by Rehfeldt *et al.* (1999). For example, they found that populations in the northern portion of its range were occupying sites 5–6°C cooler (mean annual temperature) than is optimal for those populations.

To our knowledge, this is the first application of thin-plate smoothing splines to continuous mapping of forest SI. This methodology was developed for modelling climatic surfaces (e.g. Tchebakova *et al.*, 1994; Monserud and Tchebakova, 1996), but it generalizes quite well to other ecological surfaces such as SI. The result is a map that continuously predicts lodgepole pine SI across all possible areas in the Province. Judgement must be used to limit the predictions to the current distribution limits or perhaps to the immediate vicinity. Predictions appear reasonable, primarily because of the locally adaptive nature of the spline-fitting methods. These are strongest where the data are strongest. Caution should be exercised in areas where data are sparse, such as the northern half of the Province. Predictions outside the range of the data, especially into agricultural land, should not be used.

This application is largely possible because of the availability of a very large sample of destructively sampled trees. Stem analysis is quite labour-intensive and expensive. It is rare that an organization will go to such great expense to obtain comprehensive productivity data on a regional scale. Although we might appear to be highlighting the interpolation and mapping technology (ANUSPLIN and ArcView), the real star is the enormous and comprehensive database on productivity and growth that the Province of Alberta has amassed over several decades.

This study is a pilot of sorts. Future work will expand this to all commercial species in Alberta. We will then replace SI with m^3/ha/year, which is more directly related to net primary production. The thin-plate smoothing splines methodology can also be applied to many other variables. We plan to incorporate climate into our analyses, including climate-change scenarios, to assess the impact of climate change on site productivity and forest growth and yield in general.

Acknowledgements

We are very grateful to the many people and organizations that supported this project. In particular, we thank Yuqing Yang, Tammy Kobliuk, Leonard Barnhardt, Narinder Dhir, Christine Hansen, Thomas Braun, Bob Held, Greg Behuniak, Dave Morgan, Grady Ung, Barb Osterhout, Frank Liu, Olenka Bakowsky, Yanguo Qin and Daryl Gilday for their tremendous help on data management, data analysis and data mapping. We thank Leonard Barnhardt and Andy Hudak for their excellent review comments. Because of his vast experience in the forests of Alberta, Leonard Barnhardt served ably as our Jedi Knight. We greatly appreciate the contributions of Weyerhaeuser Canada, Sunpine Forest Products, The Forestry Corp. and, particularly, Weldwood of Canada, for providing the data and/or supporting the data collection and analyses. We are also grateful to Alberta Departments of Environment and

Sustainable Resource Development for providing part of the funding necessary to carry out the work described here. Robert Monserud was supported by the Rocky Mountain Research Station (USDA Forest Service) for his contribution to this research.

References

Alberta Environmental Protection (1994) *Natural Regions and Subregions of Alberta*. Alberta Environment, Publication No. I/531, Edmonton, Alberta, 18 pp.

Alberta Forest Service (1988) *Alberta Phase 3 Forest Inventory: Tree Sectioning Manual*. Forest Measurement Section, Edmonton. Publication No. T/168. (Revised 1988 (formerly ENF Rep. Dep. 56).)

Craven, P. and Wahba, G. (1979) Smoothing noisy data with spline functions. *Numerische Mathematik* 31, 377–403.

Forestry Corp. (2002) *Site Index Increment Study*. Prepared for Weldwood of Canada (Hinton Division) by the Forestry Corp., Edmonton, Alberta, Canada, 70 pp.

Huang, S. (1997) Development of compatible height and site index models for young and mature stands within an ecologically based management framework. In: Amaro, A. and Tomé, M. (eds) *Empirical and Process-based Models for Forest Tree and Stand Growth Simulation*. September 21–27, Portugal, pp. 61–98.

Huang, S., Titus, S.J. and Klappstein, G. (1997) *Development of a Subregion-based Compatible Height–Site Index–age Model for Young and Mature Lodgepole Pine in Alberta*. Forest Management Research Note No. 6, Land and Forest Service, Technical Publication No. T/353.

Huang, S., Price, D., Morgan, D.J. and Peck, K. (2000) Kozak's variable-exponent taper equation regionalized for white spruce in Alberta. *Western Journal of Applied Forestry* 15(2), 75–85.

Huang, S., Morgan, D.J., Klappstein, G., Heidt, J., Yang, Y. and Greidanus, G. (2001) *GYPSY: a Growth and Yield Projection System for Natural and Regenerated Lodgepole Pine Stands Within an Ecologically Based, Enhanced Forest Management Framework: Yield Tables for Seed-origin in Natural and Regenerated Lodgepole Pine Stands*. Alberta Sustainable Resource Development, Technical Report Publication No. T/485, Edmonton, Alberta, 193 pp.

Hutchinson, M.F. (1991) The application of thin plate smoothing splines to continent-wide data assimilation. In: Jasper, J.D. (ed.) *BMRC Research Report No. 27, Data Assimilation Systems*. Bureau of Meteorology, Melbourne, pp. 104–113.

Hutchinson, M.F. (1993) On thin plate splines and kriging. In: Tarter, M.E. and Lock, M.D. (eds) *Computing and Science in Statistics* 25. Interface Foundation of North America, University of California, Berkeley, pp. 55–62.

Hutchinson, M.F. (2002) *ANUSPLIN Version 4.2*. Centre for Resource and Environmental Studies, The Australian National University, Canberra. Website: cres.anu.edu.au/outputs/anusplin.html

Hutchinson, M.F. and de Hoog, F.R. (1985) Smoothing noisy data with spline functions. *Numerische Mathematik* 47, 99–106.

Hutchinson, M.F. and Gessler, P.E. (1994) Splines: more than just a smooth interpolator. *Geoderma* 62, 45–67.

Little, E.L. Jr (1971) *Atlas of United States Trees: Volume 1, Conifers and Important Hardwoods*. US Department of Agriculture Miscellaneous Publication 1146, 9 pp., 200 maps.

Monserud, R.A. and Tchebakova, N. (1996) A vegetation model for the Sayan Mountains, Southern Siberia. *Canadian Journal of Forest Research* 26, 1055–1068.

Rehfeldt, G.E., Ying, C.C., Spittlehouse, D.L. and Hamilton, D.A. (1999) Genetic responses to climate in *Pinus contorta*: niche breadth, climate change, and reforestation. *Ecological Monographs* 69, 375–407.

Rudolph, T.D. and Yeatman, C.W. (1982) *Genetics of Jack Pine*. USDA Forest Service, Research Paper WO-38, Washington, DC, 60 pp.

Tchebakova, N.M., Monserud, R.A. and Nazimova, D. (1994) A Siberian vegetation model based on climatic parameters. *Canadian Journal of Forest Research* 24, 1597–1607.

Wahba, G. (1979) How to smooth curves and surfaces with splines and cross-validation. In: *Proceedings of the 24th Conference on the Design of Experiments*. US Army Research Office 79-2, Research Triangle Park, North Carolina, pp. 167–192.

Wahba, G. (1983) Bayesian confidence intervals for the cross-validated smoothing spline. *Journal of the Royal Statistical Society, Series B*, 45, 133–150.

Wahba, G. (1990) *Spline Models for Observational Data*. CBMS-NSF Regional Conference Series in Applied Mathematics 59, SIAM, Philadelphia, Pennsylvania.

Wheeler, N.C. (1981) Genetic variation in *Pinus contorta* Dougl. and related species of the subsection *Contortae*. PhD thesis, University of Wisconsin, Madison, Wisconsin, 196 pp.

Wheeler, N.C. and Critchfield, W.B. (1985) The distribution and botanical characteristics of lodgepole pine: biographical and management implications. In: Baumgartner, D.M. (ed.) *Lodgepole Pine: the Species and its Management*. Washington State University, Pullman, Washington, pp. 1–13.

3 Growth Modelling of *Eucalyptus regnans* for Carbon Accounting at the Landscape Scale

Christopher Dean,[1] Stephen Roxburgh[1] and Brendan Mackey[1]

Abstract

Modelling biomass and carbon pool fluxes at the landscape scale allows ecosystem carbon-carrying capacity to be estimated and provides a baseline for evaluating effects due to disturbance and climate change. We present a new biomass model 'CAR4D' for *Eucalyptus regnans*-dominated forests, an important forest type in Australia. CAR4D simulates changes in carbon stocks and fluxes over time, and can also incorporate spatial data in GIS format.

We adopted new modelling methods and acquired appropriate data for modelling *E. regnans* growth up to 450 years of age, well beyond the usual age of 100 years previously considered in most *E. regnans* growth models. In this chapter we report results from two scenarios: (i) stand replacement fires at 321 years and (ii) oldgrowth logging at 321 years followed by reforestation and harvesting on an 80-year cycle. CAR4D was also applied to a landscape case study.

Magnitudes of significant carbon pools and fluxes are shown for durations over 100 years. For example, an initial increase in the carbon stored in *E. regnans* biomass, until 215 years (553 t C/ha) was followed by a decrease due to stand thinning and decay of living trees. If undisturbed, the developing rainforest understorey in these forests compensates for this decay, with the total carbon levelling off at 1500 t C/ha near 375 years. The soil carbon levelled off at 670 t C/ha near 120 years. Changes in net biome productivity (per cycle) for the stand replacement fire and harvesting scenarios were 0.80 and 2.52 t C/ha/year, respectively. Mean values of soil carbon and total carbon for the stand replacement fire cycle were 654 t C/ha and 1230 t C/ha, respectively. Equivalent values for the harvesting cycle were: for soil carbon 97 t C/ha and for total carbon (including wood products) 387 t C/ha (once dynamic equilibrium was achieved, i.e. after about five cycles).

Introduction

The tree species *Eucalyptus regnans* is regarded as important in Victoria and Tasmania, Australia for pulpwood, sawlogs, fauna habitat and urban mains water supply. Our work arises from its potential at the landscape scale for high carbon sequestration. Biomass and carbon pool fluxes must be modelled at the landscape

[1] CRC for Greenhouse Accounting, Australian National University, Australia
Correspondence to: cdean@rsbs.anu.edu.au

scale to establish carbon-carrying capacity and a baseline for evaluating effects due to disturbance and climate change. In this chapter, we present a new biomass model for *E. regnans* and its associated carbon stocks.

S. Roxburgh and B. Mackey (unpublished results) used a terrestrial carbon budget model based on that described by Klein Goldewijk *et al.* (1994) to generate a carbon budget for an *E. regnans*-dominated landscape in Victoria. A generalized (non-spatial) model describing oldgrowth dynamics was parameterized using published data, and that model was extended spatially by varying forest growth and decomposition rates with topography, and by incorporating a spatially explicit database of historical fire events. That work provided the precursor to the study reported here.

Our model, CAR4D, is based on an approach using the most abundant type of data: the width of trees at 1.3 m (diameter at breast height or DBH). Height data for *E. regnans* are sometimes reported but often using methodologies that cannot be compared between different studies. In addition, some older or more disturbed stands suffer crown loss, but there is also mention (Mount, 1964) of a genetic disposition for crown retention in *E. regnans*. Statistically unbiased data for *E. regnans* over its habitat range and life cycle are scarce. Most data and models are dominated by volume data on stems less than 100 years old. In addition, stands over 110 m tall have been milled for pulp and lumber, burned (Galbraith, 1939), or cleared for farming. In the 1960s in Tasmania, specimens up to 98 m tall were recorded in logging records (Australian Newsprint Mills, *c.* 1960). Large areas of *E. regnans* in Tasmania were felled for newsprint manufacture (Helms, 1945; Australian Newsprint Mills, *c.* 1960) and afterwards for photocopy quality paper and lumber. Recently the tallest reported *E. regnans* was 91 m in Victoria (Mace, 1996) and 92 m in Tasmania (Hickey *et al.*, 2000). Only about 13% of the pre-1750 area of oldgrowth *E. regnans* in Tasmania remains and about 94% has been severely disturbed (Law, 1999). Overall, the retention of a natural range of sizes of the more mature *E. regnans* in any one logging district, or even in a state, is rare. The tallest *E. regnans* grow on specifically good soil and in preferred elevation and latitude niches. With time, voluminous and sound *E. regnans* can exist again and therefore such trees need to be accommodated in forecasting models for carbon sequestration in these forests. (For example, the most common age cohort of stands in the O'Shannessy and Maroondah catchments in Victoria is only 60 years old; many of these are in the high site index localities, and the catchment is reserved for water supply and consequently reserved from logging.)

In Australian wet sclerophyll forests and mixed forests, logging is usually by clearfelling followed by a high intensity burn (Bassett *et al.*, 2000). This process collapses or burns the habitat of most of the individual marsupials, reptiles and birds occupying the area logged, prior to logging (e.g. Mooney and Holdsworth, 1991). In this study, detailed measurements of older forests, to provide essential information on their growth and decay processes, were able to be taken during logging operations in Tasmania. This was advantageous because these measurements require destruction of the habitat (e.g. soil and rainforest tree removal from *E. regnans* buttresses to determine taper, felling of *E. regnans* to evaluate hollow content, and disc extraction from rainforest species for dendrochronology). In Victoria, sampling of older, single-aged stands of *E. regnans* in a way that significantly disturbs the habitat contravenes conservation protocols, although some valuable information can be acquired with minimal habitat disruption (e.g. stand level DBH and stocking rate). The oldest single-aged stand in Victoria (300 ± 50 years) is in the Otway Ranges' Big Tree Flora Reserve. Many older but less decayed stands exist in Tasmania and the largest, contiguous volumes per hectare of timber in Australia also exist in Tasmania

(Mount, 1964). Our fieldwork was mostly undertaken in logging coupes in Tasmania and in forest reserves in Victoria.

Experimental

Data acquisition and parameter estimation

The model comprises diverse components such as trunk morphology, leaf biomass and decay rates for coarse, woody debris. Consequently the sources of data were also diverse, ranging from fieldwork to published literature. The reserves used for data collection in Victoria were the O'Shannessy catchment and the Ada Tree reserve, both in the Central Highlands, and the Big Tree Flora Reserve in the Otway Ranges. Additionally, individual Victorian DBH data (1998 trees) and understorey data were kindly supplied by D.H. Ashton (Melbourne, 2001, personal communication). Tasmanian data were from logging coupe registers (Australian Newsprint Mills, *c.* 1960) and our field work in coupes AR023B and SX004C. Each major component of the model is described below in a separate section. Corresponding equations are presented in Table 3.1. For some of the model components the derived parameters were estimated by non-linear least squares minimization using SYSTAT (SYSTAT, 1992). Other components were estimated from published data and our fieldwork. These latter analyses involved numerous, complex steps, the details of which will be published separately.

Variation and distribution of DBH with age

The Victorian and Tasmanian data were combined to yield DBH_{ob} (DBH over bark) as a function of stand age (Equation 1 in Table 3.1 and Colour Plate 1). These data form the most comprehensive DBH_{ob} data available for *E. regnans* single-aged stands grown without silvicultural thinning. The graph (Colour Plate 1) shows DBH_{ob} for individual trees within a stand plotted against the age of the stand. The age of the stand in coupe SX004C was estimated to be at least 321 years based on the methodology of Hickey *et al.* (1999), using discs from celery-top pine (*Phyllocladus aspleniifolius*). The full set of DBH_{ob} data were also used to formulate frequency distributions of DBH_{ob} for any age stand, i.e. although stands are assumed to be even aged, the distribution of tree sizes within an even-aged stand is typically variable. The log-normal distribution curve provided the best fit, its parameters were the mean and standard deviation of $\ln(DBH_{ob})$; and the variables were DBH_{ob} and stand age. Each year the DBH_{ob} limits depended upon the widest tree that fell in the previous year and on reported DBH_{ob} limits. Also, for any particular DBH_{ob}, any trees with a frequency of <3% of that for trees with the mean DBH_{ob} were discarded.

Taper curves

The majority of an *E. regnans* tree's biomass is in its stem. Other than by destructive sampling, the stem volume is best obtained by generating volumes from taper curves. There was a paucity of data in the existing literature for DBH_{ob} greater than 3 m, even though the largest reported DBH_{ob} for *E. regnans* is 10.8 m (Ashton, 1975a). Galbraith (1937) reported 15 smoothed taper curves for trees with DBH_{ub}

Table 3.1. Derived equations. Units used: distance – metres, area – hectares, volume – metres^3, time (age) – years, rate – years^{-1}, weight – tonnes, density – tonnes/metre^3, stocking rate – stems/hectare.

Eqn no.	Equation	Variables, parameters and constants	Error analysis
1	$DBH_{ob} = a - (b \times e^{\,c \times stand\_Age})$	DBH_{ob} = diameter over bark at 1.3 m of one *E. regnans*, $stand\_Age$ = stand age, $a = 3.53(6)$, $b = 3.59(6)$, $c = -0.00323(8)$	$R^2 = 0.82$, MSE= 0.095, $N = 2184$
2	$stem\_volume = Vol\_Max \times \left(1 - \dfrac{1}{1 + \left(\dfrac{DBH_{ob} + 0.01}{DBH_{ob}\_Mid} \right)^{k}} \right)$	$stem\_volume$ = stem volume under bark of one *E. regnans*, $Vol\_Max = 380$, $DBH_{ob}\_Mid = 4.3$, $k = 2.57$	*
3	$bark\_volume =$ $Vc + Vol\_Max \times \left(1 - \dfrac{1}{1 + \left(\dfrac{DBH_{ob} + 0.01}{DBH_{ob}\_Mid} \right)^{k}} \right)$	$bark\_volume$ = bark volume of one *E. regnans*, DBH_{ob} = diameter at 1.3 m of one *E. regnans*, $Vol\_Max = 6.5$, $DBH_{ob}\_Mid = 3.7$, $k = 1.8$, $Vc = -5.2 \times 10^{-6}$	*
4	$\ln(stocking\_rate) = a - (b \times \ln(Age))$	$stocking\_rate$ = number of *E. regnans* per hectare, Age = stand age, $a = 10.6(2)$, $b = 1.27(4)$	$R^2 = 0.92$, MSE = 0.39, $N = 82$
5	$height = 75 \times 1 - (e^{-1.49556 \times DBH_{ob}})$	$height$= height of one *E. regnans*, DBH_{ob} = diameter at 1.3 m	*
6	$stem\_b\_density = a \times (1 - (b \times e^{-c \times Age}))$	$stem\_b\_density$ = basic density of *E. regnans* stem, Age = tree age, $a = 0.5124$, $b = 0.5809$, $c = 0.2$	*
7	$\ln(root\_biomass / above\_ground\_biomass) =$ $\ln(a \times e^{-b \times Age} + c)$	$root\_biomass$ = (dry) mass of roots of one *E. regnans*, $above\_ground\_biomass$ = (dry) mass of above ground components of one *E. regnans*, Age = tree age, $a = 3.3(9)$, $b = 0.78(9)$, $c = 0.13(2)$	$R^2 = 0.95$, MSE = 0.075, $N = 10$
8	$fraction\_decayed = \left(\dfrac{1}{1 + \left(\dfrac{Age}{Age\_Mid} \right)^{k}} \right)$	$fraction\_decayed$ = fraction of the total biomass decayed of one standing *E. regnans*, Age = tree age, $k = -6$, $Age\_Mid = 400$	*
9	$Er\_biomass =$ $fraction\_decayed \times$ $(stem+bark+branch+leaf+root)\_biomass$	$Er\_biomass$ = (dry) mass of one standing *E. regnans* after growth and decay, $fraction\_decayed$ (as above), $(stem+bark+branch+leaf+root)\_biomass$ = mass of the tree without decay	*

Continued

Table 3.1. *Continued.*

Eqn no.	Equation	Variables, parameters and constants	Error analysis
10	$f\_Er\_decay\_rate_{Age} =$ $a + (b \times e^{-c \times Age\_according\_to\_DBH_{ob}})$	$f\_Er\_decay\_rate_{Age}$ = rate at which a fallen *E. regnans* decays according to its age, $Age\_according\_to\_DBH_{ob}$ = age of a *E. regnans* judging from its DBH_{ob}, $a = 0.020(3)$, $b = 0.170(6)$, $c = 0.061(3)$	$R^2 = 0.98$, MSE = 0.002, $N = 10$
11	$f\_Er\_pool_{Age} =$ $f\_Er\_pool\_last\_year_{Age} \times e^{-f\_Er\_decay\_rate}$	$f\_Er\_pool_{Age}$ = mass of fallen, decaying *E. regnans* from a particular stand age, $f\_Er\_pool\_last\_year_{Age}$ = mass of the same pool last year, $f\_Er\_decay\_rate_{Age}$ (as above)	*
12	$u\_tree\_biomass =$ $Mass\_Max \times \left(1 - \dfrac{1}{1 + \left(\dfrac{DBH_{ob} + 0.01}{DBH_{ob}\_Mid}\right)^{k}}\right)$	$u\_tree\_biomass$ = above ground (dry) mass of one standing understorey tree, DBH_{ob} = diameter at 1.3 m of one understorey tree, $Mass\_Max = 45$, $DBH_{ob}\_Mid = 1.48$, $k = 2.8$	*
13	$ag\_u\_biomass =$ $Bio\_Max \times \left(1 - \dfrac{1}{1 + \left(\dfrac{Age}{Age\_Mid}\right)^{k}}\right)$	$ag\_u\_biomass$ = [dry] mass of standing understorey per hectare, Age = time since last major understorey fire, $Bio\_Max = 1600$, $Age\_Mid = 350$, $k = 5$	*

Numbers in parentheses after the last significant digit are the parameters' SE, for that last significant digit.
* The parameters for these formulae were derived as described in the corresponding sections in the text.

(DBH under bark) of 0.3 to 3.0 m. Data for larger trees were from Helms (1945) and A.M. Goodwin (Hobart, 2002, personal communication), $DBH_{ub} = 6.31$ m and 5.32 m, respectively. Additionally, we measured the taper of several trees by geometric correction (rectification) of photographs. The image rectification facility in ERDAS/IMAGINE, which included a camera model designed for use in remote sensing, was used for our ground-based photographs. The positions of various points on the tree were recorded using tapes, quadrats and a clinometer; these were the equivalent of 'geometric control points' in the remote-sensing context. The sides of the trunk were digitized on screen, as lines, over the rectified photographs. Horizontal lines, intersecting the sides of the trunk, were added at 0.5 m intervals up the trunk (Colour Plate 2). The widths of these lines were assumed to be diameters, and thereby yielded taper curves. Other measurements, e.g. branch diameters, can be read off the rectified imagery but only if they are in a plane perpendicular to the camera. The calculation of above-ground, woody volume from rectified photographs appears to be novel and will be detailed by the current authors in a separate publication. Other applications of measurement from ground-based, single photographs are given in Criminisi *et al.* (1999).

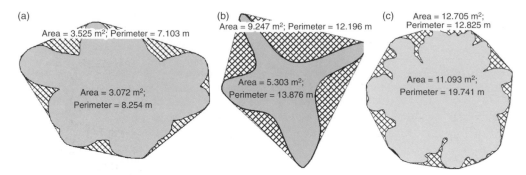

Fig. 3.1. Cross-sections showing perimeters corresponding to both diameter tape measurement (outer) and accounting for area deficit (inner). Areas were calculated for the perimeters using GIS software. Diameter, age and circular area (from diameter tape) are: (a) DBH = 2.29 m, ~321 years old, 4.02 m^2; (b) DBH = 3.86 m, ~321 years old, 11.84 m^2; (c) DBH = 4.08 m, ~400 years old, 13.09 m^2.

Buttress detail

The deviation of the trunks from circular cross-section reduces the cross-sectional area (compared with that of a circle, e.g. Fig. 3.1). The reduction is mostly due to the flutes between the spurs and to the outer perimeters not being circular. Little work has been published on cross-sectional area deficit in large eucalypts. We measured some cross-sections after they had been felled (from rectified photographs) and in some standing trees, both in Tasmania and in Victoria. The data processing and formula development representing the morphology of the buttress were complex and that work will be presented in a separate publication; a summary of the findings is mentioned here for completeness. The fractional area deficit varies with DBH_{ob} (at breast height of 1.3 m), as the tree ages. It had a peak of 0.6 (i.e. 40% less cross-sectional area than a circle of the same perimeter) at about DBH_{ob} = 3.5 m, within the range of DBH_{ob} from 0.6 to 11.5m.

The fractional area deficit due to buttressing and non-circular cross section also varies with height up the tree. It approaches zero higher up the trunk, but closer to the ground, as the buttress spurs spread apart, the area deficit is higher. Without a buttress, the trunk shape would be a conical section. It was assumed that the fractional increase in area between this cone and the buttress was equal to the fractional increase in the area deficit. Consequently, the area deficit was calculated as a function of height up the buttress and of DBH_{ob}. From the taper curves of Galbraith plus that of the Helms tree, equations were developed to describe: the taper of the buttress, slope of the theoretical cone, and buttress height. The theoretical buttress diameter was used in a relative manner, to calculate the relative cross-sectional area deficit, and directly, to calculate the diameter at ground level (e.g. Fig. 3.2).

Stem volume as a function of DBH$_{ob}$

While integrating the volume under the taper curves, the relative area deficit was multiplied by the empirical diameter to obtain the actual area deficit at various heights up the buttress. For trees not much taller than 1.3 m, the stem volume was calculated from the basal diameters and heights given by Ashton (1975a), assuming the trunks to be conical. A proxy DBH_{ob} was assigned to those smaller sizes. Stem volume was formulated as a function of DBH_{ob} (Equation 2 in Table 3.1 and Fig. 3.3).

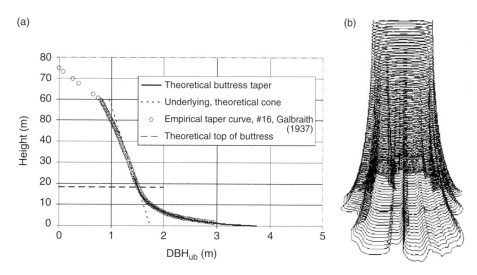

Fig. 3.2. (a) Example taper from Galbraith (1937), DBH$_{ub}$=3.03 m and theoretical components from our formulae. (b) Virtual reality modelling language (VRML) wireframe model created using the buttress diameter and fractional area deficit formulae.

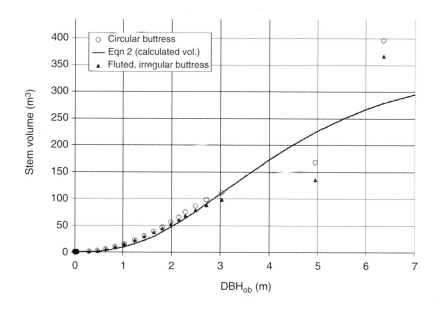

Fig. 3.3. Stem volume vs. DBH$_{ob}$ from the taper curves of Galbraith (1937), tree 495 (current work) and Helms (1945). Circular points were calculated ignoring area deficit; triangular (lower) points include deficit.

A conservative value was selected for *Vol_Max* (the maximum volume) – only 10 m^3 more than the 370 m^3 calculated for the tree of Helms. Values for bark thickness of 1.58 cm in the buttress zone and 0.45 cm higher up were derived from field data. The equation for bark volume (Equation 3 in Table 3.1) was the same type (a sigmoid) as for stem volume but with an additional constant added so that the bark volume for seedlings and thicket stage trees stayed close to zero. The convolution of the buttress was not modelled for bark and therefore the bark volumes are conservative.

Standing *E. regnans* biomass per hectare

Data on stocking rates without silvicultural thinning were collated from many literature sources plus the present field work to yield stocking rate as a function of stand age (Equation 4 in Table 3.1). From the graph in Galbraith (1937), we formulated *E. regnans* height as a function of DBH_{ob} (Equation 5 in Table 3.1). The stem biomass is the stem volume multiplied by the basic density, for which literature values ranged from 0.4 to 0.8 t/m^3 (tonnes per metre cubed). In the present work, samples from one oldgrowth Tasmanian buttress (300 ± 50 years) yielded a basic density of 0.596 t/m^3. We used an average of values from Ilic *et al.* (2000) of 0.5124 t/m^3 (low in order to yield a conservative mass). Density of younger trees is less than for mature trees so we formulated basic density as a function of tree age (Equation 6 in Table 3.1). From the present work, we found the majority of bark to have a basic density of ~0.495 t/m^3. Throughout our model the carbon in biomass and litter was approximated as 50% of (dry) biomass by weight.

The proportion of biomass allocated to branches, leaves and roots was estimated from the component proportions reported in Ashton (1975a, b, 1976), Feller (1980) and Grierson *et al.* (1991). For example, the formula for root biomass (Equation 7 in Table 3.1) is a function of above-ground biomass and age.

Examples of 400-year-old *E. regnans* with negligible pith decay are the tree of Helms (1945) and the log shown in Fig. 3.1c and Colour Plate 3c. Some other 400-year-old trees have substantial buttress or crown rot (Colour Plate 3a and b). This reduction in stem volume and consequent eventual decline in stocking rate was accounted for by a sigmoidal function (Equation 8 in Table 3.1). The parameter *Age_Mid* in Equation 8 can be reduced for stands in difficult environments or increased for stands in ideal conditions. This standing-tree fractional decay function was used as a multiplier of the live tree biomass (due to growth), to yield the biomass present for each tree (Equation 9 in Table 3.1).

Litter accumulation and decay

The current year's stocking rate minus the previous year's stocking rate was the annual tree fall, i.e. the majority of the annual coarse woody debris (CWD). The fallen trees were assigned to numerous pools with decay rates dependent on their DBH_{ob} (Equation 10 in Table 3.1). The (dry) mass remaining in each of the fallen tree pools in any particular year was therefore a function (Equation 11) of the mass in those pools in the previous year and the decay rate for the corresponding stand ages. This method of calculation of the remaining mass, by using a decay rate, follows that of Olson (1963) and is employed throughout our model. Some example half-lives are: 1-year-old fallen tree – 4 years, 50-year-old tree – 25 years, 350-year-old tree – 35 years. The accumulation of fallen bark and leaves was modelled from the data given by Polglase and Attiwill (1992) (multiplied by the fractional tree decay), and their decay was modelled on data from Ashton (1975b).

Each year 60% of the carbon in the rotted biomass was distributed to the atmosphere and 40% to the humus layer (Klein Goldewijk *et al.*, 1994). We used two soil pools (humus and slow soil carbon) with 45% of humus entering the slow pool and 55% entering the atmosphere. Humus had a half-life of 2 years and the slow pool a half-life of 693 years (from the turnover time of 1000 years in Grierson *et al.*, 1991). Soil was treated as being homogeneous with no division of carbon content between different depths.

Net biome productivity (NBP) was calculated as the per hectare carbon mass in all of the pools minus the per hectare carbon mass in those pools in the previous year. Carbon released to the atmosphere, by the decay processes (natural or anthropogenic) and fires, was subtracted from the running NBP total.

Understorey

Understorey DBH data were converted to biomass using the equations for temperate rainforest of Keith *et al.* (2000) with an adjustment for higher DBH values above 1 m. The adjustment was necessary as their allometrics were for DBH_{ob} <1 m and therefore did not reflect the more conical nature of wider trees. (The largest myrtle beech (*Nothofagus cunninghamii*) we encountered had DBH_{ob} = 2.6 m.) Ours is a simple sigmoidal function (Equation 12 in Table 3.1) and errs on the conservative side by giving a biomass of 37 t for a myrtle of DBH_{ob} = 2.6 m, compared with 104 t if using the formula of Keith *et al.* (2000). Understorey biomass per unit area is highest for the rainforest type but varies with location, natural fire history, disturbance and age of the *E. regnans* stand, to a short, scrubby understorey, as observed in one stand at 840 m in the Victorian Central Highlands. An average understorey was modelled, as a function of understorey age, using data from numerous literature sources and our fieldwork (Equation 13 in Table 3.1, where age refers to the time since the last major understorey fire). The curve was placed conservatively low at the high end because many of the larger myrtle trunks were substantially decayed, even though many had additional mass in the form of large burls. Other understorey tree species showed little trunk rot. Our formula was also conservative in the mid-range because, during least squares refinement, the parameter *Age_Mid* converged at 324, but we chose to use *Age_Mid* = 350, thereby reducing the calculated understorey biomass at age 300 years from 650 t/ha to 500 t/ha. Understorey root biomass was approximated to be 10% of the above-ground biomass. The understorey litter biomass was assumed to be zero in the first 2 years and thereafter modelled on data for twigs and leaves (Ashton, 1975b; Polglase and Attiwill, 1992). (Values for understorey CWD were unavailable.) The decay rate used was that for cool temperate rainforest: 0.409 (Turnbull and Madden, 1986).

Temporal and spatial scenarios

Stand age was limited to 450 years (beyond which we had insufficient information on forest succession and the understorey). CAR4D can handle environmental data in GIS format and thereby allows temporal and spatial analysis of carbon sequestration. It was run for the O'Shannessy catchment in the central highlands of Victoria.

Two aspatial (i.e. per hectare amounts, with temporal variation) scenarios were run. Before each scenario, ten cycles of 450 years were run (with a forest fire at 450 years) in order for carbon pools to approach dynamic equilibrium values. The first scenario was a sequence of stand replacement fires at 321 years. The second scenario was clearfell logging at 321 years (as in coupe SX004C) followed by repeated reforestation and harvesting for sawlogs every 80 years. Stands corresponding to these scenarios are shown in Colour Plate 4.

For the first scenario an estimated 20% of the *E. regnans* stem biomass, 50% of *E. regnans* branches, 100% of *E. regnans* leaves, 40% of the understorey biomass and 50% of decaying timber were emitted to the atmosphere during the stand replacement fire. The remaining biomass was committed to the decay pools.

The reduction in soil carbon used was 32% from the data in Polglase *et al.* (1994) after clearfelling of *E. regnans*. Clearfelling of wet forests such as *E. regnans* often results in substantial soil disturbance, with soil inversion down to 0.3 m common on snig tracks (log-dragging paths) and near landings (sorting, trimming, storage and loading zones). Stand replacement fires incur less soil disturbance: we assumed a drop in soil carbon of 10%. Typical values were used for mass extracted upon logging and for its partitioning between pulp and sawlog. For example, in oldgrowth logging 85% of the *E. regnans* stems were extracted and 2% of the understorey biomass (with the remainder left as debris and burnt). The fraction of removed stem pulped was 84% for oldgrowth logging and 55% for reforestation harvesting. After the hot burn, 50% of the debris remains (Stewart and Flinn, 1985) and no live trees remain. Typical values were used for half-lives of wood products (e.g. from Grierson *et al.*, 1991).

Results

The main trends following a stand replacement fire at 321 years were an initial increase in the carbon stored in standing *E. regnans*, until 215 years (553 t C/ha) (tonnes of carbon per hectare) followed by a decrease due to stand thinning and decay. The mass from self-thinning had a peak of ~16 t C/ha/year at 215 years. The remaining non-soil carbon from the fire (635 t C/ha) quickly decayed, causing a rise in the humus and slow soil pool. The soil pool loses about 70 t C/ha due to the fire then increases and levels off at 667 t C/ha at ~120 years, followed by a slight decline to 290 years, then a slight increase due to senescing *E. regnans* (Fig. 3.4). The growth of the undisturbed rainforest understorey compensates for the decay of the ageing *E. regnans* and the total carbon content of the forest levels off at 1500 t C/ha near 375 years.

The average changes (per cycle) in NBP for the stand replacement and logging scenarios were 0.80 and 2.52 t C/ha/year, respectively (with 5.90 t C/ha/year^2 after the oldgrowth logging event). Mean values of soil carbon and total carbon for the stand replacement fire cycle were 654 t C/ha and 1230 t C/ha, respectively. The logging scenario took longer to reach dynamic equilibrium (i.e. after about five cycles), after which its mean soil carbon and total carbon (including wood products) were 97 t C/ha and 387 t C/ha, respectively (Fig. 3.5).

Discussion and Conclusion

The modelled *E. regnans* stem carbon (Fig. 3.4b) falls between the values of Feller (1980) (for a 40-year-old stand) and Goodwin (1999) (generated using a site index of 45). The curve of Grierson *et al.* (1991) for *E. regnans* biomass is slightly higher than our curve but otherwise is similar. The data points of Feller are higher for each biomass pool, possibly because we chose to be conservative in our formulation and because his site was ideal for *E. regnans*. CAR4D illustrates that carbon sequestration

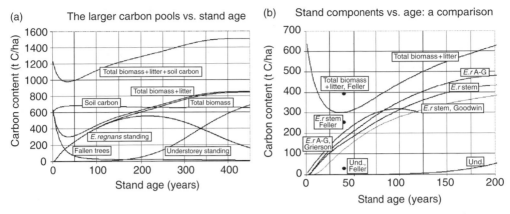

Fig. 3.4. Progress after a stand replacement fire at 321 years (without logging). Und., understorey biomass; A-G, above ground. In (b) data from: Feller (1980), Grierson *et al.* (1991) and Goodwin (1999).

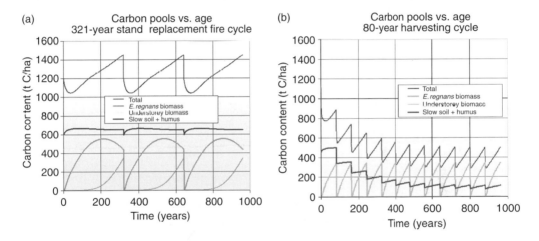

Fig. 3.5. Carbon pools. Starting from ten cycles of 450 years then a further 321 years of growth followed by (a) further stand replacement fires every 321 years (without logging); and (b) clearfelling for pulp followed by reforestation and harvesting for sawlog every 80 years. Total includes wood products.

increases beyond 100 years up to 200 years (for *E. regnans* alone), and at 400 years the total can approach ~1500 t C/ha, with an undisturbed rainforest type understorey.

Polglase *et al.* (1994) analysed soil carbon in two 53-year-old stands to about 1 m depth and found an average of about 646 t C/ha. It increased with *E. regnans* stand age, up to 60 years of age. The soil carbon forecast by CAR4D shows very similar behaviour. Grierson *et al.* (1991) modelled an 80-year reforestation and harvesting cycle at dynamic equilibrium but forecast slightly lower soil carbon than those calculated by CAR4D.

Carbon pools for the harvesting cycle were low because: (i) relatively frequent fires emit gaseous carbon; (ii) wood products were not returned to the soil; (iii) short half-life of the pulp component; (iv) high soil disturbance; (v) *E. regnans* were grown only to 63% of optimum size; and (vi) the understorey was underdeveloped. The sequestered carbon might be even lower if the following landscape scale effects were considered: escaped burns into forested areas, effect of logging roads on fire frequency, duration between felling and germination (1 to 2 years), and the diesel fuel used by the log skidders and loaders. These influences can be included in GIS data for landscape-scale analyses using CAR4D and visualized from animated output such as in Colour Plate 5.

Acknowledgements

For assistance with data acquisition we thank: Prof D.H. Ashton (Melbourne), Forestry Tasmania (Head and District Offices and field contractors), Parks Victoria (Woori Yallock and Apollo Bay), Dept NRE Victoria (Research Branch), and the CRC for Greenhouse Accounting for funding.

References

Australian Newsprint Mills (*c.* 1960) Unpublished coupe registers, Library of Forestry Tasmania, Hobart, Tasmania.

Ashton, D.H. (1975a) The root and shoot development of *Eucalyptus regnans* F. Muell. *Australian Journal of Botany* 23, 867–887.

Ashton, D.H. (1975b) Studies of litter in *Eucalyptus regnans* forests. *Australian Journal of Botany* 23, 413–433.

Ashton, D.H. (1976) The development of even-aged stands of *Eucalyptus regnans* F. Muell. in Central Victoria. *Australian Journal of Botany* 24, 397–414.

Bassett, O.D., Edwards, L.G. and Plumpton, B.S. (2000) A comparison of four-year-old regeneration following six silvicultural treatments in a wet eucalypt forest in southern Tasmania. *Tasforests* 12, 35–54.

Brown, B. (2001) *The Valley of the Giants: a Guide to Tasmania's Styx River Forests.* Bob Brown, Hobart, Tasmania, 66 pp.

Criminisi, A., Reid, I. and Zisserman, A. (1999) Single view metrology. In: *Proceedings of International Conference on Computer Vision,* 20–25 September 1999, Corfu, Greece. IEEE Computer Society Press, Los Alamitos, California, Vol. 1, pp. 434–441.

Feller, M.C. (1980) Biomass and nutrient distribution in two eucalypt forest ecosystems. *Australian Journal of Ecology* 5, 309–333.

Galbraith, A.V. (1937) *Mountain Ash* (Eucalyptus regnans *F. von Mueller): a General Treatise on its Silviculture, Management, and Utilization.* H.J. Green, Government Printer, Melbourne, 51 pp.

Galbraith, A.V. (1939) Proceedings of the Victorian Bush Fire Conference, August 1939. *Australian Forestry* 4, 79–81.

Goodwin, A. (1999) *Generalised Yield Curves for Single-aged Tasmanian Eucalypt Forest.* Forestry Tasmania, Hobart, Tasmania.

Grierson, P.F., Adams, M.A. and Attiwill, P.M. (1991) *Carbon Sequestration in Victoria's Forests* and *Carbon Storage in Soil and in Forest Products.* Reports commissioned by the State Electricity Commission of Victoria.

Helms, A.D. (1945) A giant eucalypt. *Australian Forestry* 9, 24–28.

Hickey, J.E., Su, W., Rowe, P., Brown, M.J. and Edwards, L. (1999) Fire history of the tall wet eucalypt forests of the Warra ecological research site, Tasmania. *Australian Forestry* 62(1), 66–71.

Hickey, J.E., Kostoglou, P. and Sargison, G.J. (2000) Tasmania's tallest trees. *Tasforests* 12, 105–121.

Ilic, J., Boland, D., McDonald, M., Downes, G. and Blakemore, P. (2000) *Woody Density Phase 1 – State of Knowledge.* Australian Greenhouse Office, National Carbon Accounting System Technical Report No. 18.

Keith, H., Barrett, D. and Keenan, R. (2000) *Review of Allometric Relationships for Estimating Woody Biomass for New South Wales, the Australian Capital Territory, Victoria, Tasmania and*

South Australia. Australian Greenhouse Office, National Carbon Accounting System Technical Report No. 5B.

Klein Goldewijk, K., van Minnen, J.G., Kreileman, G.J.J., Vloebeld, M. and Leemans, R. (1994) Simulating the carbon flux between the terrestrial environment and the atmosphere. In: Alcamo, J. (ed.) *IMAGE 2.0 Integrated Modelling of Global Climate Change.* Kluwer Academic, London, pp. 199–230.

Law, G. (1999) *Tasmania's Magnificent Tall Forests and the RFA.* Available at: www.greens.org.au/bobbrown/tall.htm

Mace, B. (1996) Mueller, champion of Victoria's giant trees. *The Victorian Naturalist* 113(4), 199–207.

Mooney, N. and Holdsworth, M. (1991) The effects of disturbance on nesting wedge-tailed eagles (*Aquila audax fleayi*) in Tasmania. *Tasforests* 3, 15–31.

Mount, A.B. (1964) Three studies in forest ecology. MSc thesis, University of Tasmania, Hobart, Australia.

Olson, J.S. (1963) Energy storage and the balance of producers and decomposers in ecological systems. *Ecology* 44, 322–331.

Patton, R.T. (1919) On the growth, treatment and structure of some common hardwoods. *Proceedings of the Royal Society of Tasmania* 31(II), 394–411.

Polglase, P.J. and Attiwill, P.M. (1992) Nitrogen and phosphorous cycling in relation to stand age of *Eucalyptus regnans* F. Muell. I. Return from plant to soil in litterfall. *Plant and Soil* 142, 157–166.

Polglase, P.J., Adams, M.A. and Attiwill, P.M. (1994) *Measurement and Modelling of Carbon Storage in a Chronosequence of Mountain Ash Forests: Implications for Regional and Global Carbon Budgets.* State Electricity Commission, Victoria, Australia.

Stewart, H.T.L. and Flinn, D.W. (1985) Nutrient losses from broadcast burning of eucalypt debris in north-east Victoria, *Australian Forest Research* 15, 321–332.

SYSTAT (1992) *SYSTAT for Windows: Statistics, Version 5 Edition.* SYSTAT, Evanston, Illinois, 750 pp.

Turnbull, C.R.A. and Madden, J.L. (1986) Litter accession, accumulation and decomposition in cool temperate forests of southern Tasmania. *Australian Forest Resources* 16, 145–153.

4

Spatial Distribution Modelling of Forest Attributes Coupling Remotely Sensed Imagery and GIS Techniques

Gherardo Chirici,[1] Piermaria Corona,[2] Marco Marchetti,[3] Fabio Maselli[4] and Lorenzo Bottai[5]

Abstract

Needs for accurate information about forest resources can only partly be met by conventional inventories based on ground sampling. Earth observation (EO) techniques are a valuable source of information for several forest attributes which are linked to relevant spectral responses (tree species composition, stand biomass, stand density, etc.). In particular, EO can be effective for propagating forest inventory plot sample values through the landscape: sample values can be assigned to non-sampled locations according to the similarity of certain spectral features among the sampled and the non-sampled plots. A spatial modelling based on the integration of remotely sensed images and sample field measurements targeted to produce forest attributes maps is presented for a site in central Italy with more than 300 geocoded sampling field plots. Plot data of tree stemwood volume and other non-wood forest attributes came from a single-stage cluster design with 58 primary sampling units (clusters). Landsat 7 ETM+ images are used with two classification techniques (k-NN and fuzzy classifiers) to model the spatial distribution of stemwood volume (m³/ha) and stem density (n/ha). Modelling and mapping results are discussed.

Introduction

Forest monitoring programmes are recognized as a main information support for terrestrial natural renewable resource modelling and management. Distinctively, the sustainable management of forest resources requires a large amount of supporting information. Traditional ground-sampling-based forest inventories provide summary statistics of forest attributes. However, given the usual sampling intensities, inventory information cannot directly support an estimation and visualization of the measured attributes (e.g. tree species composition, stand density, stand height,

[1] geoLAB, Dipartimento di Scienze e Tecnologie Ambientali Forestali, Università di Firenze, Italy
Correspondence to: gherardo.chirici@unifi.it
[2] Dipartimento di Scienze dell'Ambiente Forestale e delle sue Risorse, Università della Tuscia, Italy
[3] Dipartimento di Colture Arboree, Università di Palermo, Italy
[4] IATA, Consiglio Nazionale delle Ricerche, Italy
[5] LAMMA, Regione Toscana, Italy

stand volume, etc.) at the local level. A complete mapping of forest variables and associated characteristics would greatly benefit forest management planning. A complete mapping of forest attributes also provides a basic information source for many modelling tasks such as animal habitat suitability assessment, hydrological applications, and so on.

Earth observation (EO) techniques at various scales (Barbati *et al.*, 2000) play a distinctive role in the effective development of environmental mapping programmes. In the specific sector of natural resource surveys, it is important to evaluate change. Consequently, frequent coverage is essential, and satellite remote sensing can provide data which are cost-effective and not available from any other source. In particular, EO can be effective for propagating forest plot inventory values through the landscape; sampled values can be assigned to the remaining non-selected plot locations according to their spectral features (Franco-Lopez *et al.*, 2001). Non-parametric classification methods are intrinsically suited to this objective, since they allow a flexible treatment of the remotely sensed data without assuming pre-determined statistical distributions which are unrealistic in many forest environments (Richards, 1993; Fazakas *et al.*, 1999).

The present research aims at assessing the potential of known non-parametric classification methods to model the spatial distribution of forest attributes relying on Landsat 7 ETM+ data. In particular, the investigation was carried out in a forested area in central Italy for which a routine ground forest inventory had recently produced more than 300 sampling points grouped in 58 clusters. Inventory data were used for both calibrating and verifying the performances of two EO classification methods, based respectively on *k*-NN and fuzzy classifiers.

Experimental Test Site

The test site is located in the Municipality of Acquapendente, within the Mediterranean biogeographical region in central Italy. The site is about 8600 ha, and completely includes the Riserva Naturale di Monte Rufeno, a regional designated conservation area of about 3000 ha. The site has been chosen both for its representativeness of Mediterranean hilly conditions and for the availability of recent studies regarding several environmental aspects. The area is characterized by a fairly smooth morphology with slopes ranging around 30–35%.

The most common vegetation types are: forests (oak woods with *Quercus cerris* prevailing), shrubs (mantle communities and bushes), grassland, pasture and garigue, riparian shrub and tree vegetation (with alder, poplar and willow), agricultural fields and forest plantations (various species of fast-growing conifers). Forests within the Riserva Naturale di Monte Rufeno are in conversion to high forest with the application of selective thinning cuts. Most coppices outside the Riserva are still managed for the production of fuelwood, with a 15-year rotation. The most common silviculture system is coppice with standards. Some coppices are abandoned and revert to their natural evolution. Conifer plantations were established between 1950 and 1970, using various species: including *Pinus pinea*, *Pinus excelsa*, *Pinus nigra* and *Cupressus arizonica*. Some of the plantations have been thinned, whereas others still have a very high density. The Mediterranean macchia is well represented in the area.

For the surveyed forestland, the average forest coverage is about 65%, the average stand density is close to 3500 stems/ha, the average basal area is 18 m^2/ha and the average stand volume is 128 m^3/ha.

Data

Ground sampling data

Ground sampling was configured as a single-stage systematic cluster design. On the basis of the UTM geographical reference system, a kilometric sample grid was generated to locate plot clusters. Two different kind of clusters were displaced on the grid: 'dense' clusters, made of nine sample plots in a 3 × 3 configuration, and 'sparse' clusters, made of five sample plots in an L-shaped configuration. Each sample plot has a circular shape, with a 10-m radius. In dense clusters, sample plots are 25 m apart, while in sparse clusters they are 100 m apart (Fig. 4.1).

A first selection of forested clusters was carried out on the basis of a photo-interpretation of digital aerial orthophotographs from 1996. Photographs were rendered in an 8-bit greyscale with a spatial resolution of 1 m. After this selection, 31 cluster sampling points (i.e. the origin of each cluster) were classified as non-forested by photo-interpretation and excluded from further analysis. Only the remaining 58 clusters (38 sparse and 20 dense) were used for analyses (Fig. 4.2).

Satellite data

Satellite data consisted of a Landsat 7 ETM+ scene taken on 31st July 2000. The scene was orthocorrected using more than 50 ground control points and a 75-m resolution digital elevation model. The global positional accuracy was less than the nominal side length of a pixel. Subsequently the scene was superimposed on the ground sampling grid which reported the 370 sample plots. No topographic normalization was performed because the spectral effect of topographic irregularities was considered to be negligible in the gently rugged terrain.

Spectral signatures of all forest sample plots were extracted from ETM+ bands 3, 4 and 5 as recommended by Horler and Ahern (1986).

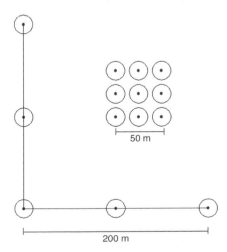

50 m

200 m

Fig. 4.1. Configuration of the sparse (bottom left) and dense (upper right) primary sampling units (clusters).

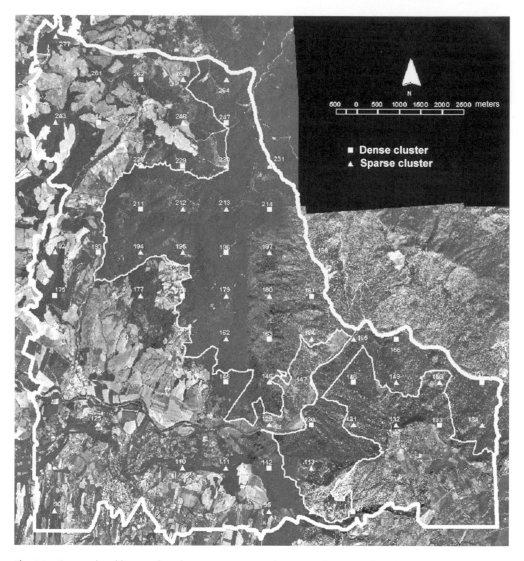

Fig. 4.2. Geographical layout of single-stage systematic cluster sampling carried out by the forest inventory (Acquapendente, central Italy). Cluster locations are displayed on the mosaic of digital orthophotographs: triangular dots show the locations with the sparse configuration of sample plots, while square dots locate the dense configuration of sample plots. Boundaries of the study site are shown by the wider white line, while boundaries of the Riserva Naturale Regionale di Monte Rufeno are indicated by the thinner white line.

Methods

Tested classifiers

The spectral signatures of all sample plots were used within two popular non-parametric classification techniques (Fazakas *et al.*, 1999; Katila and Tomppo, 2001; Maselli, 2001; Maselli *et al.*, 2001). Both classifiers are flexible, and allow a series of options which can adapt their application to the specific environmental situation and data used.

The first classifier is a k-nearest neighbour (k-NN) procedure, which has been used by Finnish and Swedish forest experts for updating their national inventories (Fazakas *et al.*, 1999; Katila and Tomppo, 2001). In the k-NN estimation procedure, the variables for a specified pixel are predicted as weighted averages of the most spectrally similar reference pixels (the k nearest neighbours). Several options are allowed regarding the form of spectral distance computed and the thresholding of spatial and altitudinal distances (Franco-Lopez *et al.*, 2001; Katila and Tomppo, 2001). Our version of k-NN used the spectral Euclidean distance as the discriminating feature, with the number k determined in each single case (see below). Due to the limited size and topographic variability of the study area, no horizontal or vertical thresholding was applied.

The second classifier is a simplified version of the fuzzy discrimination procedure recently proposed by Maselli (2001). This uses the same Euclidean distance as before to compute for each estimation point spectral weights with respect to all reference pixels. These spectral weights W_s are defined as:

$$W_s = \exp^{(-E_d/R_d)} \tag{1}$$

where E_d is the spectral Euclidean distance between the estimation and measured points, and R_d is a constant (range) which regulates the weighting of the spectral distances. A very large R_d effectively assigns a weight of 1 to all observations, while a small R_d assigns a weight that decreases rapidly with the spectral distance. The main difference between the two estimation methods is the inclusion of all reference points in the fuzzy classifier (even though a spectral threshold can also be applied). As can easily be understood, this approach is conceptually similar to kriging, which uses estimates of the covariance between the location to be estimated and all sampled locations to compute a prediction that is a weighted averages of all reference points for each point examined (Davis, 1973). Similarly to kriging, the determination of the spectral range is a crucial step in the procedure, which was currently carried out for each parameter as described below.

Forest attributes and classifiers configuration

The modelling experimentation focused on the identification of the best configuration of the two classifiers for tree stemwood volume (m^3/ha) and tree stem density (n/ha). The fine-tuning of the classifers was done by a *leave-one-out* strategy (Franco-Lopez *et al.*, 2001), which allowed the consideration of almost all reference pixels: it must be recalled that both classifiers, being intrinsically non-parametric, are very sensitive to the representativeness of the training sample, which must therefore be as large as possible for a correct accuracy assessment. Application of the *leave-one-out* strategy allowed the identification of the k and R_d values that minimize the average prediction error for a single location.

Results

k-NN classifier

The results of the *leave-one-out* application of the k-NN classifier are summarized in Fig. 4.3. As can be seen, attractive configurations of the classifier are found with

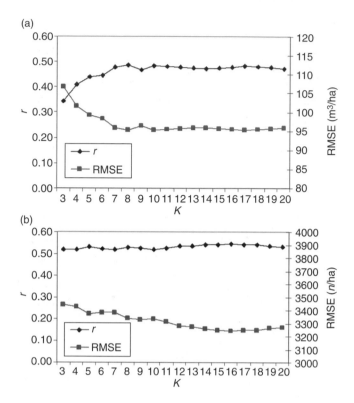

Fig. 4.3. Accuracy statistics (*r* and RMSE) obtained by the *leave-one-out* application of the *k*-NN classifier for the estimation of stemwood volume (a) and stem density (b).

medium numbers of points considered; specifically 13 for stemwood volume and 18 for stem density. These numbers are of the same order as those reported in the litera-ture from the Finnish and Swedish investigations (Fazakas *et al.*, 1999; Katila and Tomppo, 2001). The prediction accuracy achieved with the final parameter choice is actually not high, as testified by a correlation coefficient < 0.5. Thanks to the high number of sample plots, however, these correlations are statistically significant for both parameters ($P < 0.01$). Similar indications are provided by the root mean square errors (RMSEs), which are actually medium-high with respect to the observed vari-ability of the two forest attributes.

Fuzzy classifier

As regards the fuzzy classifier, its best configurations were found with very small values of R_d (2 and 3 digital counts for stemwood volume and stem density, respec-tively; Fig. 4.4). This indicates that the most influential plots are spectrally very close to each estimation point. The best prediction accuracy (correlation and RMSE) achieved with the fuzzy classifier was very similar to what could be achieved with the *k*-NN. As before, the highest achieved correlation was statistically significant for both study attributes ($P < 0.01$).

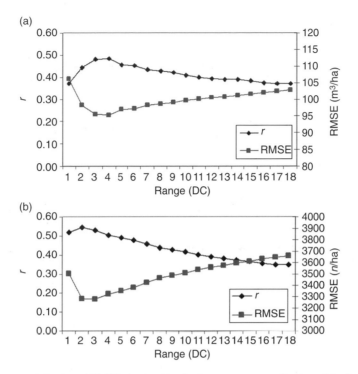

Fig. 4.4. Accuracy statistics (*r* and RMSE) obtained by the *leave-one-out* application of the fuzzy classifier for the estimation of stemwood volume (a) and stem density (b).

Spatial predictions of stemwood volume

Spatial predictions of stemwood volume for the forested locations not included in the inventory sample are shown in Fig. 4.5. In these maps, all non-forest pixels are masked out by the application of a threshold to the band subset considered (ETM+ bands 3, 4 and 5). The two stemwood volume maps are quite similar, and differences in stemwood volume estimates are hardly visible.

Discussion and Conclusions

The current investigation presented the results obtained by applying two non-parametric classification procedures to Landsat 7 ETM+ data for modelling the spatial distribution of two major forest attributes; stemwood volume and tree stem density. The tested modelling techniques are 'transparent' and easy to understand, while allowing flexibility in the incorporation of ancillary information. In addition, a number of forest variables can be estimated at the same time. Both techniques can easily be integrated into existing forest monitoring programmes, with the added potential for combining different sources of information not only from outside a region of interest, but even from different forest inventory designs.

The similar results obtained by the two tested non-parametric classifiers are in accordance with their substantially similar nature: both of them use Euclidean distances to compute spectral weights for the final parameter estimation. Some differences are, however, evident in the trends of the classifiers on the *k* numbers and

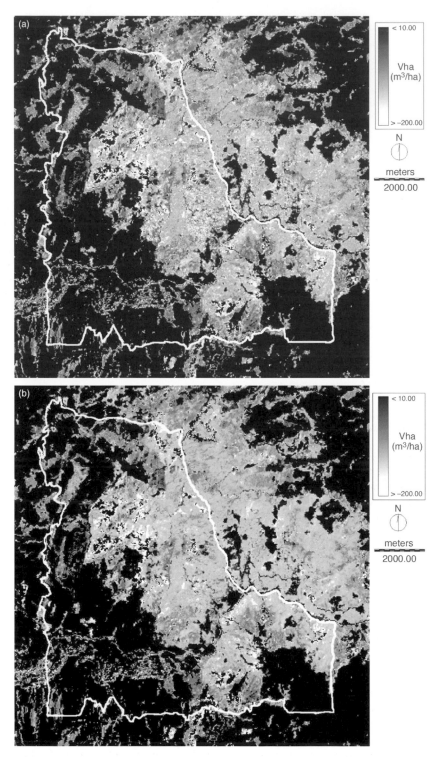

Fig. 4.5. Maps of stemwood volume obtained by the *k*-NN (a) and the fuzzy (b) classifiers (non-forest pixels are masked out). The white line delineates the study site.

ranges considered. While the sensitivity of the k-NN to the number of points considered is rather low, the fuzzy classifier presents a very clear accuracy maximum in correspondence to very small R_d values (2–4 digital counts).

The quality of the estimation by the two classifiers relies on the correlation among the variables and the image features. The key to success is therefore having spectral data correlated to the examined forest attributes and enough ground samples to cover all relevant variations in the study area (Franco-Lopez *et al.*, 2001; Maselli *et al.*, 2001). Similar procedures are being tested in northern Europe and the USA, where the spatial variability in forest composition and structure are frequently lower. The results obtained from this experimentation are therefore significant in showing that non-parametric spatial smoothers may produce useful per-pixel estimates of forest attributes even in complex Mediterranean environments.

In this regard, the statistical significance of the correlations found is a first indication that the ETM+ data considered are actually informative on the two forest attributes and that the two non-parametric classifiers are capable of extracting at least part of this information. It must, however, be remembered that the accuracies achieved were not high. This can be explained, in part, by considering the complexity of the study environments – especially the fragmentation and heterogeneity of local forest coverage – as well as the difficulty in estimating the forest attributes considered, which are probably only indirectly linked to relevant spectral responses. Furthermore, when assessing the low correlation coefficients, it should be kept in mind that they should not be compared with those of ideal cases but with correlations that could be expected from conventional methods (i.e. estimating stand volume by photo-interpretation). In this respect, even the relatively low accuracies currently achieved can improve operational forest monitoring and management in complex Mediterranean areas.

From a methodological viewpoint, it can be expected that notable improvements could be obtained by further investigations using both the current and other data sets. First, only a subset of TM bands was used for this analysis, and some improvement could be expected by the utilization of all ETM+ bands, and, even more, of multi-temporal satellite acquisitions. Different spectral distances, such as that of Mahalanobis, which accounts for the intercorrelations of the satellite sensor bands but not for the spatial intra-cluster correlations, should also be tested. Modified forms of such distances could be defined for each considered attribute in order to give more importance to the most informative bands (Maselli, 2001).

Another option to consider is the effect of differential thresholding and weighting. These factors could assume a particular relevance in relation to the spatial structure of ground sampling design. The accuracy assessment based on the *leave-one-out* strategy is in fact sensitive to the presence of autocorrelated ground attributes when some reference points are spatially close, as in cluster sampling. In these cases the presence of spectrally similar plots within the same cluster leads to the preferential inclusion of plots within the same cluster. This fact could partly explain the previously mentioned different sensitivities of the two classifiers to changes in the k and R_d values. The implications of both these behaviours must be quantified in order to progress further with these spatialization methods.

Acknowledgements

The work, carried out with equal contributions by the authors, was partially supported by the Italian Ministero delle Politiche Agricole e Forestali – RISELVITALIA funds. We are grateful to two anonymous reviewers for their helpful comments.

References

Barbati, A., Corona, P., De Natale, F., Marchetti, M. and Tosi, V. (2000) Forest remote sensing in Italy in the framework of FRA2000. In: Zawila-Niedzwiecki, T. and Brach, M. (eds) *Remote Sensing and Forest Monitoring, EUR 19530*. Office for Official Publications of the European Communities, Luxemburg, pp. 284–299.

Davis, J.C. (1973) *Statistics and Data Analysis in Geology*. Wiley, New York.

Fazakas, Z., Nilsson, M. and Olsson, H. (1999) Regional forest biomass and wood volume estimation using satellite data and ancillary data. *Agricultural and Forest Meteorology* 98–99, 417–425.

Franco-Lopez, H., Ek, A.R. and Bauer, M.E. (2001) Estimation and mapping of forest stand density, volume, and cover type using the *k*-nearest neighbours method. *Remote Sensing of Environment* 77, 251–274.

Horler, D.N.H. and Ahern, F.J. (1986) Forestry information content of Thematic Mapper data. *International Journal of Remote Sensing* 8, 1785–1796.

Katila, M. and Tomppo, E. (2001) Selecting estimation parameters for the Finnish multisource National Forest Inventory. *Remote Sensing of Environment* 76, 16–32.

Maselli, F. (2001) Extension of environmental parameters over the land surface by improved fuzzy classification of remotely sensed data. *International Journal of Remote Sensing* 22, 3597–3610.

Maselli, F., Bonora, L. and Battista, P. (2001) Integration of spatial analysis and fuzzy classification for the estimation of forest parameters in Mediterranean areas. *Remote Sensing Reviews* 20, 71–88.

Richards, J.A. (1993) *Remote Sensing Digital Image Analysis: an Introduction*, 2nd edn. Springer-Verlag, Heidelberg, Germany.

5

Algorithmic and Interactive Approaches to Stand Growth Modelling

Michael Hauhs,[1] Falk-Juri Knauft[1] and Holger Lange[1]

Abstract

The rationale for stand growth modelling is often grounded either in a search for improved scientific understanding or in support for management decisions. The ultimate goal under the first task is seen in mechanistic models, i.e. models that represent the stand structure realistically and predict future growth as a function of the current status of the stand. Such mechanistic models tend to be over-parameterized with respect to the data actually available for a given stand. Calibration of these models may lead to non-unique representations and unreliable predictions. Empirical models, the second major line of growth modelling, typically match available data sets as well as process-based models do. They have fewer degrees of freedom, and thus mitigate the problem of non-unique calibration results, but they often employ parameters without physiological or physical meaning. That is why empirical models cannot be extrapolated beyond the existing conditions of observations. Here we argue that this widespread dilemma can be overcome by using interactive models as an alternative approach to mechanistic (algorithmic) models. Interactive models can be used at two levels: (i) the interactions among trees of a species or ecosystem and (ii) the interactions between forest management and a stand structure, for instance in thinning trials. In such a model, data from a range of sources (scientific, administrative, empirical) can be incorporated into consistent growth reconstructions. Interactive selection among such growth reconstructions may be theoretically more powerful than algorithmic automatic selection. We suggest a modelling approach in which this theoretical conjecture can be put to a practical test. To this end, growth models need to be equipped with interactive visualization interfaces in order to be utilized as *input* devices for silvicultural expertise. Interactive models will not affect the difficulties of predicting forest growth, but may be at their best in documenting and disseminating silvicultural competence in forestry.

Introduction

Modern information technology (IT) has made modelling of many real world systems possible that were not long ago thought to be too complicated for the application of formal methods of prediction. Distributed workstations and PC-based

[1] BITÖK, University of Bayreuth, Germany
Correspondence to: michael.hauhs@bitoek.uni-bayreuth.de

computing extended modelling also to managed ecosystems, including the assessment of forest growth. Yet the predictive power of growth models has not lived up to the often high expectations derived from other applications (Vanclay and Skovsgaard, 1997). Forest growth modelling is one example that may even be typical of the general difficulty in linking and integrating research and management of (forest) ecosystems. In an exhaustive poll among German foresters in public service, science and research came last when they were asked to rank external factors potentially significant for sustainable forest management (Schanz, 1995). Scepticism of managers towards models seems to be related to the generally low reputation of formal methods and reasoning in silviculture practice.

Modelling is not only a method to bridge this gap between science and management, it can also help to analyse its causes. Are growth models under-appreciated by practitioners of silviculture or do models suffer from conceptual and technical limitations that render them unsuitable for practical management? In any modelling project, several aspects of the posed problem need to be reconciled: conceptual, mathematical, engineering and ecological aspects. The sceptical attitude of German foresters towards new growth models is the background for the question: Is there anything that makes forestry qualitatively distinct from other application areas of modern IT? In these other fields, weather prediction is such a case; computer models have had a large and lasting impact on the routines of managers and the reliability of services provided to customers. Historically, weather prediction mainly rested on an empirical (and local) basis. Predictions were derived from past experiences, rather than from the solution of well-understood equations describing non-linear atmospheric transport processes. These days, such equations can be fed with sufficiently accurate, actual data and solved computationally such that the corresponding predictions now outcompete the crude empirical models of the past in most cases. Limits in the time horizon of weather predictability are understood as an inevitable feature of a complex dynamic system. In forestry, however, the yield table, a form of 'tabulated experience', still dominates prediction under operational conditions. Growth equations have a quite different character from transport equations; they are heuristic generalizations rather than applications of an underlying theory. Practical management seems to work, yet often remains based on empirical predictions that are sometimes outdated by the current context of forestry. Why is it so hard for rigorous objective observations in forest ecosystems to provide knowledge that is capable of substituting for empirical expertise? Here it is suggested that the answer is interaction. Unlike the links to the weather system, any ecosystem management includes *interactive* relationships at the scales relevant for managers. The following discussion will show how it is technically possible to reflect interactive relationships in modern software technology, and an application of a prototype of such a model will be sketched out.

We use a specific task typical of German forestry, in order to analyse and discuss the challenges and opportunities that forest growth modelling offers as an application area of IT. The example is a series of thinning operations that typically occur over the extended rotation periods for managed tree species of central Europe. For Norway spruce, the economically most important species, the total timber yield from such thinnings is about the same as the final harvest. Expertise in selective thinning, the ability to regulate the growth competition among individual trees or small groups of trees, can be regarded as a paradigmatic example of practical silvicultural knowledge and expertise. Of course, the economic and ecological rationale on which thinning decisions were based will have changed over the last rotation period (80–120 years) due to the changing social, economic and environmental boundary conditions of forestry. At the same time the regional, and sometimes even local, experience about growth responses to thinning has accumulated within forestry. We will focus the following discussion on this knowledge about the perceived technical choices that form the basis of each thinning from the perspective of a forester. What are the silvicultural options, for example, when the task at hand is the thinning of a 60-year-old spruce stand in southern Bavaria? In order to answer this question, modelling forest growth has two options.

- Use the most recent and complete pool of relevant management experience, e.g. the most appropriate yield table, local expertise, etc. (typically this approach is currently used and organized within forestry); and/or
- use the best process-level understanding that is relevant to describe tree growth, e.g. from plant ecology, biogeochemistry of the nutrient turnover, etc. (typically applied for regions or future scenarios where expertise is scarce).

From a modelling perspective, it has been customary to separate the dynamics of forest growth from the occasional evaluation and (thinning) interference imposed by the forester. The former is represented as an autonomous set of algorithms that can be run on a computer, the second can either be represented as an extra (goal) function acting upon the growth simulation or by a real forester acting through an appropriate (visual) interface (see Knauft, Chapter 30, this volume). In order to be useful in an economic (or ecological) evaluation of thinning operations, the human decisions are provided to the growth model through a visual interface in the form of an *external actor*.

Interaction in Information Science

Among the various aspects of IT, the vastly increased possibility of interacting with computer programs is one of most fascinating and relevant issues for the (cultural) impact and theoretical foundations of IT (Wegner, 1997; Stein, 1999; Wegner and Goldin, 2003). Based on our hypothesis that silviculture is an interactive enterprise at its core, we regard forest growth modelling as particularly well suited to exploiting the interactive character of today's IT techniques. The technical aspects of developing interactive growth models through visualization is the topic of a companion study (Knauft, Chapter 30, this volume); the example of applying an interactive growth simulator to data from a thinning trial in Bavaria is presented elsewhere (Hauhs *et al.*, 2001). An interactive version of this reconstruction will be used as a demonstration of the concept and is available from our website (www.bitoek. uni-bayreuth.de/mod/html/webapps).

Interaction is a widely used (technical) term in scientific writing. A number of systems from basic sciences (interacting elementary particles), engineering (interactive software) to sociology ('true' interaction among humans) might be characterized as interactive. In order to discuss links between the notions of interaction useful for growth modelling and used in information science, we use a definition that is based on the relationship between just two actors, i.e. systems that have choices and make decisions. In these sequential interactions, two decision-making systems are brought into mutual and sequential dependence for their respective choices, such as in a two-person game. We use choice as a purely phenomenological description attached to systems for their external behaviour. It does not relate to any internal feature or organization inside the system displaying choice in a given situation (see the chess example below, where a computer actually makes choices about the next move). Choice is hence attributed to a system, whereas interaction is an attribute of the relationship between two (or several) systems.

In computer networks, more powerful (non-sequential) forms of interaction have been implemented, but there are few theoretical tools available yet to study them. However, in information science the sequential form of interaction is already regarded as more powerful than algorithmic computing. 'Interactive computing can be broken into algorithmic steps, but viewing it as nothing more than that would "miss the forest for the trees"' (Goldin *et al.*, 2001).

A further difference is related to the perspective under which choices or interaction can be further classified: a decision or choice for a system can be assessed by an external observer of its behaviour who may want to analyse the choice-making system for internal mechanisms, i.e. finding out whether the choice is 'real' or virtual. For example, in the case of a chess computer, an observer might be informed about the implemented algorithm and the conditions upon which it stops and decides about the next move. However, with the quality of today's chess computer, there is no way to discover this just by playing against such an algorithm. Knowledge about the internal organization of such an actor comes from the engineers who build chess computers (Levy and Newborn, 1991). As long as we do not have the engineers who build living systems from non-living ones, any question about the nature of the choices displayed by biota remains open to speculation. It is sufficient to note that all behaviour of living entities is currently history-dependent and its memory is not fully accessible to an external observer. However, the above definition of interaction is phenomenological, including, for example, interaction with a telephone-answering machine. That is why we do not need a more sophisticated discussion about the 'true nature' of the various types of choices occurring within the systems that interact.

The notion of interaction as introduced here is actor-dependent and needs to be posed in the context of the computational and memory resources of the respective actor. This implies that asking whether any external relationship between two actors is interactive or not has to be asked and answered separately for each of them. Results of this classification will depend on the actor's respective computational capacity, as an example from chess will show. One could imagine that a powerful computer or player discovers a (conditional) winning strategy at a given point in the game and hence renders the remaining moves a simple choice problem whose outcome will be fixed. To this actor, the remaining sequence of choices is no longer fully interactive and will in fact no longer have the character of a game. An opponent who has yet to discover the full picture still regards the same situation as truly interactive. Mathematically, chess is interesting as a fully deterministic game with a finite number of simple rules, yet a game for which no winning strategy is known and may not even exist. From a forestry perspective, a thinning trial can be abstracted as a sequential game with predefined rules (at least for the forester) and the yield table as an archive or crude look-up table of played out games.

Interaction in Biology

There have been different proposals in information science about how to formalize (sequential) interaction (van Leeuwen and Wiedermann, 2001; Wegner and Goldin, 2003). For the purpose of this discussion it is sufficient to note that any capacity to engage in an interaction among programs requires persistence of data (states) between the individual decisions provided by the actor. Any of these individual decisions may in itself be algorithmic, as in the example of a chess computer. The memory on which an algorithmic decision is based is usually not available to the partner in a sequential interaction such as a game, and that is why choice and decision making seem to occur for an external observer. Different types of interactions can be distinguished by the form of persistent data, i.e. the type of memory utilized by the respective actor engaged in interaction. Here we need two versions of interaction, human interaction (i.e. interaction based on cultural forms of memory, such as yield tables or growth equations) and biological interaction with a genetically encoded memory.

Biological interaction describes the ability of organisms to engage in a mutually dependent relationship that has the character of a temporal chain of choices (in our phenomenological sense). In these mutually dependent decisions, the future choices for one organism are partly constrained by the past choices of the others and its own decisions, similar to the character of a game. In ecosystems this becomes a multi-player game in most cases. The possibility of biotic choices is an expression of phenotypic plasticity, i.e. the property of a given genotype to produce different phenotypes in response to distinct environmental conditions (Pigliucci, 2001). The strategic aspects of height growth of trees, for example, can be regarded as a behavioural phenotype evolved from past interactions with competing species. A forester has to deduce this growth potential of a specific stand from the actualized growth that can only resemble a subset of the potential behaviour.

The key assumption for the proposed interactive approach to growth modelling is to regard biological interaction as 'irreducible', i.e. at least for pragmatic reasons we will not seek to reduce it to an algorithm, or engage in discussions about the chances of doing so (Hauhs and Lange, 2003). It is important to note that the special character of interaction does not derive from internal 'mechanisms' within the considered actors, but rather derives from the richness and responsiveness of their environment. Given the computational and memory capacity of the actors involved, no algorithmic prediction can be given for the environment under which forest growth occurs.

Interaction in Silviculture

Modern forestry has to provide and manage a number of services, not just the goal of a maximum (or economically optimal) rate of fibre production. This starting point resembles the observation that 'at the heart of the new computing paradigm is the notion that a system's job is not to transform a single *static* input to an output, but rather to *provide an ongoing service*' (Goldin *et al.*, 2001, emphasis in original). Silviculture is a repetitive interference with the growth processes under a specific goal and in the face of several technical and biological options that act to influence forest growth. However, in any particular stand, the silvicultural options with respect to thinnings are not known at the time of planting or stand regeneration; they have to be inferred and updated from actual growth. This reflects the history dependence and unique character of each ecosystem (Lange, 1999). A common principle for any specific thinning operation is therefore to make only those decisions that are regarded as necessary at the current stage of growth and leave a maximum number of potential choices for the next visit, e.g. after 10–20 years. In this respect, thinning trials do resemble the chess example. Only the goals and the rules are known at the beginning, decisions can only be made by assessing each actualized situation. In information science, such a system of data supply, where each element in a *stream* of data, which only becomes available after the preceding one has been processed, is termed 'lazy evaluation'. This distinguishes streams from strings, or sequences that define the known and finite amount of input to ordinary algorithmic computation. 'Lazy evaluation' describes the attitude of practical silviculture towards stand growth quite well (Leibundgut, 1978), even including some of the prejudices acquired in the scientific study of forest growth.

Currently the empirical models of forest growth seem to match the needs of such a 'lazy evaluation' task much better than process-based models. We suggest that testing of an interactive approach in forestry may even become a paradigmatic case in ecosystem modelling, mainly because the empirical basis for management decisions is so well documented and provides stark contrast to the (surprisingly) weak performance of process-based models in this application area. Process-based models typically leave few options for adjusting predictions as new data about the actual growth responses become available. The preference of silvicultural practitioners for empirical over process-based models may thus be rooted in the distinction that only the former can currently be included easily into interactive management schemes, while the latter conceptually enforces an attitude towards the forest similar to that towards the weather and weather predictions.

The example of chess games (interaction based on cultural memory)[2] demonstrates the potential roles of theory and modelling in situations where interactive computing (the game as it is played out) can be distinguished from algorithmic computing, i.e. with each single decision leading to a next move. This corresponds to the implicit and ultimate promise of many process-level models for forest growth. In the case when a realistic state presentation of the stand growth dynamics process can be found, predictions become feasible and their precision would be limited only through uncertainties about initial or future boundary conditions (Lange, 1999). The empirical traditions and heuristics of practical decisions so prominent in silviculture would be outcompeted and forestry as an 'art' or 'game' would cease to exist in a similar manner to which chess would cease to exist with the knowledge of a practically relevant winning strategy.

What, however, would happen, if our universe is too small, too short lived, or we are too impatient to wait for a winning strategy in chess to be discovered if it exists; or if our forests are too complicated and the biotic interaction among trees too history-dependent for a realistic description by a process model to be invented/discovered in science? In this case, the only way to become a good chess player is to resort to experienced players and to study played-out games. A master player's expertise seems largely to rest on fast access to an archive of played-out games in his/her brain (Ognjen *et al.*, 2001). When this analogy is taken back to forestry, the only way to deal with novel goals and novel boundary conditions for forest ecosystems is to update and improve the practical expertise gained from interaction and to study successfully managed forests in the past and present. Hence the potential need and impact for interactive models would be large for forestry in central Europe, where a range of new boundary conditions devaluates the heuristic basis of silviculture and a large but hitherto poorly documented archive of played-out games exists only within experts' memories. A crucial difference from the chess example, however, is that in silviculture the rules or even the goals of the game are changed 'on the fly' as hidden behavioural phenotypes of trees may be evoked by the new boundary conditions (e.g. increasing atmospheric CO_2 levels) or new priorities, such as biodiversity, appear on the political agenda.

The problem of silvicultural choice does not reduce to the question of whether and when somebody acts properly as a scientist or as a forest manager, but the issue

---

[2] This is cultural interaction even though sometimes played by machines. Hence the internal configurations or mechanisms do not matter for the type of interaction. Its classification focuses on behaviour only. This example is also discussed by Dreyfus and Dreyfus (1988) and Flyvbjerg (2002).

is whether interaction can be abstracted out of a given problem in forest management or in growth modelling. If we take thinning schemes as an example, the task is the reconstruction of a specific growth process in which biological and human interaction *did take place*. As interaction is defined through the external relationship of a process, the focus in modelling shifts from the search for a realistic representation of the forest stand itself to a useful representation of an external interface. In particular the role of visualization is altered to an input interface (Hauhs *et al.*, 1999, 2001). As an illustrative example we use flight simulators.[3] Flight simulators are not intended to represent a plane as such, only its behaviour during its operation as it appears to a pilot. It is another example where software engineering is providing an ongoing service rather than transforming static input to output.

In forestry this concept takes on the character of an inverse flight simulator or the task of inverse engineering of such a program. Can we design the behavioural interface to a simulated forest in such a way that an interactive visualization will be accepted by experts in silviculture? Such an approach consists of two steps: first, an interactive growth simulator needs to be set up on the basis of the available observations. This first step does not differ from a conventional calibration exercise. Its result is typically non-unique. The reason is that (for all practical purposes) the set of observations is not large enough to uniquely identify all the parameters of a process-level model. Therefore the resulting calibrated models may still contain many *ad hoc* decisions that will lead to unreliable predictions. The calibration exercise of this first step defines the initial state for the subsequent interaction process. In our example, it simulated a given forest stand up to the age of the first thinning.

The second step consists of an interactive selection of that model behaviour that best resembles the (subjective) memories by which an experienced forester describes the responses of a thinned stand to his interference. The advantage of this coupling of two steps is that interactive selection appears theoretically more powerful than (non-interactive) algorithmic selection of data (Goldin and Keil, 2001). The disadvantage is that the reference point for successful modelling is shifted to the pool of expert knowledge in the respective realm. Without such experts, who remember and are able to agree upon proper growth behaviour and set a standard for reconstructions, the interactive model would be of no help, i.e. prediction would remain as elusive as before. However, if there is a group of experts that has something to say about a silvicultural problem, the model may greatly support and facilitate communication within this group of experts. To return to the flight simulator metaphor, if good pilots approve of a simulator, it can be used to pass their expertise on to novices and ultimately to serve as a communication tool among the experts themselves.

An Interactive Model Applied to a Thinning Trial at Denklingen (Bavaria)

We illustrate this approach with a well-documented stand from Denklingen (Bavaria). The Denklingen Norway spruce (*Picea abies* Karst.) thinning trial was situated on a glacial moraine in southern Bavaria. It is one of several trials started in the 1880s. At the time of its harvest in 1991 (due to stormfelling), it had become one of the oldest Norway spruce thinning trials and had accumulated the longest record from such trials in Bavaria (H. Pretzsch, personal comunication). It consisted of three treatments, all of which received a periodic low-thinning regime, although with varying intensity. In the

---

[3] J.K. Vanclay has already proposed this analogy before, as was pointed out to us by Paul van Gardingen.

A-grade condition, only those trees were removed that were expected to die within the next growth period. B-grade consisted of moderate and C-grade of heavy thinning (late 19th century definitions).

As a model we use the individual tree simulator TRAGIC++ described elsewhere[4] (Hauhs *et al.*, 1995). It allows us to specify a local and variable nutrient supply for one (growth-limiting) nutrient for which roots compete and the photosynthetic capacity of leaves competing for light interception. An important feature of this model is that any stand has to be started from seeds, plants or natural regeneration for which no or a narrow distribution of initial heights is specified. Here we started with identical trees representing the 1.4 × 1.4 m plantation from 1848. The simulated distributions in height, biomass, diameter, etc. are consequences of competition between individual trees (algorithmic part) and, in addition (interactive part), the external interferences imposed by the thinning regimes (Hauhs *et al.*, 2001).

The Denklingen A-grade data set comes closest to the concept of an autonomous growth process not impacted by management. That is why we used it here for the first of the two steps outlined above. From the plantation of identical trees the model is able to match observed distributions of height and diameter measurements for 108 years between the ages of 35 (when measurements started) and 143 (the last recording before the stand was felled by a storm). We found several parameterizations of this model in which the number of trees and the averages for height diameter and basal area are relatively close to observations. It is difficult for such a multi-dimensional calibration exercise to define a criterion for the best fit (Fig. 5.1). In addition, a highly parameterized model such as TRAGIC++ gives typically non-unique results in a calibration exercise.

This is the point where interactive selection provides a further and refined testing possibility for the several, equally satisfactory, calibrations of a process-based model. Any calibrated reconstruction of the A-grade can be taken as the non-interactive growth part behind the responses observed in the thinned variants B- and C-grades. An experienced forester is asked to select the trees in each thinning event based on the visualization of the simulated reconstruction (Fig. 5.2). In addition, one might provide the forester with the explicit list of felled trees with diameter and height, or alternatively just with the total basal area of the trees taken out in each thinning (a much more realistic constraint in terms of operational forestry).

The forester is confronted with this image and asked whether the presented forest is acceptable as a plausible thinning task and, if yes, the thinning can be imposed interactively at the tree level through the visual interface. After completing this thinning the resulting stand structure is passed back to the autonomous growth part of the model (TRAGIC++). That is how the algorithmic part of the simulation is extended, say for the next 10–20 years. The forester is then asked again whether he or she accepts the resulting visualized stand as a plausible result of the preceding actions and, if yes, continues with the thinning exercise. Whenever the model response is judged as unrealistic, the forester has the option of a restart with a slightly changed set of behavioural tree rules or a slightly changed nutrient availability, etc.[5]

We will not present results of such trials with experts here, as we want only to point out how this can be organized and the new role that forest growth models would play in this interactive task, which eventually becomes a communication process. The current best calibration of the A-grade is accessible through a graphical

---

[4] For a technical description of the current version see: www.bitoek.uni-bayreuth.de/mod/html/tragic/documentation/documentationnotes.pdf
[5] Currently this step will not happen automatically, a user has to mail the database administrator for this request.

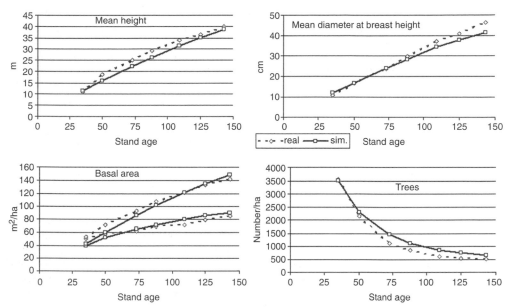

Fig. 5.1. The results of a TRAGIC++ calibration (squares) to the Denklingen A-grade data set (diamonds). The stand was planted as 5000 identical seedlings per hectare and then run with constant boundary conditions and without any interference for 143 years. Mortality up to age 40 is entirely due to competition among trees. After age 40 a random mortality of 0.5%/year has been added. The trees randomly removed by this procedure are larger than the trees selected by the (interactive) low-thinning regime that was applied at the real stand. This caused the tree number towards the end of the run to be higher and diameter to be lower in the simulation. This deviation has been accepted as we wanted to see how closely the model can match observations in the non-interactive mode. Shown are averages from five runs over a 50 × 50 m section with yearly time steps. The standard deviation is about ±3% and is not shown.

Fig. 5.2. A typical screenshot from an interactive session using the graphical interface JTRAGIC. The shown stand simulates a 20 × 20 m section of the reconstructed A-grade Denklingen stand at the age of 60 years due for thinning.

interface, JTRAGIC, where interested readers can place their own thinning attempts (www.bitoek.uni-bayreuth.de/mod/html/webapps).

It is necessary, however, to start this selection with experts only, because we want to exclude the possibility that the model was discarded due to a thinning choice imposed by a silviculturally incompetent user. The main question becomes: Are there models that are suitable to be accepted by such experts as a proper documentation and repository of their decisions, as these reside in their subjective memories? To what extent can these reconstructions be kept consistent with the objective (non-interactive) observations of such a stand?

Discussion and Outlook

Computation has become a powerful tool in the automatic generation of complex patterns from simple iterated functions. One prominent example is the implementation of grammars such as L-systems in which modular organisms can be generated from the typical building blocks. The architecture of a tree can be algorithmically derived from a set of modules and rules. When linked with a powerful visualization software, such programs have come close to photo-realism in presenting individual plants (Prusinkiewicz and Lindenmayer, 1990; Godin, 2000; see also Knauft, Chapter 30, this volume).

This phase of autonomous growth of a stand can be extended when the (seemingly) interactive aspects can be reduced to an algorithmic function. For example, interactive competition among trees for photosynthetically active radiation can be explicitly calculated from a simplified light extinction function and the actual vertical distribution of above-ground biomass of the stand. It is here that currently much of the computational power of IT has been allocated in forest growth modelling. This rigorous, but still unsuccessful, approach in terms of convincing forest managers, represents the 'brute force' method, as additional resources of computing are entirely directed towards an extension of existing *algorithmic* tasks.

The interactive approach proposed here is motivated by the observation that interactive selection among an immense number of choices is in many real life situations more powerful than an algorithmic selection. For example, in the world of cultural interaction, (interactive) job interviews, where the next question depends on the previously given answers, are obviously more powerful than questionnaires (where the sequence of answering is not prescribed). At the biological level of interaction as observed in ecosystems, the interactive nature of living organisms may itself be an irreducible feature (Hauhs and Lange, 2003). Past attempts to get rid of this feature have more often been motivated by its incompatibility with the prevailing paradigm of process-oriented modelling rather than by a proper abstraction of ecosystem behaviour. Here we propose to utilize the additional classificatory power of interaction for a faster and more complete (in terms of the available sources of knowledge) zoom on to consistent growth reconstructions.

An expansion of forest growth modelling has been sketched from serving predictive tasks based on algorithmic modelling towards also serving communicative tasks based on interactive modelling. This proposed change moves the focus in stand modelling from the search for a realistic representation of stand dynamics to the support of the documentations of silvicultural experience; in the words of our examples, creating an archive of played-out games as remembered by experts rather than looking further for a winning strategy for the whole game.

Human beings are (still) superior to computers in the area of pattern *recognition*, in fact this is the reason why humans are still able to beat a computer at chess.

Computers are much more powerful in pattern *generation*. A proper coupling of pattern generation (algorithmic) and pattern selection (interactive, experience-based) becomes possible if the generated patterns from a computer model can rapidly be visualized as a virtual forest. An integration between the two perspectives would be achieved when the unavoidable plasticity of the model (and the range of calibrated parameter sets) is sufficient to cover the range of expert behaviours encountered in an interactive session; in addition in such a framework it can be checked whether or not the subjective judgements of individual experts converge to reproducible and accepted standards in the field. Another potential outcome (although we regard it as unlikely) is that the approach indicates a lack of any standard among experienced practitioners. However, in the case where a reproducible pattern or reaction norm has been found, the scientific problem can be stated in a new way: What is the best (simplest) explanation in terms of processes for this reproducible pattern in silviculture?

We suggest that much more effort can and should be devoted to the problem of interaction of which silviculture may serve as a paradigmatic example. The 'should' is stated from the perspective of a largely empirical management regime that requires new and appropriate technology for a continuous updating of its expertise when faced by the changes in its boundary conditions, e.g. in central Europe. The 'can' is stated from the perspective of the potential that IT provides for growth modelling. Modelling becomes in this case a predominately backward-oriented exercise in time, by which valuable data and experiences from the past become more efficiently documented than by books or words. Instead of predictions that substitute for experience, interactive growth models have the potential to preserve and update the relevant experience in a transparent and efficient way.

Acknowledgements

This work was supported by the German Ministry of Science and Education (BMBF) under contract no. PT BEO 0339476 D. The manuscript was prepared during a sabbatical stay of the first author at the School of Forestry of the University of Canterbury (NZ). We thank Hans Pretzsch for providing us with the Denklingen data set.

References

Dreyfus, H. and Dreyfus, S. (1988) *Mind Over Machine: the Power of Human Mind and Expertise in the Era of the Computer.* Free Press, New York, 25 pp.

Flyvbjerg, B. (2002*) Making Social Science Matter*. Cambridge University Press, Cambridge.

Godin, C. (2000) Representing and encoding plant architecture: a review. *Annals of Forest Science* 57, 413–438.

Goldin, D. and Keil, D. (2001) Evolution, Interaction, and Intelligence: Congress on *Evolutionary Computation CEC'01*, Korea, May 2001. Available at: www.cs.umb.edu/~dqg/papers/cec01.doc

Goldin, D., Keil, D. and Wegner, P. (2001) An interactive viewpoint on the role of UML. In: *Unified Modelling Language: Systems Analysis, Design, and Development Issues*. Idea Group Publishing. Available at: www.cs.umb.edu/~dqg/papers/uml.doc

Hauhs, M. and Lange, H. (2003) Virtualities and realities of artificial life. In: Reuter, H., Breckling, B. and Mittwollen, A. (eds) *Gene, Bits und Ökosysteme, Theorie in der Ökologie*. P. Lang Verlag, Frankfurt/M.

Hauhs, M., Kastner-Maresch, A. and Rost-Siebert, K. (1995) A model relating forest growth to ecosystem-scale budgets of energy and nutrients. *Ecological Modelling* 83, 229–243.

Hauhs, M., Dörwald, W., Kastner-Maresch, A. and Lange, H. (1999) The role of visualization in forest growth modelling. In: Amaro, A. and Tomé, M. (eds) *Empirical and Process-based Models for Forest Tree and Stand Growth Simulation.* Edições Salamandra, Lisbon, pp. 403–418.

Hauhs, M., Lange, H. and Kastner-Maresch, A. (2001) Complexity and simplicity in ecosystems: the case of forest management. *InterJournal for Complex Systems* 415, 1–8 (www.interjournal.org).

Lange, H. (1999) Are ecosystems dynamical systems? *International Journal of Computing Anticipatory Systems* 3, 169–186.

Leibundgut, C. (1978) *DieWaldpflege.* Haupt Verlag, Bern.

Levy, D. and Newborn, M. (1991) *How Computers Play Chess.* W.H. Freeman, New York.

Ognjen, A., Riehle, H.J., Fher, T., Wienbruch, C. and Elbert, T. (2001) Pattern of y-bursts in chess players. *Nature* 412, 603–604.

Pigliucci, M. (2001) *Phenotypic Plasticity, Beyond Nature and Nurture.* Johns Hopkins University Press, Baltimore, Maryland.

Prusinkiewicz, P. and Lindenmayer, A. (1990) *The Algorithmic Beauty of Plants.* Springer, New York.

Schanz, H. (1995) Forstliche Nachhaltigkeit. Befragung zum Begriffsverständnis der Forstleute in Deutschland. *Allgemeine Forstzeitschrift* 50(4), 188–192.

Stein, L.A. (1999) Challenging the computational metaphor: implications for how we think. *Cybernetics and Systems* 30 (6), September 1999.

Vanclay, J.K. and Skovsgaard, J.P. (1997) Evaluating forest growth models. *Ecological Modelling* 98, 1–12.

van Leeuwen, J. and Wiedermann, J. (2001) Beyond the Turing limit: evolving interactive systems. In: Pacholski, L. and Ruzicka, P. (eds) *SOFSEM 2001: Theory and Practice in Informatics*, Proceedings of the 28th conference, Lecture Notes in Computer Science Vol. 2234. Springer-Verlag, Berlin, pp. 90–109.

Wegner, P. (1997) Why interaction is more powerful than algorithms. *Communication Association Computing Machinery* 40(5).

Wegner, P. and Goldin, D. (2003) Computation beyond Turing machines. *Communication Association Computing Machinery* (in press).

6

Linking Process-based and Empirical Forest Models in Eucalyptus Plantations in Brazil

Auro C. Almeida,[1,2] Romualdo Maestri,[1]
Joe J. Landsberg[2] and José R.S. Scolforo[3]

Abstract

The 3-PG model (Landsberg and Waring, 1997) was parameterized to predict potential productivity across 170,000 ha of *Eucalyptus grandis* hybrid plantation distributed in 19 regions in eastern Brazil. The regions were defined on the basis of meteorological measurements made by automatic weather stations. Mean annual increments estimated by the model for a 6-year rotation were compared with available observations made annually in permanent sample plots (PSPs). The goodness of fit between estimated and observed growth was determined by $R^2 =$ 0.92. Comparisons between model estimates and measurements such as basal area and total volume are presented.

An empirical model called E-GROW ARCEL was developed and fitted using PSP data from the same region. The model is based on recovering the parameters of the Weibull probability density function by matching their moments to estimated stand level variables. Stand models were fitted for projections of stand basal area, mortality, dominant height, tree height, DBH (diameter at breast height) variance and stem taper. Volume of log types in the DBH distribution can be estimated.

Mean annual increment (MAI), one of the outputs of 3-PG, was used to establish a hybrid approach, linking the two models by matching the relationship between MAI and site index from E-GROW ARCEL. Growth curves and yields are generated.

The hybrid approach is being established as a basis for decision making and management of fast-growing *E. grandis* hybrid plantations in eastern Brazil.

Introduction

The area of fast growing eucalyptus plantations in Brazil has been increasing during the last 20 years. According to the Brazilian Silvicultural Society (SBS, 2001), planted forest in Brazil in 2000 was estimated at 4.8 million ha; these plantations are the basis for the pulpwood industry in Brazil. Its high productivity is a result of ideal

[1] Aracruz Celulose S.A., Brazil
Correspondence to: aca@aracruz.com.br or rmaestri@aracruz.com.br
[2] The Australian National University, School of Resources Environmental and Society, Australia
[3] Universidade Federal de Lavras, Brazil

climatic conditions, applied research, and improvements in genetics and silvicultural practices. Many forest companies are working with a plantation rotation length of 6–7 years. With this short rotation, any change in climate, such as severe drought stress, has direct effects on productivity and wood quality.

Forest resource management requires decisions to be made on the strategy and tactics of operations. At all these levels, decision making may be supported by mathematical models (Hinssen, 1994). Demand for specific information by forest managers and planners is one of the reasons for changing this situation and moving towards a causal-oriented approach. Questions about potential productivity, the effects of climate on forest growth, and the ability to understand and analyse the effects of silvicultural practices, such as soil preparation, weed and disease control, fertilization and water management, can be answered using process-based models and in some cases integrating these with empirical models in operational systems.

The development of effective and accurate models to predict forest growth and products during the forest rotation is essential for forest managers and planners. Empirical growth and yield models, which rely on fitting functions to measurement data from a sample of the forest population of interest (Burkhart, 1997), are the tools that have mainly been used to provide decision-support information that meets basic operational needs for evaluating various forest management scenarios (Mohren and Burkhart, 1994). Growth and yield predictions are used to assess profitability, determine harvesting schedules, estimate site potential and risks involved, and evaluate silvicultural practices (Dye, 2001).

The development of process-based models to predict forest growth has been progressing rapidly in the last few years. However, operational applications in forest plantations are still at an early stage. These models focus on the representation and explanation of the processes that take place in forest ecosystems and highlight state variables that are closely associated with these processes (Larocque, 1999). However, despite the logical concepts underlying process-based models, they are not yet being used as operational tools in forest management; most of them are used as research tools only. Baldwin *et al.* (1993), MacLean (1999), Kimmins *et al.* (1999), Host *et al.* (1999), Hauhs *et al.* (1999) and Battaglia *et al.* (1999) have recently combined process-based and empirical models. Mäkelä *et al.* (2000) consider that the future lies in models that combine biophysical processes and empirical relationships – referred to as mixed or hybrid models. Interest is being shown, by some forestry companies, in the use of these models as operational tools (Almeida, 2000).

Our aim in this chapter is to demonstrate the possibility of integrating the process-based model 3-PG with the empirical growth model called E-GROW ARCEL, to produce a hybrid model developed and adjusted for *Eucalyptus grandis* in Brazil. We used a dense network of measurements of growth rates from permanent plots to evaluate the model's performance.

State of the Art

Vanclay (1999) defined stand growth models as abstractions of the natural dynamics of a forest stand, which may encompass growth, mortality, and other changes in stand composition and structure. Korzukhin *et al.* (1996) presented a detailed analysis of the relative merits of process-based and empirical forest models, which highlighted the value of both classes of models and indicated how they can be applied in forest ecosystem management.

Currently available process-based models can provide good estimates of growth and biomass productivity at various scales; combined with conventional mensuration-based growth and yield models they can provide information of the type required by managers and planners (Landsberg, 2003). Some of the most commonly cited models in the literature are: FOREST-BGC (Running and Coughlan, 1988; Running and Gower, 1991), CENTURY (Parton *et al.*, 1987), G'DAY (Comins and McMurtrie, 1993), 3-PG (Landsberg and Waring, 1997), PROMOD (Battaglia and Sands, 1997), JABOWA (Botkin *et al.*, 1972) and MAESTRO (Wang and Jarvis, 1990). All these have been used and tested as research tools in different parts of the world, with data from a range of environments. Several authors argue that the limited application of process-based models as practical tools is a consequence of the large number of parameter values required, the complexity of the models and the lack of appropriate documentation. However, despite these factors, the use of process-based models must increase our understanding of the environmental factors affecting growth, and they can be used to estimate potential productivity in areas without forest and under changing environmental conditions (Mohren and Burkhart, 1994; Korzukhin *et al.*, 1996). Models with fewer parameters that express the physiological processes in simple terms are more likely to be used in forest management.

Empirical growth models may be at different levels of detail (Maestri *et al.*, 1995). They may be size class models, single-tree models, or apply to a whole stand, depending on the detail required. These models are derived from tree size data from stands in a range of ages, site indices (SIs), stand densities and management conditions. They are widely used in forest planning activities. However, they are limited to be transportable to new areas where no measured growth data are available.

Several studies have been done using empirical growth models that include environmental variables. Hunter and Gibson (1984) used principal component analysis (PCA) to select soil characteristics and climatic variables that exerted significant effects on growth. They observed a positive relationship between SI and rainfall, nutrients, topsoil depth and soil penetrability of *Pinus radiata* stands in New Zealand. Carter and Klinka (1989) related SI of coastal Douglas-fir stands in British Columbia to available soil micronutrients and soil water deficits during the growth season. Snowdon *et al.* (1998) incorporated climatic indices derived from a process-based model, BIOMASS, into an empirical growth model, to describe stand height, basal area and volume in an initial spacing trial with *P. radiata*. These indices improved the fit compared with the basic empirical equations by 13%, 22% and 31% for mean tree height, stand basal area and stand volume, respectively. Woollons *et al.* (1997) incorporated climatic variables into a basal area model of *P. radiata* in New Zealand, improving the accuracy of the model by 10%.

A hybrid approach combining the main advantages of process-based and empirical models has been adopted in some cases. Baldwin *et al.* (1993) combined a single-tree empirical model called PTAEDA2 (Burkhart *et al.*, 1987) with a process-based model called MAESTRO (Wang and Jarvis, 1990). Using PTAEDA2 they projected to a certain age the stand variables used by MAESTRO: individual mean crown ratio, crown shape, crown length, and the vertical and horizontal distributions of foliage biomass. This information was then used by MAESTRO to calculate biomass production, which was fed back to PTAEDA2 to adjust its predictions. These steps were repeated to the end of the rotation.

Battaglia *et al.* (1999) used the process-based model PROMOD and the empirical model NITGRO developed for *Eucalyptus nitens* plantations. The resulting hybrid model was applied in 16 *Eucalyptus globulus* stands in Tasmania, Australia. PROMOD predicted the mean annual increment (MAI, m^3/ha/year) and estimated the SI applying an empirical relationship between MAI and SI.

Methodology

We used the process-based model 3-PG (physiological principles in predicting growth), (Landsberg and Waring, 1997) to predict the potential productivity of 19 regions in eastern Brazil, lying between 17° 25′ S and 20° 05′ S, covering 170,000 ha of eucalypt plantation (*E. grandis* Hill ex. Maiden hybrids) (see Fig. 6.1). The regions were defined based on meteorological measurements made by automatic weather stations. Mean annual increments estimated by the model for a 6-year rotation were compared with available observations made annually in permanent sample plots (PSPs). The forest areas used in the comparison were those planted in 1995, which had PSPs during the rotation. They represented 10 of the 19 regions and included 213 PSPs.

For the same area, using PSP measurements, we developed an empirical growth and yield model called E-GROW ARCEL. The two models have been linked to allow the prediction of forest productivity and wood products under different environmental conditions. The link is based on the MAI at age 6 years (MAI_6). There is a robust relationship between MAI_6 and SI – defined to be the mean dominant height at age 5 years – derived from E-GROW ARCEL. Once 3-PG has provided the MAI_6 values, this empirical relationship estimates the SI that is applied in E-GROW ARCEL to predict growth and yield at stand structure and product profile levels.

3-PG model structure

The 3-PG model is a simple process-based model of forest growth. It requires few parameters, climatic data inputs, a basic knowledge about the local soil water-holding capacity and an indication of soil fertility. The model consists of five sim-

Fig. 6.1. Location of Aracruz Celulose S.A. eucalypt plantation and automatic weather station network.

ple submodels: these describe the assimilation of carbohydrates, the distribution of biomass between foliage, roots and stems, the determination of stem number, soil water balance, and conversion of biomass values into variables of interest to forest managers (Sands and Landsberg, 2002). The time step is monthly and the main outputs are: gross primary productivity (GPP), net primary productivity (NPP), leaf area index (LAI), mean stem diameter at breast height (DBH), basal area (BA), standing biomass (partitioned into foliage, stem and root), MAI and stem volume.

3-PG differs from other process-based models because it incorporates important simplifications to some well-known physiological processes and can be used as a practical tool in forest management. Waring and McDowell (2002) described these simplifications as: (i) constant ratio of net primary production to gross photosynthesis (Pn/Pg); (ii) canopy conductance approaches a maximum above an LAI of 3.0; (iii) the ratio of actual/potential photosynthesis decreases in response to the most restrictive (monthly) environmental limitation; (iv) the fraction of production not allocated to roots is partitioned among foliage, stem and branches based on species-specific allometric relationships with tree diameter; (v) canopy quantum efficiency (α_c) is assumed to increase linearly with soil fertility.

3-PG has been tested for forest species in different countries (see Landsberg *et al.*, 2000, 2003; Dye, 2001; Sands and Landsberg, 2002).

E-GROW ARCEL model structure

E-GROW ARCEL is an implicit model, based on the recovery of Weibull distribution parameters by diameter class. It includes a set of stand models to estimate dominant height, mortality, basal area, DBH variance, individual tree height and stem taper. Integrating basal area generated from two levels of resolution (diameter distribution method and stand-based model) increases the system reliability; following methodology presented in Scolforo (1998).

E-GROW ARCEL allows predictions and projection estimates. Prediction refers to growth and yield based on early stage information, and projection means estimation using data from forest inventory over time. The model also provides simulation of stand growth and yield under a solid wood regime.

The basic E-GROW ARCEL inputs and outputs are:

Inputs

- Stand basic information: genetic material, management regime.
- Forest information: stand age, dominant height, mean diameter, diameter coefficient of variation, trees per hectare and basal area.
- Definition of the projection age.

Outputs

- Estimation of dominant height, stem density, basal area, diameter variance and minimum diameter.
- Weibull *a*, *b* and *c* recovered parameters and diameter distribution estimation.
- Total height estimation for each diameter class.
- Merchantable volume estimation for each log class defined by log length and small end diameter.

Data

Used in 3-pg model

The 3-PG model was rigorously tested against data from an experimental catchment site before applying it across landscapes. Measurements at the catchment site provided information about biomass allocation, monthly growth rates, plant nutrition and water balance.

The catchment area is 286 ha, of which 190 ha are covered by eucalyptus plantations and 86 ha by the original tropical rainforest (*Mata Atlântica*), the remainder being roads. The topography of the plantation area is flat, while the tropical forest covers the hilly drainage area of the system. Measurements have been made in this area since 1993 and include:

- Weekly soil moisture measurements in 16 tubes using neutron probe in the first 2.8 m.
- Weekly water table level in five piezometers.
- Monthly plantation growth in 16 PSPs including different genetic material.
- Three automatic weather stations measuring precipitation air temperature, relative humidity, global solar radiation, net radiation, PAR (photosynthetically active radiation) , and wind speed and direction.
- Monthly LAI using a plant canopy analyser instrument calibrated against destructive samples taken annually.
- Annual biomass allocation in three trees for each genetic material including stem, bark, branches, foliage and roots.
- Monthly litter fall dry mass in 35 traps.
- Annual stomatal conductance in five genotypes.

The PSPs were well distributed across all regions; however, the area planted in 1995 covered ten regions only. To calibrate 3-PG for other regions we used the available information about soil water-holding capacity and soil fertility from soil surveys at scale 1:10,000. The automatic weather station network provided the monthly rainfall, solar radiation, vapour pressure deficit and temperature data used in 3-PG.

Used in E-GROW ARCEL model

E-GROW ARCEL was developed using data from 1010 PSPs, covering the range of ages, SI and management conditions faced by *E. grandis* hybrid stands across Aracruz plantations. The measurements included upper-stem diameters at different ages in different regions, SIs and stem densities. Information about genetic material was also available. The PSPs cover an area of *c.* 170,000 ha of *E. grandis* hybrid plantations (see Fig. 6.1). Some PSPs are located inside thinned stands. As each PSP was measured more than once, there were 3281 sets of observations. Table 6.1 summarizes the measurements made. Additionally, taper functions were developed based on diameter measurements along the stem from a set of 401 felled trees.

Linking process-based and empirical models

Both 3-PG and E-GROW ARCEL can be applied to obtain predictions of growth and yield under different management practices (establishment, re-establishment or cop-

Table 6.1. Number of measurements of the PSPs used to develop E-GROW ARCEL.

Age (years)	Dominant height (m)				Total
	<15	15–20	20–25	>25	
<3	153	203	17	0	373
3–5	183	881	751	117	1932
5–7	2	90	535	329	956
>7	0	0	5	15	20
Total	338	1174	1308	461	3281

pice), genetic material and initial stem density. In this study the link between models assumed re-establishment as the standard management practice for clonal *E. grandis* hybrids with 1111 trees/ha. Under these conditions, E-GROW ARCEL estimated MAI_6 for a range of SI, resulting in a robust relationship, defined by Equation 1.

$$SI = 10.84 + 0.403 \times MAI_6 - 0.001 \times (MAI_6)^2$$
$$R^2 = 0.99 \qquad RMSE = 0.26 \text{ m}$$

(1)

Equation 1 provides the necessary link between predictions of MAI_6 from 3-PG and the SI required as an input for E-GROW ARCEL.

Results

The values of MAI estimated by 3-PG for 6-year-old stands in different regions were compared with observations made annually in PSPs in ten of 19 regions (Fig. 6.2), showing the observed MAI for each region. The nine regions without PSP data are presented in Fig. 6.2, indicating the 3-PG predictions under different environmental conditions. The estimated values obtained with 3-PG assumed the real climate conditions and the applied silvicultural practices such as fertilization, soil preparation and weed control. The histograms show that 3-PG provides an accurate estimate of

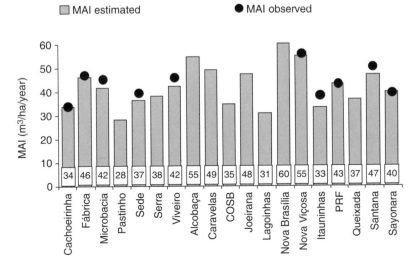

Fig. 6.2. Mean annual increment (MAI) in different regions planted in 1995, estimated using 3-PG (bars), and observed MAI (circles) at age 6 years.

the MAI likely to be obtained in any area.

Values of basal area and stand volume at 6 years of age, measured in PSPs and estimated by 3-PG, are presented in Table 6.2 for each region.

Figures 6.2 and 6.3 and Table 6.2 show that 3-PG is well calibrated for age 6-years. The resulting MAI_6 value was inserted in Equation 1 generating the SI for each region that was used in E-GROW ARCEL, to predict the volume curve of commercial products.

We selected the Nova Viçosa region as an example of the final results. 3-PG estimated a MAI_6 of 55 m^3/ha/ year, which generated an SI of 30 m, using Equation 1. Based on this site index, E-GROW ARCEL describes the volume curve and the structure of the stand at each age of the projection period. Figure 6.4 shows the volume curve by predefined log type based on its small end diameter (SED) and log length. The distribution of diameter at breast height (DBH) is also obtained. Figure 6.5 presents the log volume for each DBH class at age 7 years.

Discussion and Conclusions

Long-term planning for the production of fast-growing species is strongly affected by the productivity of the area concerned. Traditionally, potential productivity is estimated from growing history and/or local expertise, without considering the eco-

Table 6.2. Basal area and stand volume observed in PSPs and estimated by 3-PG at age 6 years.

Region	Basal area observed (m²/ha)	Basal area estimated (m²/ha)	Stand volume observed (m³/ha)	Stand volume estimated (m³/ha)
Cachoeirinha	21.1	21.2	200.3	203.5
Fabrica	28.0	25.7	296.0	276.6
Microbacia	23.3	24.2	268.2	250.2
Sede	21.3	22.3	245.6	219.9
Viveiro	24.4	24.4	274.4	254.6
Nova Viçosa	26.5	28.7	324.9	331.7
Itauninhas	22.3	21.0	241.3	200.3
PRF	23.6	24.8	258.6	260.8
Santana	26.9	26.1	298.2	282.7
Sayonara	21.5	23.6	245.8	240.2

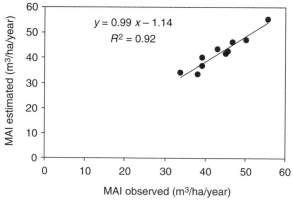

Fig. 6.3. Relationship between observed and estimated MAI by 3-PG. The goodness of fit of the regression line is given by the R^2 value of 0.92.

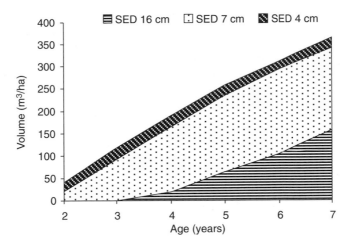

Fig. 6.4. Volume curve per predefined log type based on its small end diameter (SED).

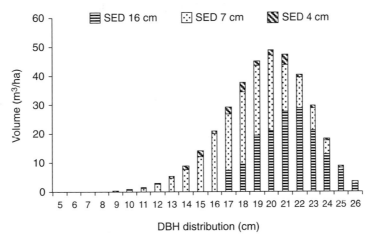

Fig. 6.5. Diameter at breast height (DBH) distribution and log volume per small end diameter (SED) at age 7 years.

physiological processes involved. This is critical when production in new areas, where no forest has been planted before, is being planned. Another important aspect for forestry industries is information about the product volumes likely to become available from any area. The hybrid approach developed in this study provides information about potential productivity and the product mix.

This study has illustrated the significant variation in MAI that occurs between regions as a consequence of environmental conditions. The 3-PG model is a powerful and practical tool that can be used to analyse and predict these effects, while its combination with E-GROW ARCEL provides enough detail in terms of forest products to meet planning needs.

This approach is being tested in operational practice by Aracruz Celulose S.A. in Brazil. The results are encouraging and appear likely to lead to continuous improvements in forest planning and management.

Acknowledgements

We acknowledge with thanks permission given by Aracruz Celulose S.A. to publish these data and the results of our analyses of them. We also acknowledge our colleagues Ricardo Penchel, Sebastião Fonseca and Simone M. Barddal for their efforts on technical support and database maintenance. Finally we thank Marcelo Wiecheteck and Natalino Calegario for their review of the manuscript.

References

Almeida, A.C. (2000) *Eucalyptus* plantations in Brazil: data from Aracruz Celulose S.A. and proposed analysis using 3-PG. In: Williams, K.J. (ed.) *3PG 2000: a Workshop on the Forest Model 3-PG: New Developments in Calibration, Performance, Spatial Inputs and Practical Applications.* Forest Ecosystem Research and Assessment Technical Papers 00-18, August 2000, DNRQ00126. Queensland Departments of Natural Resources and Primary Industries, Brisbane, pp. 82–85.

Baldwin, V.C., Burkhart, H.E., Dougherty, P.M. and Teskey, R.O. (1993) *Using a Growth and Yield Model (PTAEDA2) as a Driver for a Biological Process Model (MAESTRO).* US Department of Agricultural, Forest Service, Southern Forest Experiment Station, New Orleans, Louisiana, Research Paper SO-276, 9 pp.

Battaglia, M. and Sands, P. (1997) Modelling site productivity of *Eucalyptus globulus* in response to climatic and site factors. *Australian Journal of Plant Physiology* 24, 831–850.

Battaglia, M., Sands, P.J. and Candy, S.G. (1999) Hybrid growth model to predict height and volume growth in young *Eucalyptus globulus* plantations. *Forest Ecology and Management* 120, 193–201.

Botkin, D.B., Janak, J.F. and Wallis, J.R. (1972) Some ecological consequences of a computer model of forest growth. *Journal of Ecology* 60, 849–872.

Burkhart, H.E. (1997) Development of empirical growth and yield models. In: Amaro, A. and Tomé, M. (eds) *Empirical and Process-based Models for Forest Tree and Stand Growth Simulation.* Edições Salamandra, Oeiras, Portugal, pp. 53–60.

Burkhart, H.E., Farrar, L.R., Amateis, R.L. and Daniels, R.F. (1987) *Simulation of Individual Tree Growth and Stand Development in Loblolly Pine Plantations on Cutover, Site Prepared Areas.* Virginia Polytechnic Institute and State University, School of Forestry and Wild Life Resources, Blacksburg, FWS 1-87, 47 pp.

Carter, R.E. and Klinka, K. (1989) Relations between growing-season soil water-deficit, mineralizable soil nitrogen and site index of coastal Douglas fir. *Forest Ecology and Management* 30, 301–311.

Comins, H.N. and McMurtrie, R.E. (1993) Long-term response of nutrient-limited forests to CO_2 enrichment; equilibrium behaviour of plant–soil models. *Ecological Applications* 3, 666–681.

Dye, P.J. (2001) Modelling growth and water use in four *Pinus patula* stands with the 3-PG model. *Southern African Forestry Journal* 191, 53–63.

Hauhs, M., Dörwald, W., Kastner Maresch, A. and Lange, H. (1999) The role of visualization in forest growth modelling. In: Amaro, A. and Tomé, M. (eds) *Empirical and Process-based Models for Forest Tree and Stand Growth Simulation.* Edições Salamandra, Oeiras, Portugal, pp. 403–418.

Hinssen, P.J.W. (1994) HOPSY, a model to support strategic decision making in forest resource management. *Forest Ecology and Management* 69, 321–330.

Host, G.E., Theseira, G.W., Heim, C.S., Isebrands, J.G. and Graham, R.L. (1999) Epic–Ecophys: a linkage of empirical and process models for simulation of poplar plantation growth. In: Amaro, A. and Tomé, M. (eds) *Empirical and Process-based Models for Forest Tree and Stand Growth Simulation.* Edições Salamandra, Oeiras, Portugal, pp. 419–429.

Hunter, I.R. and Gibson, A.R. (1984) Predicting *Pinus radiata* site index from environmental variables. *New Zealand Journal of Forestry Science* 14, 53–64.

Kimmins, J.P., Scoullar, K.A., Seely, B., Andison, D.W., Bradley, R., Mailly, D. and Tsze, K.M. (1999) Forseeing and forecasting the horizon: hybrid simulation modeling of forest ecosystem sustainability. In: Amaro, A. and Tomé, M. (eds) *Empirical and Process-based Models for Forest Tree and Stand Growth Simulation*. Edições Salamandra, Oeiras, Portugal, pp. 431–441.

Korzukhin, M.D., Ter-Mikaelian, M.T. and Wagner, R.G. (1996) Process versus empirical models: which approach for forest ecosystem management? *Canadian Journal of Forest Research* 26, 879–887.

Landsberg, J.J. (2003) Modelling forest ecosytems: state-of-the-art, challenges and future directions. *Canadian Journal of Forest Research* 33, 385–397.

Landsberg, J.J. and Waring, R.H. (1997) A generalised model of forest productivity using simplified concepts of radiation-use efficiency, carbon balance and partitioning. *Forest Ecology and Management* 95, 209–228.

Landsberg, J.J., Waring, R.H. and Coops, N.C. (2000) The 3-PG forest model: matters arising from evaluation against plantation data from different countries. In: Carnus, J.M., Dewar, R., Loustau, D., Tomé, M. and Orazio, C. (eds) *Models for the Sustainable Management of Temperate Plantation Forests*. IEFC/INCRA Meeting on models for the sustainable management of temperate plantation forests. EFI Publications, Bourdeaux, France, pp. 1–15.

Landsberg, J.J., Waring, R.H. and Coops, N.C. (2003) Performance of the forest productivity model 3-PG applied to a wide range of forest types. *Forest Ecology and Management* 172, 199–214.

Larocque, G.R. (1999) Preface. *Ecological Modelling* 122, 135–137.

MacLean, D.A. (1999) Incorporating of defoliation impacts into stand growth models: empirical and process-based approaches. In: Amaro, A. and Tomé, M. (eds) *Empirical and Process-based Models for Forest Tree and Stand Growth Simulation*. Edições Salamandra, Oeiras, Portugal, pp. 443–459.

Maestri, R., Scolforo, J.R.S. and Hosokawa, R.T. (1995) Um sistema de predição do crescimento e da produção para povoamentos de acácia-negra (*Acacia mearnsii* de Wild). *Revista Árvore* 19, 358–381.

Mäkelä, A., Landsberg, J.J., Ek, A.R., Burk, E.T., Ter-Mikaelian, M., Agren, G.I., Oliver, C.D. and Puttonen, P. (2000) Process-based models for forest ecosystem management: current state of the art and challenges for practical implementation. *Tree Physiology* 20, 289–298.

Mohren, G.M.J. and Burkhart, H.E. (1994) Contrasts between biologically-based process models and management-oriented growth and yield models. *Forest Ecology and Management* 69, 1–5.

Parton, W.J., Schimel, D.S., Cole, C.V. and Ojima, D.S. (1987) Analysis of factors controlling soil organic matter levels in Great Plains grasslands. *Soil Science Society of America Journal* 51, 1173–1179.

Running, S.W. and Coughlan, J.C. (1988) A general model of forest ecosystem processes for regional applications. I. Hydrologic balance, canopy gas exchange and primary production processes. *Ecological Modelling* 42, 125–154.

Running, S.W. and Gower, S.T. (1991) FOREST-BGC: a general model of forest ecosystem processes for regional applications. II. Dynamic carbon allocation and nitrogen budgets. *Tree Physiology* 9, 147–160.

Sands, P.J. and Landsberg, J.J. (2002) Parameterisation of 3-PG for plantation grown *Eucalyptus globulus*. *Forest Ecology and Management* 163, 273–292.

SBS (2001) Area plantada com pinus e eucaliptos no Brasil (ha) – 2000. SBS, São Paulo/ Available at: www.sbs.org.br/area_plantada.htm

Scolforo, J.R.S. (1998) *Modelagem do Crescimento e da Produção de Florestas Plantadas e Nativas*. UFLA/FAEPE, Lavras, Brazil, 441 pp.

Snowdon, P., Woollons, R.C. and Benson, M.L. (1998) Incorporation of climatic indices into models of growth of *Pinus radiata* in a spacing experiment. *New Forests* 16, 101–123.

Vanclay, J.K. (1999) *Modelling Forest Growth and Yield: Applications to Mixed Tropical Forests*. CAB International, Wallingford, UK, 312 pp.

Wang, Y.P. and Jarvis, P.G. (1990) Description and validation of an array model – MAESTRO. *Agricultural and Forest Meteorology* 51, 257–280.

Waring, R.H. and McDowell, N. (2002) Use of physiological process model with forestry yield tables to set limits on annual carbon balances. *Tree Physiology* 22, 179–188.

Woollons, R.C., Snowdon, P. and Mitchell, N.D. (1997) Augmenting empirical stand projection equation with edaphic and climatic variables. *Forest Ecology and Management* 98, 267–275.

7

A Strategy for Growth and Yield Research in Pine and Eucalypt Plantations in Komatiland Forests in South Africa

Heyns Kotze[1]

Abstract

In Komatiland Forests the objectives of growth and yield research are to develop a range of empirical stand-level growth and yield models for thinned pine stands. The models are used for resource description, growth and yield predictions for the purpose of management planning and for the development of optimum stand-level management regimes.

The growth and yield research programme consists of four main components, namely a field programme, long-term data management, growth modelling and the development of simulation systems. For each of these components a strategy has been developed. In the field, programme data are collected for each of the target model components from a number of sources, such as spacing trials, permanent sample plots (PSPs), silvicultural response trials, plantation inventories and destructive sampling for volume and taper data.

The modelling strategy is to develop multi-component model architectures for modelling stand-level growth, namely dominant height, unthinned basal area and survival. For thinned stands, additional functions for the modelling of the basal area thinning ratio and the basal area response after thinning are required. Depending on the data, both the Chapman–Richards-type and the Schumacher-type functions are used for modelling unthinned stand-level basal area. Various methods are used for modelling basal area response after thinning, e.g. the index of suppression or the age-adjustment method. A major challenge in growth modelling is to develop models for thinned stands based on PSP data only, while conforming to the hypothesis of thinned stand growth. Stand structure is modelled using the Weibull function in conjunction with the method of moments approach for recovering the relevant Weibull parameters.

Finally, the models are built into simulation systems for stand-level scenario analysis, e.g. the development of rule-based thinning and pruning systems, and also forestry scenario analysis. However, the main function of stand-level growth and yield models is still to provide for more accurate company planning systems.

Introduction

The Komatiland Forests plantations are situated on the great north-eastern escarpment of South Africa (Fig. 7.1). The company is a recently created subsidiary of the

[1] Komatiland Forests, Research Division, South Africa
Correspondence to: hkotze@safcol.co.za

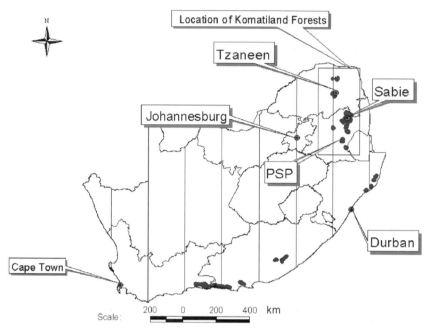

Fig. 7.1. Location of Komatiland Forests in South Africa.

South African Forest Company Limited (SAFCOL). The plantations cover approximately 130,000 ha and comprise various pine species mainly grown for sawtimber. A small area of eucalyptus is used for the production of special products for niche markets. The plantations are spread over a number of physiographical regions with varying geology and climate.

The objectives of the growth and yield research programme in Komatiland Forests are to develop a set of growth and yield models that can be used for resource description, growth and yield predictions for management planning purposes and the analysis and development of optimum stand-level management regimes.

Figure 7.2 shows the components in the growth and yield cycle. Models and simulation systems are the end-products of the growth and yield programme. Therefore, the field programme has been designed to meet the requirements of models for stand growth, tree volume and taper, stand structure, response to silviculture and site quality.

The purpose of this chapter is to describe the strategy for growth and yield research for Komatiland Forests by describing the past and current situation and indicating how the programme should develop towards the future.

The Field Programme

Due to its government-based background, with intense involvement in past growth and yield research in South Africa, Komatiland Forests inherited most of the historical data on long-term spacing and thinning trials, which started in 1936. More recently, since the late 1980s, a series of new spacing trials has been established and, since 1995, a stratified permanent sample plot (PSP) programme has been started. The field programme has been designed to gather data for the requirements of

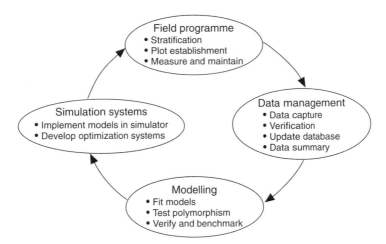

Fig. 7.2. Components of the growth and yield cycle.

stand-level modelling. New data augment existing historical data in order to provide for more accurate growth models. Table 7.1 shows how the various types of data are used to develop the required model components.

The field programme consists of long-term spacing trials, PSPs, silvicultural response trials and *ad hoc* destructive volume and taper sampling. Inventory data are also used for some model components. In a field programme, long-term trials and PSPs should be well stratified over the resource base. Stratification factors of importance are species, working circle, site types, productivity classes and age classes. A detailed 1:10,000 site classification system is available for bottom-up stratification purposes. This classification provides detailed descriptions of climate class, soil group, soil depth, depth limiting material, geology and other terrain descriptive variables. A top-down stratification framework was also developed by the forest

Table 7.1. Model components and their different data sources.

Model components	Spacing trials	PSPs	Thinning control data	Management inventory data	Silviculture response trials	Destructive sampling
Stand growth						
Dominant height	✓	✓				
Average height from dominant height						
Basal area	✓	✓				
Survival	✓	✓				
Growth response						
Thinning response					✓	
Thinning						
BA thinning ratio			✓			
Stand structure						
DBH distribution (Dmin, Sdev)	✓	✓	✓	✓		
Average height by DBH class	✓	✓				
Products						
Tree volume and taper						✓
Live crown height	✓	✓				✓

industry for the whole summer rainfall area (Kunz and Pallet, 2000). This classifies the forest area in terms of physiographical regions, i.e. groupings of geology and climate classes. PSPs are located so that they represent the various strata proportional to the species area.

The aim of the field programme is to establish at least one spacing trial per species per physiographical region. The historic spacing trials were mostly of the correlated curve trend (CCT) (O'Connor, 1935) and Nelder 1a (Nelder, 1962) types, with the current spacing trials being of the Triple-S CCT type (Bredenkamp, 1990). It is also aimed to establish at least 40 PSPs per species per physiographical region. Currently there is about one PSP for every 400 ha. A PSP is approximately 40 m × 40 m in size and contains at least 40 trees at rotation age. PSPs are established inside existing stands, about 40 m away from the roadside. Du Plessis *et al.* (1997) summarized the guidelines for the establishment and maintenance of the field programmes. For minor species, with small areas, additional data are sourced from neighbouring companies, belonging to the industry-based Mensuration and Modelling Research Consortium (MMRC). A biennial measurement cycle is followed, using the same measurement team as far as possible. Table 7.2 shows the current extent of the field programme and indicates data available from historic sources and the MMRC. The field programme continues to develop along with our understanding of site complexity and the requirements for developing each model component.

Data Management

Research data follow three distinct steps from the field, through to the database and to the models. In the field, trees are measured on demarcated plots. Tree diameter at breast height (DBH) is measured with a DBH tape and all heights with a Vertex hypsometer. Individual tree measurements are verified against the previous set of measurements. Measurement data are recorded on paper field sheets, or on hand-held electronic data-capturing devices. In the office, data are captured (downloaded) on electronic media to a standardized format. Data are again verified by range checking and distribution plots. For growth modelling purposes, the data are summarized to stand-level variables. At the end of each measurement cycle, a measurement report is compiled, which summarizes all relevant parameters.

Table 7.2. Extent of the field programme.

Variable	P. patula	P. elliotii	P. taeda	P. ellxcar hybrid	E. grandis	Other (7)	Total
% of area (130,000 ha)	46.4	26.5	6.7	5.3	4.1	11.0	100
Spacing trials							
Current	2	1	0	1	3	2	9
Historic	3	1	1	0	0	1	5
MMRC	2	0	0	2	0	0	6
PSPs							
Current	169	109	0	18	27	28	351
MMRC	52	66	42	0	50	61	271
Sample trees for taper							
Historic	358	180	0	100	174	0	812
MMRC	170	100	100	0	0	420	790

Data must flow through the above steps speedily and without error. As measurement data are also collected over a very long period of time, continuous and careful data management is of the utmost importance. In the MMRC, this has been realized to be an important need for the whole industry, and a joint general growth and yield database is under development. In order to ensure consistency in the industry, great efforts were made to develop common standards for the database (Kotze, 1995). The data-management process for spacing trials has also been well described in Kotze and Groenewald (1995).

Growth Modelling

Komatiland Forests adopted a stand-level modelling approach because all our forests are planted, even-aged, single-species stands. Stand-level models can be classified under stand models, which is a general term to denote equations or systems of equations developed for providing information on forest growth and yield. Stand models are essential for decision making under intensive forest management, provide a means for evaluating treatment alternatives, and are but one component in comprehensive systems for the management of forests. Stand-level models predict stand growth and yield as a function of stand age, site index and stand density. They can generally be applied with existing inventory data and are computationally simple. They can be combined with procedures for predicting diameter distribution to estimate product mixes. Stand-level models are somewhat inflexible for analysing a wide range of stand treatments and are not likely to give reliable results outside the range of conditions included in the sample plots on which they are based. There are a number of essential properties which stand-level models must exhibit. These are analytical compatibility between growth and yield, invariance for projection length, and numeric equivalency between alternative applications of the equations (Pienaar *et al.*, 1988).

Various yield prediction models, with different input requirements, can be used for management planning purposes under varying circumstances, e.g. the classical biological stand-level growth models (Chapman–Richards type) and the stand-level yield prediction and projection models (Schumacher type). Remeasured data from spacing trials are suitable for the Chapman–Richards-type growth models, and PSP data are, in turn, more suited for the Schumacher-type growth models. Table 7.1 indicates the model components required and which data can be used to develop them.

Growth and yield research has been carried out in South Africa since the early 1920s. A number of researchers – O'Connor (1935), Craib (1939) and Marsh and Burgers (1973) – did pioneering work in stand-density management, especially with the CCT trials. Many others developed models or model components from trial, inventory and other data. The theory and technology pertaining to growth and yield models have changed over time, and numerous function types and model architectures have been used over the years. Kotze (2001) has summarized all the available models and their coefficient sets pertaining to South Africa.

Many of the existing models used in Komatiland Forests are based on data from the old CCT trials started in 1936. They are still the most suitable models, because there are no models based on more recent data. The purpose of a growth and yield research programme is to ensure that relevant data become available for the development of more accurate models. That is why a series of new spacing trials was installed in the late 1980s and a PSP programme was started in 1995.

The best existing models and their coefficient sets are selected for each species and region combination. These are packaged into a model configuration for ease of

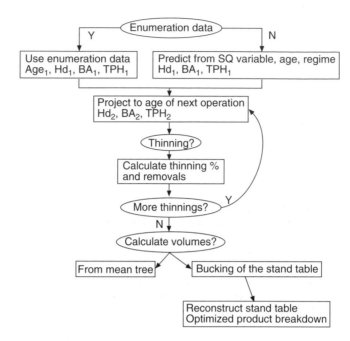

Fig. 7.3. Multi-component growth and yield model framework.

use and reference. These functions can work together in a simulation framework (Fig. 7.3). Stand growth and yield are normally projected from a calibration point where inventory data provide the calibration age (Age_1), dominant height (Hd_1), basal area per hectare (BA_1) and stems per hectare (TPH_1). If such inventory data do not exist, then a calibration point is created before the first thinning age. Each stand has a default site index value at base age 20 years (SI_{20}) and this is used to estimate height at Age_1. TPH_1 is estimated by multiplying planting density (TPH_0) with a default survival percentage (Surv%). BA_1 is estimated with a Schumacher-type basal area function from Age_1, TPH_1 and Hd_1.

The most problematic part of growth and yield prediction is to project basal area over time in thinned stands. An example is provided, showing the current models in use. The following multi-step process is used:

1. Project Hd, TPH and BA from calibration age to thinning age. The most used functions are given below.
1a. The Chapman–Richards-type height growth function (Richards, 1959):

$$HD_2 = HD_1 \left[\frac{\left(1 - e^{\left(\beta_1\left(Age_2 + \beta_2\right)\right)}\right)}{\left(1 - e^{\left(\beta_1\left(Age_1 + \beta_2\right)\right)}\right)} \right]^{\beta_3} \tag{1}$$

where HD_1 is the dominant height at Age_1, the calibration age; HD_2 is the projected height at Age_2; and β_1 to β_3 are parameters to be estimated.
1b. The Clutter–Jones survival function (Clutter and Jones, 1980):

$$TPH_2 = \left[TPH_1^{\beta_1} + \beta_2 \cdot \left[\left(\frac{Age_2}{100} \right)^{\beta_3} - \left(\frac{Age_1}{100} \right)^{\beta_3} \right] \right]^{\frac{1}{\beta_1}} \tag{2}$$

where TPH_1 is the stems per hectare at Age_1, the calibration age; and TPH_2 is the projected stems per hectare at Age_2.

1c. The Schumacher-type stand-level basal area function for unthinned stands (Schumacher, 1939; Pienaar and Harrison, 1989):

$$BA_2 = \exp\left[\begin{array}{l}\ln(BA_1) + \beta_1 \cdot \left(\dfrac{1}{Age_2} - \dfrac{1}{Age_1}\right) + \beta_2 \cdot \left(\ln(TPH_2) - \ln(TPH_1)\right) \\ + \beta_3 \cdot \left(\ln(HD_2) - \ln(HD_1)\right) + \beta_4 \cdot \left(\dfrac{\ln(TPH_2)}{Age_2} - \dfrac{\ln(TPH_1)}{Age_1}\right) \\ + \beta_5 \cdot \left(\dfrac{\ln(HD_2)}{Age_2} - \dfrac{\ln(HD_1)}{Age_1}\right)\end{array}\right] \quad (3)$$

where BA_1 is the basal area per hectare at Age_1, the calibration age; and BA_2 is the projected basal area per hectare at Age_2.

1d. The Chapman–Richards-type basal area function for unthinned stands (Pienaar and Turnbull, 1973):

$$BA_2 = A \cdot \left[1 - \exp\left(k \cdot \left(Age_2 + t_0\right)\right)\right]^m \quad (4)$$

where A, k, t_0 and m are parameters to be estimated.

2. Reduce the TPH according to the thinning schedule. A typical thinning schedule for pine sawtimber is as follows: plant at 1372 trees/ha, a survival percentage of 89% is the default. Thin at age 8 years to 650 stems/ha, at age 13 years to 400 stems/ha, at age 18 years to 250 stems/ha and clearfell at 35 years.

3. Estimate the BA remaining after thinning with Field's thinning ratio function (Field *et al.*, 1978):

$$BA_t = BA_b \cdot \left(\frac{TPH_t}{TPH_b}\right)^{\beta_0} \quad (5)$$

where BA_t is the basal area per hectare removed in the thinning; BA_b is the basal area before the thinning; TPH_t is the stems per hectare removed in the thinning; and TPH_b is the stems per hectare before the thinning.

4. Advance the calibration point to an age just after the thinning (Age_1):

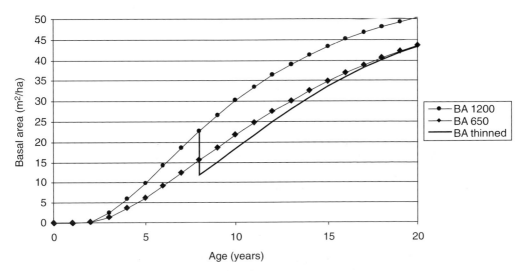

Fig. 7.4. Basal area after thinning using the index of suppression.

- Hd$_1$:   Hd after thin = Hd before thin;
- TPH$_1$:   TPH after thin = TPH remaining as prescribed in the thinning regime;
- BA$_1$:   BA after thin = BA before thin − BA removed.

5. Project dominant height as indicated in Equation 1 above from Age$_1$ to the new Age$_2$.

6. The assumption is made that there is no mortality between intense and frequent thinnings. Therefore, TPH remains constant until the next thinning.

7. Basal area response after thinning is one of the most difficult model components to develop and to explain. The basic assumption is that a stand thinned from, say 1200 stems/ha to 650 stems/ha, will have a lower basal area than an equivalent unthinned stand standing at 650 stems/ha (Fig. 7.4). The hypothesis for basal area growth after thinning is that the thinned stand will eventually catch up with the unthinned stand (Marsh and Burgers, 1973). This type of growth response, in which data are available for both the unthinned and thinned curves, can only be obtained from formal thinning response trials as, for example, in the old CCT trials.

Three approaches have been defined to model this basal area response, namely:

7a. The index of suppression, used in conjunction with the Chapman–Richards-type models (Pienaar and Shiver, 1984).

$$BA_2 = BA_u \left(1 - IS_2\right) \tag{6}$$

$$IS_2 = IS_1 \cdot e^{-\beta_0 \left[\frac{(AGE2 - AGE1)}{Age_1} \right]^{\beta_1}}$$

$$IS_1 = 1 - \left(\frac{BA_a}{BA_{ut}} \right)$$

where BA$_a$ is the basal area per hectare after thinning at Age$_2$; BA$_u$ is the basal area per hectare of the unthinned stand growing at the same stand density to which the stand is thinned to at Age$_2$; Ba$_{ut}$ is the basal area per hectare of the unthinned stand at the age of thinning; IS$_1$ is the index of suppression at the age of thinning; and IS$_2$ is the index of suppression at Age$_2$.

7b. The stand-level thinning response term used with the Schumacher-type functions (Pienaar and Harrison, 1989).

$$BA_2 = \exp \left[\begin{array}{l} \ln(BA_1) + \beta_1 \cdot \left(\dfrac{1}{Age_2} - \dfrac{1}{Age_1} \right) + \beta_2 \cdot \left(\ln(TPH_2) - \ln(TPH_1) \right) \\[2ex] + \beta_3 \cdot \left(\ln(HD_2) - \ln(HD_1) \right) + \beta_4 \cdot \left(\dfrac{\ln(TPH_2)}{Age_2} - \dfrac{\ln(TPH_1)}{Age_1} \right) \\[2ex] + \beta_5 \cdot \left(\dfrac{\ln(HD_2)}{Age_2} - \dfrac{\ln(HD_1)}{Age_1} \right) + \beta_6 \cdot \dfrac{TPH_t}{TPH_a} \cdot \left[\left(\dfrac{Age_t}{Age_2} \right)^{\beta_7} - \left(\dfrac{Age_t}{Age_1} \right)^{\beta_7} \right] \end{array} \right] \tag{7}$$

7c. The age-adjustment method, when only unthinned curves are available. This is an algorithmic method and is based on observations by Marsh on South African CCTs (Marsh and Burgers, 1973). Marsh's hypothesis states that the increment of a thinned stand equals that of an unthinned stand of the same density and basal area, but at an earlier age. The effect is that the growth rate of the thinned stand after the thinning is initially faster than that of an unthinned stand until the difference in basal area is made up. This hypothesis works well, except where excessive suppression has been allowed to develop between thinnings. Alder (1980) and Crockford (1995) have described the yield prediction procedure.

8. A number of other variables can be derived from the basic growth curves, namely

quadratic mean DBH, mean tree volume, mean tree-based whole stand volume, mean annual increment, stock tables with a DBH distribution and average height by DBH class and product volumes. The details of these variables are not discussed.

Simulation Systems

The growth and yield research cycle is completed when the model configurations have been implemented in growth and yield simulators. These simulators are forestry planning tools and have different roles, for instance they are used in management planning systems (e.g. the G&Y Shell; Nel, 2002) or as stand-alone systems for forestry scenario analyses (FORSAT; Kotze *et al.*, 2001), the Pruning Scheduler (Kotze and Vonck, 1997) or for optimal bucking of stands (WINCUTUP; Vonck, 1997).

Recommendations

The key for success is to find an effective methodology for the development of each component in the growth and yield cycle and to maintain continuity over a long period of time. Each aspect of the growth and yield research programme requires its due attention. The emphasis is on an efficient and cost-effective field programme. Data are sourced from a well-stratified field programme for each model component. Industry interaction is aligned with the company objectives as far as possible to enable maximum benefit.

The empirical stand-level modelling approach is recommended, because these models can be easily implemented in simulators and can be calibrated with data that are readily available in the planning database. However, not all the basal area and basal area response after thinning model types as illustrated in the paragraph on growth modelling can be directly applied to (for example) the PSP data from thinned stands. It is recommended that other modelling approaches be applied for this specific data source.

It is recommended that Komatiland Forests supports other research institutions in South Africa currently working on other modelling strategies, such as process-based models, until they can be applied in a practical manner. It is also recommended that joint modelling exercises be coordinated between member companies in the MMRC for the processing of pooled data. This will be beneficial, especially for minor species covering small areas.

References

Alder, D. (1980) *Forest Volume Estimation and Yield Prediction*, Vol. 2. FAO Forestry Paper 22/2, Rome.

Bredenkamp, B.V. (1990) The triple-S CCT design. In: von Gadow, K. and Bredenkamp, B.V. (eds) *Proceedings of a Symposium arranged by the Forest Mensuration and Modelling Working Group in Collaboration with the Southern African Institute of Forestry and the Eucalyptus grandis Research Network on 'Management of Eucalyptus grandis in South Africa'*, pp. 98–205.

Clutter, J.L. and Jones, E.P. Jr (1980) *Prediction of Growth after Thinning in Old-field Slash Pine Plantations*. US Department of Agriculture, Service Research Paper SE-217, pp. 1–8.

Craib, I.J. (1939) *Thinning, Pruning and Management Studies on the Main Exotic Conifers Grown in South Africa*. Department of Agriculture and Forestry, Government Printer, Pretoria, 179 pp.

Crockford, K.J. (1995) Yield models and tables for *Pinus patula, Pinus elliottii* and *Pinus taeda* in Zimbabwe. *Zimbabwe Bulletin of Forestry Research* 12, Forestry Commission, 51 pp.

Du Plessis, M., Chiswell, K.E. and Morley, T. (1997) *Criteria for the Inclusion of Permanent Sample Plots in the Mensuration and Modelling Research Consortium, Version 1.* Mensuration and Modelling Research Consortium, South Africa, 30 pp.

Field, R.C., Clutter, J.L. and Jones, E.P. Jr (1978) Predicting thinning volumes for pine plantations. *Southern Journal of Applied Forestry* 2(2), 59–61.

Kotze, H. (1995) *Standards and Formats for Data and Databases.* CSIR Report FOR-C-299, 47 pp.

Kotze, H. (2001) *A Summary of Growth and Yield Models in South Africa.* SAFCOL Research, South Africa, 29 pp.

Kotze, H. and Groenewald, W.H. (1995) *Progress Report: a Data Manipulation System for Growth and Yield Trials.* CSIR Report FOR-I-616, 74 pp.

Kotze, H. and Vonck, D.I. (1997) A growth simulator and pruning scheduler for *Pinus patula* in Mpumalanga–North Province, South Africa. In: Amaro, A. and Tomé, M. (eds) *Empirical and Process-based Models for Forest Tree and Stand Growth Simulation.* 21–27 September, Portugal, pp. 205–221.

Kotze, H., Kassier, H.W. and van der Merwe, D.G. (2001) *FORSAT Specification, Design and Implementation Document.* SAFCOL Research, South Africa, 90 pp.

Kunz, R. and Pallet, R. (2000) *A Stratification System Based on Climate and Lithology for Locating Commercial Forestry Permanent Sample Plots.* ICFR Bulletin Series 01/2000, 17 pp.

Marsh, E.K. and Burgers, T.F. (1973) The response of even-aged pine stands to thinning. *Forestry in South Africa* 14, 103–110.

Nel, J.H. (2002) *A Generic Shell for Growth and Yield Models.* AfriGis Environmental Solutions, South Africa, 69 pp.

Nelder, J.A. (1962) New kinds of systematic designs for spacing experiments. *Biometrics* 18, 283–307.

O'Connor, A.J. (1935) Forest research with special reference to planting distances and thinning. *British Empire Forestry Conference*, South Africa, 30 pp.

Pienaar, L.V. and Harrison, W.M. (1989) Simultaneous growth and yield prediction equations for *Pinus elliottii* plantations in Zululand. *South African Forestry Journal* 149, 48–53.

Pienaar, L.V. and Shiver, B.D. (1984) An analysis and models of basal area growth in 45-year-old unthinned and thinned slash pine plantation plots. *Forest Science* 30, 933–942.

Pienaar, L.V. and Turnbull, K.J. (1973) The Chapman–Richards generalization of Von Bertalanffy's growth model for basal area growth and yield in even-aged stands. *Forest Science* 19, 2–22.

Pienaar, L.V., Harrison, W.M. and Bredenkamp, B.V. (1988) *Yield Prediction Methodology for E. grandis Based on Langepan CCT Data.* University of Georgia, 36 pp.

Richards, F.J. (1959) A flexible growth function for empirical use. *Journal of Experimental Botany* 10, 290–300.

Schumacher, F.X. (1939) A new growth curve and its application to timber yield studies. *Journal of Forestry* 37, 819–820.

Vonck, D.I. (1997) *WINCUTUP User's Manual.* SAFCOL Planning and Development, South Africa, 6 pp.

Part 2

Mathematical Approaches and Reasoning

All models reflect scale in some way. Models may be developed at one particular scale and applied at that scale. Other models may be developed and applied at multiple scales. Some models may be developed at one scale and applied at another. Scale may be an explicit component in a model or merely implied by the structure and nature of the model. Regardless of how scale is used or functions within a model, a model cannot be assessed or evaluated without some consideration of scale. As models become more 'realistic', it is likely that issues revolving around scale will become more acute. Thus, it is important to continue to develop and refine tools for modelling within and across spatial, temporal and hierarchical scales.

Most models of forest systems that have been developed are state models. They describe or pertain to the conditions of the system of interest. Most process models are state models, as are some empirical models. An interactive model is sensitive to changing conditions and allows the user to become a part of the modelling exercise, from the initial conceptual and design phase of model development to final use and application. It is likely that an increasingly sophisticated public will expect models of all types to be more interactive in nature.

Society's expectations of how forest models can be used and what information they will provide can be incongruous with the design and intent of developed models. Sometimes this occurs because society is asking different questions or perhaps broader questions than the models are capable of addressing. At other times there is inadequate continuity between the scientific work of model development and what society needs or wants. Additionally, sometimes society's questions or priorities change. Therefore, it is important to develop modelling methods or approaches that address the current issues that society is facing as well as being flexible enough to accommodate future concerns.

Usually, the development of a forest model is initiated by developing a simple submodel or component that will form the basis of the full model. From this point forward, the modeller confronts the tension that exists between simplification and aggregation. On the one hand, forest models are simplifications of forest reality. Thus, models that become too complex lose their usefulness. On the other hand, the need and desire for more realistic models tend to promote the aggregation of submodels and components into more complex model systems. Clearly, modellers must use a development process that aggregates submodel components such that the proper balance between model simplicity (for the sake of usefulness) and complexity (for the sake of reality) is achieved.

Many of the traditional modelling tools, such as regression and other statistical techniques, continue to provide satisfactory results for many modelling tasks. Their refinement continues to be an important area of research. There are also some relatively new modelling tools that have not been fully developed or applied frequently. Markov and Bayesian techniques are examples of tools that utilize the temporal nature of permanent plot histories to maximize the use of prior plot information. Another tool, just in its infancy, is cellular automata. In fact, even its potential as a forest-modelling tool has not yet been evaluated.

With regard to future work, two questions arise. The first is, how good are our models for communication? While a modeller may be satisfied that a model meets the needs of targeted end-users, those same end-users may not fully agree. This can occur for a variety of reasons including misperception of the needs of the end-user by the modeller and a failure of the modeller and end-users to adequately communicate and understand the modelling goals and objectives. Developing new and better techniques for assessing how successful modellers are at communicating should be an important area of research.

A second question pertains to the extension of models for addressing larger questions: can traditional process and statistical models be combined into a more general heuristic system of problem-solving tools? This question implies that, beyond the purpose for which they were developed, forest models may have additional value as information sources for larger, more comprehensive systems. This may be a fertile field for future research.

The contributions that follow address these and other issues encountered in the choice of modelling approach and the development of the rationale used in selecting modelling approaches for various purposes. The editors wish to acknowledge the efforts of Ralph Amateis (USA) and Paula Soares (Portugal) in coordinating this section, and thank them for their contributions to this volume.

8

Quantitative Tools and Strategies for Modelling Forest Systems at Different Scales

Ralph L. Amateis[1]

Abstract

Decisions regarding the management and monitoring of forests are often predicated on information obtained from models. The nature, scope, scale and purpose of models used for assessing forests are many and vary widely. Correspondingly, the mathematical and statistical tools used to develop and test such models are equally prodigious and diverse. Choosing a particular quantitative tool depends on a host of factors including the purpose for which the model is being developed, available data for parameter estimation, model scale (both spatial and temporal), inherent assumptions in the model, background and philosophical bent of the modeller, and potential users of the model. While some modelling tasks seem tailor-made for the application of a certain mathematical or statistical technique, other tasks seem not to lend themselves so obviously to the use of a particular quantitative tool. One arduous task for which adequate quantitative tools seem to be lacking is spanning scales at various resolutions. This chapter considers the task of modelling across scales and discusses some of the quantitative tools useful for that task. Particular attention is paid to the use of simulation and experimentation.

Introduction

Every professional carries a bag of tools to practise their trade: hammers, saws and chisels for carpenters; beakers, bottles and balances for chemists; and ANOVA, regression and experimental designs for statisticians. As a growth and yield modeller entering the forestry profession in the 1970s, my bag of tools contained a few statistical and sampling procedures, a rudimentary knowledge of Fortran and Basic, and some experience of measuring field plots. As a beginner's toolbox, it proved adequate for addressing a number of modelling problems. Additions (and a few lost items) have occurred as the years have come and gone following the general rule that problems arise and then tools are developed to solve those problems.

[1] Department of Forestry, Virginia Tech, USA
Correspondence to: ralph@vt.edu

Since the 1970s, I have witnessed two important changes in the field of forest biometrics and modelling that have significantly affected 'tool development' in forest modelling. The first is a blurring of the lines between different levels, or resolutions, of forest modelling. My training and early work were almost exclusively in growth and yield modelling. In recent years, however, I have worked on a number of other forest modelling projects including response to various silvicultural practices, physiological processes of tree growth, population cycles of forest insects, decay rates of woody debris, aspects of climate change as they affect forest development, and wood quality characteristics, to name but a few. While it is tempting to focus attention on the differences between these hierarchies or resolutions, their similarities are probably more important (Robinson and Ek, 2000). In particular, I found that many of the quantitative tools used by growth and yield modellers could be applied to other levels of forest structure and function. In some cases, tools had to be refined prior to application; in other cases, new and unfamiliar tools were borrowed from elsewhere or created.

Each of the many resolutions of forest structure functions at a particular spatial and temporal scale. The disparity of these scales is great, ranging from micrometres and seconds for physiological processes to hectares and millennia for stand and landscape changes (Reed, 1999). Modern society has come to expect a comprehensive accounting of the forest's structure, function, products, services, sustainability and response to changing conditions. This 'full-scale' accounting includes inventorying, monitoring, studying and modelling across the full range of spatial and temporal scales to which it belongs.

It has been no surprise, therefore, to witness a second important change emerging, which is the need to develop methods and construct models that are effective across spatial and temporal scales. In the 1970s, when my career began, attempts to model across a range of scales were nascent. Since then, an awareness of how scale affects our modelling efforts has grown to the point that scaling issues have become an explicit consideration in most modelling projects. Overall, I have seen that considerable progress has been made with regard to modelling across scales, with the implicit understanding that much work still remains to be done.

In this chapter, I discuss the development and application of two modelling tools – simulation and experimentation – and consider how they might be used to link different levels of forest systems and model forest systems across spatial and temporal scales. Woven through the discussion of scale and how to tackle its modelling challenges through simulation and experimentation, I present some of the other quantitative tools that have proved useful for modelling forest systems at different levels of resolution. I have tried to note in passing those tools that have been well developed and fully presented in the literature and to linger over those tools that seem promising but are new or have not been as well developed or extensively applied.

Tools for Modelling Across Spatial and Temporal Scales

Simulation

A number of models have been developed over the years to quantify forest systems at different levels of resolution and encompassing different spatial and temporal scales. Some, such as growth and yield models, are useful tools for predicting the growth and development of tree and stand dimensions for making predictions of wood yield. Other models are developed to characterize biological and physiological processes of trees and particular hierarchies of forest systems (Mäkelä, 2001). Tools, such as aggregation, have also been developed that can translate knowledge about forest attributes across hierarchies, usually from finer to coarser (Rastetter *et al.*, 1992; Baldocchi, 1993; Jarvis, 1993). Spatialization tools that map forest characteristics and climate variables across the landscape are necessary for applying models to broad geographical areas (Franco-Lopez *et al.*, 2001). Additional work at the interface of management-oriented models and physiological process models has brought the physiological processes of forest systems into the manager's arena (Battaglia and Sands, 1998; Johnsen *et al.*, 2001).

No single model will be best for all purposes, and I believe it is prudent not to expect otherwise. There are several reasons for this. First, models are simplifications of reality and reflect the inherent inclinations, limitations, assumptions, biases and purposes of the modeller. Secondly, models contain collective knowledge gained from previous experience along with current information. As such, models are dynamic and change as new knowledge and more relevant information become available. On the other hand, there may be times when jettisoning the old paradigm is unwarranted even in the face of new knowledge. We do not use quantum physics to design bridges; Newtonian mechanics works quite well enough. Thirdly, it is unreasonable to expect any single model to be effective at characterizing the richness, complexity, scale and dynamic nature of forest systems. Instead, it seems likely that greater advances will be achieved when models of various types and resolutions can be integrated into systems of models. Information can be passed between models of the system reflecting how different levels of resolution and scale interact and affect each other. This approach was used to link two simulators for loblolly pine plantations; a growth and yield model (PTAEDA) with a light-interception process model (MAESTRO) (Baldwin *et al.*, 1993, 2001). In these studies, the process model and the growth and yield model were linked so that output from one model became input for the other. Thus it was possible to assess the effects of crown canopy light-interception parameters on the competitive relationships and wood production capacity of loblolly pine plantations.

In a larger and more comprehensive study, an approach termed signal transfer modelling was used by Luxmore *et al.* (2000) to assess responses to environmental stress across a number of forest types in the south-eastern USA. Beginning with a set of environmental parameters and initial conditions that affect tree growth, computer simulation results (response signals) from ecophysiological process models were 'scaled up' to generate stemwood increment responses. These responses became inputs for other models, such as growth and yield simulators. In this way, a grid of responses was created for a wide range of environmental parameters and initial conditions. Monte Carlo simulation techniques were used throughout the model propagation process, creating a frequency distribution of outputs at each level of resolution that were analysed with appropriate statistical techniques. Inferences were then made about the effects of changing environmental conditions on south-eastern US forests. From the simulation results, GIS databases were assembled for plant, soil and climate attributes across the region. Maps were then generated depicting vegetation and land-use changes for particular simulation scenarios. Forest production, evapotranspiration, carbon storage and other forest vegetation parameters could then be evaluated.

No model can be considered 'finished' without a careful process of testing and evaluation. Some of the most important testing criteria include logical and biological consistency, statistical properties and a characterization of model error (Vanclay and Skovsgaard, 1997). Some obvious questions and concerns arise when assembling models into simulators and linking together simulators of various types. How should the uncertainty associated with model predictions due to imprecisely known input variables be evaluated? How should appropriate sensitivity analyses (contribution of individual input variables to the uncertainty of model predictions) be conducted? What methods should be used to address the error propagating through a system of models? This last question becomes especially important when output from one simulator is used as input to another, and estimates and predictions span several levels of resolution. Gertner and Dzialowy (1984) used Monte Carlo methods to empirically approximate the variance of predictions in a model. The advantage of the Monte Carlo method is that it obviates the need for an independent set of data. The disadvantage is that it is computationally laborious and time consuming. In subsequent work, Gertner (1987) developed an error propagation method that was much more efficient computationally than the Monte Carlo method.

Additional tools developed recently hold great promise for model testing and evaluation. Saltelli *et al.* (1999) produced a method for conducting sensitivity analysis of model output based on the Fourier amplitude sensitivity test (FAST). FAST allows the total contribution of each input variable to the output's variance to be computed. While the motivation for this and other similar work (Helton, 1993) comes from a need to evaluate engineering models, the application to forest models is clear.

Some of the most recent work on model evaluation, termed Bayesian synthesis and Bayesian melding (a modification of Bayesian synthesis), is a probabilistic approach that takes into account both parameter and model uncertainty. Theoretical work by Raftery *et al.* (1995) and Poole and Raftery (2000) has been applied to a mechanistic forest model (PIPESTEM) (Green *et al.*, 1999). Subsequently, the technique has been refined and applied to another forest model (PnET-II) by Radtke *et al.* (2002). These diagnostic tools allow complex models and simulators to be subjected to rigorous testing procedures so that outputs and predictions can be evaluated for reliability.

Experiments

Experiments are fundamental to elucidating the cause and effect relationships that govern the development of forest systems and are essential for advancing the state of forest science. Experiments are usually established at the same spatial and temporal scale as the forest system under investigation. In this way, hypotheses concerning the system can be tested directly and data from the experiment can be used to develop formal (mathematical) models at the same scale as the population of interest. A formal model is a set of scale-relative statements that can be applied to the material system of interest. These scale-relative statements typically consist of the mathematical and statistical equations and parameters that characterize the system. Encoding and decoding occurs between the material system and the formal model as information and data from the material system are used to construct the formal model, and predictions and estimations from the formal model are used to characterize the material system. Forest experiments have been established to evaluate a wide range of management regimes including planting density, site–species interactions, genetic trials, thinning, pruning, fertilizer use and competition control, to name but a few. Large-scale, multi-disciplinary silvicultural and ecological experiments such as the Long-Term Ecological Research (LTER) network and the Forest Ecosystem Research Network of Sites (FERNS) are important for understanding forest systems and how they respond to human manipulation (Monserud, 2002).

Regression techniques are among the most commonly applied statistical tools for describing forest experiments. Regression equations are developed from experimental data collected in the material systems and comprise the formal model that describes the structures and functions of the material systems. As such they reflect the spatial and temporal scale of the material systems from which they have been derived, although they may span multiple levels, or hierarchies, of resolution. Defining and parameterizing regression models was at one time a rather straightforward procedure. A suitable model was derived from underlying mechanistic assumptions or from correlative relationships found in the data (or perhaps both). Model parameterization using ordinary least squares was the norm. In recent years this procedure has been advanced and refined to accommodate a wide range of data structures and modelling situations for both mechanistic and empirical models (e.g. Lappi and Bailey 1988; Gregoire and Dyer, 1989; Gregoire *et al.*, 1995). Many of the initial advancements in this area have been for linear models. Similar tools for the case of non-linear models are now becoming available.

Two other tools that have been applied to modelling forest systems at a given scale are neural networks and fractal analysis. Neural networks utilize data to define a particular functional relationship between variables rather than a specified model (Warner and Misra, 1996). Since no predetermined model is needed, neural networks are particularly useful when the underlying functional relationships are not known. Recently, neural networks have been used in cover type mapping and remote sensing applications (Gong *et al.*, 1997; Blackard and Dean, 1999), forest resource management (Gimblett and Ball, 1995), modelling wood characteristics (Schultz *et al.*, 1999) and assessing uncertainty in growth models (Guan *et al.*, 1997).

There are many tree and forest characteristics that can be quantified using Euclidean dimensions. Tree stem volume, for example, is Euclidean (because volume does not change with the units used to measure it). Other characteristics, however, are not Euclidean in their morphology, but fractal; their determination depends on the particular units used to measure them. Tree crowns and their composite, the forest canopy, are examples of fractals (mass fractals in this case). Processes such as respiration will also scale as fractal rather than Euclidean because they are a function of surface area, which is fractal. An awareness of the fractal dimension of some tree, forest and landscape characteristics coupled with the pioneering work of Mandelbrot (1983) has spawned the development of fractal analysis techniques in forestry. Zeide and Pfeifer (1991) developed a technique for determining the fractal dimension of tree crowns that confines the fractal dimension to lie between 2 and 3. They note that the calculated fractal dimension has a natural interpretation related to the distribution of foliage mass in the crown and perhaps its shade tolerance. In another study, Zeide and Gresham (1991) related the fractal dimension of tree crowns with two forest stand characteristics associated with growth: site index and thinning intensity. Clearly, fractal analysis techniques can be attractive tools for modelling the rich morphological complexity of certain tree and stand structures.

While current tool development promises incremental forward progress in modelling success, I believe that greater strides are possible by reducing the scale of experiments themselves in order to bring the spatial and temporal scales of various hierarchies of forest systems closer together. That is, by utilizing an experimental physical model of a forest system established at a smaller scale than the system of interest, a reduction in the disparity of spatial and temporal scales between hierarchies of forest systems is achieved. This is well understood and practised with physical systems, such as the dynamics of aeroplane wing construction, where the relationship between material systems at two scales are analogues of each other. In such systems the formal model, which is digital in nature, encodes the structure and function of the material systems and provides a pathway to mathematically relate the analogues. Scaling functions are part of the formal model and relate phenomena occurring in the small-scale analogue with its full-scale counterpart (Allen, 2001).

With some exceptions (Landsberg and Ludlow, 1970), the use of analogue models in forestry has not been widely applied. This type of experimentation, however, has a long history in the physical sciences, and the mathematical theory used to construct formal models for these systems has been well developed (Gukhman, 1965; Schuring, 1977; Sedov, 1993). It appears that much of this mathematical theory, which arises from the principles of similarity and dimensional methods, can be applied to experimentation with forest systems. Schneider (1994) makes a similar case for the application of this technology and its underlying theoretical principles to ecological research and experimentation. Ecologists have advocated using experimental model systems, called mesocosm or microcosm experiments, to elucidate mechanisms underlying ecological processes (Fraser and Keddy, 1997; Wiens, 2001). It should be possible to use these techniques to study mechanisms underlying forest systems as well.

Discussion

The progression of science seems to be that questions and problems arise, followed by the development of the analytical tools necessary to address those questions and problems. Forest systems are unique because the questions and problems that arise are complex and span a range of scales at various resolutions. When modelling forest systems we typically array before us a set of mathematical and statistical tools to build and test equations, systems of equations and simulators that represent the reality of the systems of interest. Many of these tools are satisfactory and quite appropriate for use with specific applications. Thus, one line of work has been, and probably will continue to be, refining our existing tools. We will benefit in even greater measure as we successfully integrate simulators into model systems capable of analysing the complex relationships across the broad range of scales we find in forestry.

Society has high expectations for its forests: wood supplier, repository for flora and fauna, recreation paradise, storehouse for carbon, and environmental protection, to name but a few. Likewise, society has high expectations for forest modellers. We are increasingly expected to develop models that are more realistic and produce more precise estimates of important forest characteristics in a shorter amount of time with fewer resources. To accomplish this we will need new tools that allow us to make greater leaps forward. One of the most promising may be experimentation.

Usually experiments are conducted directly on the system of interest so that inferences can be readily made. Problems arise, however, because the experiment spans considerable spatial and temporal scales at different resolutions. One solution may be to conduct the experiment on a small-scale analogue of the larger experiment and so reduce the range of scale of the experiment. This brings phenomena at different scales closer together. In a preliminary study of this type of experiment, it appears that planting loblolly pine trees at a 1/16 spatial scale reduces rotation length by approximately one order of magnitude (Amateis *et al.*, 2001).

Experiments, of course, are useful only if their results can be extrapolated to the scale of the systems of interest. In some cases, experiments may be conducted to test key concepts, examine basic principles of forest development or to look for generality. In other cases, the primary goal of experiments may be to estimate parameters and make direct predictions for the full-scale system of interest. When the latter is the goal, principles from dimensional and similarity analyses appear to offer a rigorous, mathematical framework upon which extrapolation can be based.

The whole concept of extrapolation brings to mind a host of important questions. Are there scale dependencies in forest systems that do not extrapolate well and, if so, what are they? Can the physiological processes of tree seedlings be scaled to their mature tree counterparts? Do allometric relationships differ with small-scale studies and, if so, can they be scaled up? While mean values may be extrapolated, what about variance? These and many other questions must be considered as we delve into the arena of small-scale experimentation. Experimentation is a tool that, with a good deal of sharpening and refining, holds considerable promise for helping us with our modelling tasks.

References

Allen, T.F.H. (2001) The nature of the scale issue in experimentation. In: Gardner, R.H., Kemp, W.M., Kennedy, V.S. and Petersen, J.E. (eds) *Scaling Relations in Experimental Ecology*. Columbia University Press, New York.

Amateis, R.L., Sharma, M. and Burkhart, H.E. (2001) Using miniature scale plantations as experimental tools for assessing sustainability issues. In: Lemay, V. and Marshall, P. (eds) *Forest Modeling for Ecosystem Management, Forest Certification, and Sustainable Management*. Vancouver, BC, Canada, pp. 147–156.

Baldocchi, D.D. (1993) Scaling water vapour and carbon dioxide exchange from leaves to a canopy: rules and tools. In: Ehleringer, J.R. and Field, C.B. (eds) *Scaling Physiological Processes: Leaf to Globe*. Academic Press, San Diego, California, pp. 77–114.

Baldwin, V.C. Jr, Burkhart, H.E., Dougherty, P.M. and Teskey, R.O. (1993) *Using a Growth and Yield Model (PTAEDA2) as Driver for a Biological Process Model (MAESTRO)*. USDA Forestry Service Research Paper SO-276, 9 pp.

Baldwin, V.C. Jr, Burkhart, H.E., Westfall, J.A. and Peterson, K.D. (2001) Linking growth and yield and process models to estimate impact of environmental changes on growth of loblolly pine. *Forest Science* 47, 77–82.

Battaglia, M. and Sands, P.J. (1998) Process-based forest productivity models and their application in forest management. *Forest Ecology and Management* 102, 13–32.

Blackard, J.A. and Dean, D.J. (1999) Comparative accuracies of artificial neural networks and discriminant analysis in predicting forest cover types from cartographic variables. *Computers and Electronics in Agriculture* 24(3), 131–151.

Franco-Lopez, H., Ek, A.R. and Bauer, M.E. (2001) Estimation and mapping of forest stand density, volume, and cover type using the k-nearest neighbors method. *Remote Sensing of Environment* 77, 251–274.

Fraser, L.H. and Keddy, P. (1997) The role of experimental microcosms in ecological research. *Trends in Ecology and Evolution* 12, 478–481.

Gertner, G. (1987) Approximating precision in simulation projections: an efficient alternative to Monte Carlo methods. *Forest Science* 33, 230–239.

Gertner, G. and Dzialowy, P. (1984) Effects of measurement errors on an individual tree-based growth projection system. *Canadian Journal of Forestry Research* 14, 311–316.

Gimblett, R.H. and Ball, G.L. (1995) Neural network architectures for monitoring and simulating changes in forest resource management. *Artificial Intelligence Applications* 9, 103–123.

Gong, P., Pu, R. and Chen, J. (1997) Mapping ecological land systems and classification uncertainties from digital elevation and forest cover data using neural networks. *Photogrammetric Engineering and Remote Sensing* 62, 1249–1260.

Green, E.J., MacFarlane, D.W., Valentine, H.T. and Strawderman, W.E. (1999) Assessing uncertainty in a stand growth model by Bayesian synthesis. *Forest Science* 45, 528–538.

Gregoire, T.G. and Dyer, M.E. (1989) Model fitting under patterned heterogeneity of variance. *Forest Science* 35, 105–125.

Gregoire, T.G., Schabenberger, O. and Barrett, J.P. (1995) Linear modelling of irregularly spaced unbalanced, longitudinal data from permanent-plot measurements. *Canadian Journal of Forestry Research* 25, 137–156.

Guan, B.T., Gertner, G.Z. and Parysow, P. (1997) A framework for uncertainty assessment of mechanistic forest growth models: a neural network example. *Ecological Modelling* 98, 47–58.

Gukhman, A.A. (1965) *Introduction to the Theory of Similarity*. Academic Press, New York, 256 pp.

Helton, J.C. (1993) Uncertainty and sensitivity analysis techniques for use in performance assessment for radioactive waste disposal. *Reliability Engineering and System Safety* 42, 327–367.

Jarvis, P.G. (1993) Prospects for bottom-up models. In: Ehleringer, J.R. and Field, C.B. (eds) *Scaling Physiological Processes: Leaf to Globe*. Academic Press, San Diego, California, pp. 115–126.

Johnsen, K., Samuelson, L., Teskey, R., McNulty, S. and Fox, T. (2001) Process models as tools in forestry research and management. *Forest Science* 47, 2–8.

Landsberg, J.J. and Ludlow, M.M. (1970) A technique for determining resistance to mass transfer through the boundary layers of plants with complex structure. *Journal of Applied Ecology* 7, 187–192.

Lappi, J. and Bailey, R.L. (1988) A height prediction model with random stand and tree parameters: an alternative to traditional site index methods. *Forest Science* 34, 907–927.

Luxmore, R.J., Hargrove, W.W., Tharp, M.L., Post, W.M., Berry, M.W., Minser, K.S., Cropper, W.P. Jr, Johnson, D.W., Zeide, B., Amateis, R.L., Burkhart, H.E., Baldwin, V.C. Jr and Peterson, K.D. (2000) Signal-transfer modelling for regional assessment of forest responses to environmental changes in the southeastern United States. *Environmental Modelling and Assessment* 5, 125–137.

Mäkelä, A. (2001) Modelling tree and stand growth: towards a hierarchical treatment of multi-scale processes. In: LeMay, V. and Marshall, P. (eds) *Forest Modelling for Ecosystem Management, Forest Certification and Sustainable Management*. University of British Columbia, Vancouver, pp. 27–44.

Mandelbrot, B.B. (1983) *The Fractal Geometry of Nature*. W.H. Freeman, New York, 468 pp.

Monserud, R.A. (2002) Large-scale management experiments in the moist maritime forests of the Pacific Northwest. *Landscape and Urban Planning* 59, 159–180.

Poole, D. and Raftery, A.E. (2000) Inference for deterministic simulation models: the Bayesian melding approach. *Journal of the American Statistical Association* 95, 1244–1255.

Radtke, P.J., Burk, T.E. and Bolstad, P.V. (2002) Bayesian melding of a forest ecosystem model with correlated inputs. *Forest Science* 48, 701–711.

Raftery, A.E., Givens, G.H. and Zeh, J.E. (1995) Inference from a deterministic population dynamics model for Bowhead whales. *Journal of the American Statistical Association* 90, 402–430.

Rastetter, E.B., King, A.W., Cosby, B.J., Hornberger, G.M., O'Neill, R.V. and Hobbie, J.E. (1992) Aggregating fine-scale ecological knowledge to model coarser-scale attributes of ecosystems. *Ecological Applications* 2, 55–70.

Reed, D.D. (1999) Ecophysiological models of forest growth: uses and limitations. In: Amaro, A. and Tomé, M. (eds) *Empirical and Process-based Models for Forest Tree and Stand Growth Simulation*. 21–27 September Portugal, pp. 305–311.

Robinson, A.P. and Ek, A.R. (2000) The consequences of hierarchy for modelling in forest ecosystems. *Canadian Journal of Forestry Research* 30, 1837–1846.

Saltelli, A., Tarantola, S. and Chan, K.P.S. (1999) A quantitative model-independent method for global sensitivity analysis of model output. *Technometrics* 41, 39–56.

Schneider, D.C. (1994) *Quantitative Ecology: Spatial and Temporal Scaling*. Academic Press, San Diego, California, 395 pp.

Schultz, E.B., Matney, T.G. and Koger, J.L. (1999) A neural network model for wood chip thickness distributions. *Wood and Fiber Science* 31, 2–24.

Schuring, D.J. (1977) *Scale Models in Engineering*. Pergamon Press, Oxford, 299 pp.

Sedov, L.I. (1993) *Similarity and Dimensional Methods in Mechanics*. CRC Press, Boca Raton, Florida, 479 pp.

Vanclay, J.K. and Skovsgaard, J.P. (1997) Evaluating forest growth models. *Ecological Modelling* 98, 1–12.

Warner, B. and Misra, M. (1996) Understanding neural networks as statistical tools. *Journal of the American Statistical Association* 50, 284–293.

Wiens, J.A. (2001) Understanding the problem of scale in experimental ecology. In: Gardner, R.H., Kemp, W.M., Kennedy, V.S. and Petersen, J.E. (eds) *Scaling Relations in Experimental Ecology*. Columbia University Press, New York, 373 pp.

Zeide, B. and Gresham, C.A. (1991) Fractal dimensions of tree crowns in three loblolly pine plantations of coastal South Carolina. *Canadian Journal of Forestry Research* 21, 1208–1212.

Zeide, B. and Pfeifer, P. (1991) A method for estimation of fractal dimension of tree crowns. *Forestry Science* 37, 1253–1265.

9

GLOBTREE: an Individual Tree Growth Model for *Eucalyptus globulus* in Portugal

Paula Soares[1] and Margarida Tomé[1]

Abstract

This chapter presents the development of a tree survival probability equation and a tree diameter increment submodel, of the type potential function × modifier function. These two submodels are components of a wider modelling project to obtain an individual tree growth model for first-rotation *Eucalyptus globulus* Labill. plantations located in the north and central coastal regions of Portugal. The submodels are based on data from permanent plots and trials. The effects of competition on stand structure and tree growth were analysed, leading to the definition of different stages of stand development, according to the mean stand crown ratio. The hypothesis that different submodels are required to adequately describe growth at different stages of stand development was tested. This chapter also describes the full model, which also includes the following submodels: a dominant height growth equation, a tree crown ratio prediction equation, a tree height–diameter equation and a tree volume prediction equation.

Introduction

Eucalypt plantations are mainly used by the pulp industry and are intensively managed as a short-rotation coppice system. As a consequence of the relatively simple eucalypt system characteristics, whole-stand growth models have been applied with success in Portugal. However, the development of tree growth models can be justified when these are incorporated into decision-support systems. In fact, maximization of volume per unit area could be achieved by the use of closer spacings, as supported by previous studies suggesting site suboccupancy of the eucalypt plantations (Soares and Tomé, 1996).

However, spacing effects are mainly visible on tree diameters resulting in positively skewed distributions with a high number of trees in the lowest classes. Limits defined by the pulp companies for merchantable volume and the exploitation operations are compatible with the use of either diameter distribution models or individual tree models. The GLOBTREE, an individual tree growth model, was developed in this context.

[1] Department of Forestry, Instituto Superior de Agronomia, Portugal
Correspondence to: paulasoares@isa.utl.pt

Data

Data from permanent plots, several spacing trials and a fertilization and irrigation trial of *E. globulus* in first rotation located in the central and northern coastal regions of Portugal were used (Soares, 1999). The permanent plots were remeasured at approximately annual intervals: diameter of each tree, a sample of heights and/or dominant height were obtained in each measurement; in some cases the height to the base of the live crown was also registered. The spacing trials and the fertilization and irrigation trial were intensively remeasured monthly or once every 3 months. Tree coordinates were measured in all plots.

Table 9.1 presents a summary of the principal characteristics of the 154 plots selected corresponding to 984 remeasurements and 829 growth periods. The data set resulted in 20,060 trees measured in a total of 128,493 observations at tree level.

Model Structure

Several submodels developed in previous work and two submodels presented in this chapter comprise the GLOBTREE model. Table 9.2 presents a list of variables used in the present work.

Submodels employed from the literature

Tree crown ratio prediction equation

Soares and Tomé (2001), see Table 9.2:

Table 9.1. Characterization of the 154 plots used in the modelling.

Variables	Minimum	Mean	Maximum	SD
Plot area (m^2)	243.0	980.3	2487.4	573.3
Year of plantation	1965	–	1994	–
Age at first measurement (years)	0.5	3.1	10.1	2.0
Age at last measurement (years)	2.0	9.1	24.7	5.7
Number of measurements per plot	2.0	6.3	27.0	4.0
Number of trees planted (per ha)	500	1520	5000	933
Site index at base age 10 years (m)	12.4	21.3	28.4	3.5

SD is the standard deviation of the indicated variables.

Table 9.2. List of symbols used in this work.

Symbols	Variables	Symbols	Variables
cr	Tree crown ratio	hdom	Dominant height (m)
d	Diameter measured at 1.30 m of tree height (cm)	n	Number of trees in each plot
		N	Stand density (per ha)
dg	Quadratic mean diameter (cm)	ndom	Number of dominant trees
dmax	Maximum tree diameter (cm)	nreg	Parameter defined for different ecological regions of Portugal
h	Tree total height (m)		
t	Age (years)	G	Basal area (m^2/ha)
v	Tree total volume (m^3)	SI	Site index (m)

$$cr = \dfrac{1}{\left[1 + e^{-(-5.76111 + 12.33413\,1/t - 0.27179\,N/1000 - 0.17543\,\text{hdom} + 0.20559\,d)}\right]^{1/6}}$$

Tree height–diameter equation

Soares and Tomé (2002), see Table 9.2: version (A) for young plantations (age <4 years); version (B) for use in commercial forest inventory where trees smaller than 4 cm diameter are not measured.

(A) $\quad h = 1.30 + \text{hdom}\left(1 + \left(-0.43487 - 0.0108t + 0.09772\text{hdom} - 0.06021\text{dg}\right)e^{-0.04864\,\text{hdom}}\right)$

$$\left(1 - e^{-1.58926\frac{d}{\text{hdom}}}\right)$$

(B)

$$h = \text{hdom}\left(1 + \left(0.10694 + 0.02916\frac{N}{1000} - 0.00176\,\text{dmax}\right)e^{0.03540\,\text{hdom}}\right)\left(1 - e^{-1.81117\frac{d}{\text{hdom}}}\right)$$

Tree volume equation (with bark)

Tomé (1990), see Table 9.2:

$$v = 0.00003739 d^{1.8150696} h^{1.1454998}$$

Dominant height growth equation

Tomé *et al.* (2001), see Table 9.2:

$$\text{hdom}_2 = 61.1372\left(\frac{\text{hdom}_1}{61.1372}\right)^{\left(\frac{t_1}{t_2}\right)^{\text{nreg}}}$$

New Submodels

Tree diameter increment equation

Tree diameter increment is predicted on the basis of a potential growth function multiplied by a modifier equation, which is expressed as a function of distance-independent and/or dependent competition indices. This formulation has frequently been used by modellers (e.g. Ek and Monserud, 1974; Vanclay, 1994; Soares and Tomé, 1999b).

The hypothesis that different submodels are required to adequately describe growth at different stages of stand development was tested.

Different stages of stand development

Competition processes have been defined according to two basic models: symmetrical/asymmetrical and one-sided/two-sided competition. In two-sided competition, resources are shared (equally or proportionally to size) by all the trees, while in one-sided competition larger trees are not affected by their smaller neighbours. When there is perfect sharing relative to size, competition is symmetrical. In this study, one-sided competition is considered as an extreme case of asymmetrical competition and two-sided competition is considered as being symmetrical or asymmetrical according to whether or not the sharing of resources is proportional to the size of the individuals. Based on previous studies (Soares and Tomé, 1996; Soares, 1999) it was decided that two stages of development for *E. globulus* stands would be considered: (i) the first stage, where the effects of asymmetrical competition are not evident; and (ii) the second stage, where asymmetrical competition is evident. Research focused on the definition of a variable (or an index) that expressed, in a realistic but easy way, these two stages. The following variables were considered (see Table 9.2):

- stand parameters: dominant height, $\dfrac{\sum_{i=1}^{ndom} h_i}{ndom}$; mean height/quadratic mean

  diameter, $\dfrac{\left.\sum_{i=1}^{n} h_i \right/ n}{\sqrt{\sum_{i=1}^{n} \dfrac{d_i^2}{n}}}$;

- relative density measures: $\dfrac{100}{\sqrt{N} \times hdom}$; $\dfrac{100 \times G}{\sqrt{N} \times hdom}$; $\dfrac{100}{G \times hdom}$; $\dfrac{100}{G \times SI}$; $\dfrac{100}{N \times \dfrac{dg}{100}}$;

- crown parameters: crown depth, crown ratio, leaf area, leaf area index.

For the two spacing trials that were measured starting from 1.5 years, the variables were computed and the temporal evolution was analysed. The crown ratio was the variable whose value most consistently tended to decrease over time. To define the value that allowed distinction between the two stages, a classificatory discriminant analysis was performed using PROC DISCRIM (SAS, version 6.12). As a consequence, all the measurements of the two spacing trials were classified according to the two stages previously defined. When the correlation coefficient between the relative growth rate in diameter and the diameter was significantly different from zero and either negative or positive, the stand measurement was classified as part of the first stage or second stage, respectively, of stand development.

Competition indices

Distance-dependent competition indices can be classified into distance weighted size ratio functions (DR), area potentially available (APA), point density measures (PD) and area overlap indices (AO) (Table 9.3). In this work, and based upon previous studies on *E. globulus* in Portugal (Soares and Tomé, 1999a), the AO indices were not considered. The DR and PD indices are typically two-sided, while the APA can be regarded as assuming a two-sided asymmetrical competition, with the level of asymmetry depending on the weight given to tree size in the definition of the area potentially available. The DRU and PDU indices and the modified version of the DR indices developed by Tomé and Burkhart (1989) reflect one-sided competition. The modified indices give an indication of the dominance of the tree in relation to its closest neighbours.

Based on principal component analysis (PCA), interactions and similarities between different types of indices were analysed, permitting the definition of three groups: (i) DR, DRU, PD, PDU; (ii) DR modified; and (iii) APA. This result showed that it is reasonable for a diameter growth model to include more than one competition index, yielded by different groups.

Table 9.3. Distance-dependent competition indices.

Type of index	Version	Mathematical formulation
Distance weighted size ratio functions	Traditional (DR)	$\displaystyle\sum_{j=1}^{n}\frac{d_j}{d_i}\times f(\text{dist}_{ij})$
	Unilateral (DRU)*	$\displaystyle\sum_{j=1}^{n_1}\frac{d_j}{d_i}\times f(\text{dist}_{ij}), d_j > d_i$
	Modified (DRM)**	$\displaystyle\sum_{j=1}^{n_1}\left(\frac{d_i}{d_j}\times f(\text{dist}_{ij})\right)-\sum_{j=1}^{n_2}\left(\frac{d_j}{d_i}\times f(\text{dist}_{ij})\right)-\sum_{j=1}^{m}\left(\frac{d_i}{d_{j0}}\times f(\text{dist}_{ij})\right)$ (dominant neighbours)−(dominated neighbours)− (dead neighbours)
	Modified (DD)**	$\displaystyle\sum_{j=1}^{n}(d_j-d_i)\times f(\text{dist}_{ij})+\sum_{j=1}^{m}(d_{j0}-d_i)\times f(\text{dist}_{ij})$ (dominant neighbours)−(dominated neighbours)− (dead neighbours)
Point density measures	Traditional (PD)	$\displaystyle\frac{2500}{n}\times\left(\sum_{j=1}^{n}(j-0.5)\times\left(\frac{d_j}{\text{dist}_{ij}}\right)^2\right)$, not considering the subject tree
		$\displaystyle\frac{2500}{n}\times\left(\sum_{j=1}^{n}(j+0.5)\times\left(\frac{d_j}{\text{dist}_{ij}}\right)^2\right)$, considering the subject tree
	Unilateral (PDU)*	$\displaystyle\frac{2500}{n}\times\left(\sum_{j=1}^{n_1}(j-0.5)\times\left(\frac{d_j}{\text{dist}_{ij}}\right)^2\right)$, not considering the subject tree $, d_j > d_i$
		$\displaystyle\frac{2500}{n}\times\left(\sum_{j=1}^{n_1}(j-0.5)\times\left(\frac{d_j}{\text{dist}_{ij}}\right)^2\right)$, considering the subject tree
Area overlap indices	Traditional (AO)	$\displaystyle\sum_{j=1}^{n}\left(\frac{a_{ij}}{A_i}\times\left(\frac{d_j}{d_i}\right)^{k_1}\right)$
	Unilateral (AOU)*	$\displaystyle\sum_{j=1}^{n_1}\left(\frac{a_{ij}}{A_i}\times\left(\frac{d_j}{d_i}\right)^{k_1}\right), d_j > d_i$
	Modified (AOM)**	$\displaystyle\sum_{j=1}^{n_1}\left(\frac{a_{ij}}{A_i}\times\left(\frac{d_j}{d_i}\right)^2\right)-\sum_{j=1}^{n_2}\left(\frac{a_{ij}}{A_i}\times\left(\frac{d_i}{d_j}\right)^2\right)-\sum_{j=1}^{m}\left(\frac{a_{ij}}{A_i}\times\left(\frac{d_i}{d_{j0}}\right)^2\right)$ (dominant neighbours)−(dominated neighbours)− (dead neighbours)
Area potentially available	Traditional (APA)	$\displaystyle w_k=\frac{d_i^k}{d_i^k+d_j^k}$, in its weighted version with $k=2$ (APA2) and $k=4$ (APA4)

*Larger trees are not affected by smaller neighbours; **neighbours larger than the subject tree place it at a competitive disadvantage and smaller neighbours place it at a competitive advantage (Tomé and Burkhart, 1989); d, dimension; n, total number of competitors ($n = n_1+n_2$); n_1, number of dominant neighbours; n_2, number of dominated neighbours; m, number of dead neighbours; i, subject tree; j, competitor; j_0, dead neighbour; $f(\text{dist}_{ij})$, distance function between the subject tree i and the competitor j; a_{ij}, overlap area between the subject tree i and the competitor j; A_i, area of influence of the subject tree defined as a function of its dimension.

Border trees were selected from the trees inside the plots. Rules for selecting border trees as well as competitor trees were defined as asymptotically restricted, non-linear functions of tree size (Soares, 1999; Soares and Tomé, 1999a).

Three distance-independent competition indices were also tested (see Table 9.2): a measure of dominance $\left(RBM = \dfrac{\pi N d^2}{40000G} \right)$, a measure of stand density $(100/N)$ and a variable to express past competition (crown ratio).

Potential × modifier model

In this study, the mean growth of dominant trees was used as an indicator of the potential growth of a specific stand, dominant trees being defined as the 100 thickest trees per hectare. For practical purposes the potential growth was estimated by fitting a growth model to the growth data from the dominant trees. Using this model, the potential growth for each plot was computed as the growth of the tree with a diameter (d) equivalent to the quadratic mean diameter of the respective dominant trees. To estimate the dimension of the dominant trees at age some time after $t1$ (called $t2$), several functions were proposed (Table 9.4). The selection of the potential growth function was based on the residual sum of squares (RSS), the asymptote values, the normality of the studentized residuals and absence of heteroscedasticity associated with the error term of the models. The analyses of the last two were done graphically.

Table 9.4. Candidate functions to modelling tree potential diameter growth.

Function	Parameter	Equation
Lundqvist–Korf (Stage, 1963; Korf, 1973)	k	$d_{t2} = A \left(\dfrac{d_{t1}}{A} \right)^{(t1/t2)^n}$
McDill–Amateis (Amateis and McDill, 1989; McDill and Amateis, 1992)	–	$d_{t2} = \dfrac{A}{1 - \left(1 - \dfrac{A}{d_{t1}} \right)\left(\dfrac{t1}{t2} \right)^n}$
Richards (1959)	k	$d_{t2} = A \left\{ 1 - \left[1 - \left(\dfrac{d_{t1}}{A} \right)^{1-m} \right]^{t2/t1} \right\}^{1/(1-m)}$
	m	$d_{t2} = A \left(\dfrac{d_{t1}}{A} \right)^{\frac{\ln\left(1 - e^{-kt2}\right)}{\ln\left(1 - e^{-kt1}\right)}}$
Schumacher (1939)	k	$d_{t2} = A \left(\dfrac{d_{t1}}{A} \right)^{t1/t2}$

d_{t1} and d_{t2}, diameters (cm) at ages $t1$ and $t2$ (years); $t1$ and $t2$ defined as $(t - t0)$ with $t0 = c/SI$; A, asymptote defined as $A = a + b \times SI$; SI, site index defined as dominant height at base age 10 years; k, m, n and c, function parameters.

The modifier function (mod) was selected from a set of combinations of the type $d_{t2} = d_{t1} + \text{ipot} \times \text{mod}$, where the modifier function was based on the logistic or exponential functions. The best modified functions were selected on the basis of the residual mean of squares (RMS), non-existence of collinearity between variables, significance of the parameters, normality of the studentized residuals and absence of heteroscedasticity associated with the error term of the equations. Dominant trees were eliminated from the data set, as they were assumed to attain the potential growth.

During the evaluation stage, bias and precision of the selected functions were analysed based on the prediction residuals (true value − predicted value). Bias was assessed from the mean of the prediction residuals. Precision was expressed by the interquantile range of the prediction residuals (Q99–Q1) and by computation of the mean of the absolute value of the prediction residuals. The model efficiency (ME) was computed; this statistic provides a simple index of performance on a relative scale, where 1 indicates a perfect fit, 0 reveals that the model is no better than a simple average, and negative values indicate a very poor model (Vanclay and Skovsgaard, 1997).

To develop the tree diameter increment equation, the total data set was randomly split into two subsets and both were used in order to fit, select and validate the equations. To ensure that the data splitting was not affected by systematic influences, the equations selected in one data subset were evaluated using the other subset, and vice versa.

Tree survival probability model

The probability of tree survival (P) is simulated with a logistic function fitted with the maximum likelihood method using PROC LOGISTIC (SAS, version 6.12). This function is one of the most widely employed to express the probability of tree survival (Vanclay, 1991; Zhang et al., 1997; Monserud and Sterba, 1999). The logistic function is limited to the interval [0, 1] and the probability of death is given by $(1 - P)$. The dependent variable is a binary variable assuming 0 for dead trees and 1 for live trees. The submodel to be fitted (logit transformation) is of the type:

$$f(x) = \sum_{i=1}^{n} b_i x_i = \ln \frac{p(S=1)}{p(S=0)} = \ln \frac{p(S=1)}{1 - p(S=0)},$$ where S is a binary variable, x is the vector of the independent variables, b is the vector of the parameters associated with the variables, and $p(S=1)$ is the probability of survival in a defined period, being expressed by $p(S=1) = (1 - p(S=0)) = \dfrac{e^{f(x,b)}}{1 + e^{f(x,b)}}$.

The final model is the result of the stepwise logistic selection, which allows one to define, based on the likelihood ratio χ^2 test, the most significant independent variables for the expression of the dependent variable. The tested tree and stand variables were selected based on the knowledge of the eucalypt system and the behaviour modelled in the tree diameter growth equation. The practical non-existence of density-dependent mortality in the E. globulus stands, suggesting the sub-occupancy of the site (Soares and Tomé, 1996), resulted in a reduced number of observations in the subset 'dead trees'. Consequently, distance-dependent competition indices, which imply a reduction in the number of useful trees, were not considered as independent variables. The qualitative analysis of the model was based on standard tests and statistics for logistic regression (see, for example, Hosmer and Lemeshow, 1989): likelihood ratio test − to analyse the overall significance of the model; Wald's test − to analyse the significance of one specific variable when the other variables of the model are present; odds ratio − calculated for each independent

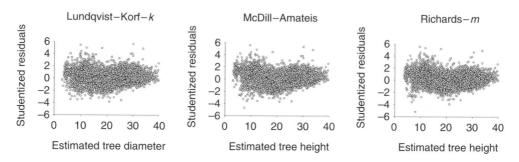

Fig. 9.1. Graphical relationship between the studentized residuals and the diameter values estimated by weighted regression with the Lundqvist–Korf – k, McDill–Amateis, and Richards – m functions.

Table 9.5. Estimated parameters of the potential growth functions fitted with the dominant trees data set ($n = 6495$).

Function	$A = a + b \times SI$		$t = t - c/SI$	n	m	k	$A_{max.}$	RSS	$R^2_{adj.}$
	a	b	c*						
Lundqvist–Korf – k	23.6257	1.2478	−1.7990	0.5591	–	–	–	2708.7	0.992
	22.7147	1.3848	–	0.5286	–	–	62.0	2712.0	0.992
McDill–Amateis	18.4375	0.9822	6.2257	1.0456	–	–	46.3	3154.2	0.991
Richards – k	20.0977	0.9773	−3.7188	–	−0.0984	–	–	4445.6	0.987
	21.4167	0.9170	–	–	−0.1575	–	–	4244.2	0.988
Richards – m	5.3007	1.4616	11.7149	–	–	0.0689	46.8	3203.2	0.991
Schumacher – k	22.3453	0.6657	−26.4407	–	–	–	–	2979.9	0.991
	8.5371	0.9757	–	–	–	–	36.2	4868.3	0.986

A, asymptote (cm); t, age (years); a, b, c, n, m and k, function parameters; RSS, residual sum of squares; *c parameter should assume positive values.

variable and usually referred to as n units of the independent variable; and concordance analysis – based on the analysis of the correspondence between real and predicted answers, thereby giving an indication of the predictive capacity of the model.

Tables were constructed with classification error rates for varying cut-off points, with the objective of selecting the optimal cut-off point. Cut-off points are used to convert probability of survival to dichotomous (0,1) data: trees with estimated probabilities above the cut-off are positive diagnostics to being considered alive. Two statistics were used for each considered cut-off point: sensitivity and specificity. Sensitivity is the proportion of true positives that were predicted as events; specificity is the proportion of true negatives that are predicted as non-events.

Results And Discussion

Different stages of stand development

The classification of all the measurements of the two spacing trials, based on the relationship between the relative growth rate in diameter and the diameter (d), resulted in 40 observations classified as part of the first stage and 144 as part of the second stage. The classificatory discriminant analysis defined a value of 0.69 for the mean crown ratio for distinguishing between the two stages of stand development.

1. First stage: effects of asymmetric competition are not evident – mean crown ratio > 0.69

2. Second stage: where asymmetric competition is evident – mean crown ratio ≤ 0.69.

Tree diameter increment equation

Potential growth function

Based on the results presented in Table 9.5, Lundqvist–Korf – k, McDill–Amateis and Richards – m functions were selected in a first analysis as potential growth functions. In the three functions it was observed that the variance increased as the predicted value of the response variable decreased, reflecting the evidence of heteroscedasticity associated with the error term of the models. Weighted regression was used to circumvent this problem (Fig. 9.1). The weight was selected from $\{(d_{t1})^{1/2}, d_{t1}, \ln(d_{t1})\}$, based on asymptote parameters and the values obtained during the fitting stage (as logarithms of negative values). The factor $(d_{t1})^{1/2}$ was chosen.

The Lundqvist–Korf – k function was selected as the tree potential diameter growth function.

$$d_{t2} = A\left(\frac{\text{ddom}_{t1}}{A}\right)^{(t1/t2)^n} = (31.6761 + 1.2067\ SI)\left(\frac{\text{ddom}_{t1}}{31.6761 + 1.2067\ SI}\right)^{(t1/t2)^{0.4905}}$$

where ipot – ddom$_{t2}$ – ddom$_{t1}$. Asymptote is a function of site index assuming high values in more productive sites.

Modifier functions

To develop the tree diameter increment equation, the total data set was divided according to the mean stand crown ratio value of 0.69 (subset 1 and subset 2). Each subset was randomly split into two other subsets (subsets 1_1, 1_2, 2_1 and 2_2) and both were used to fit, select and validate the equations.

Distance-independent modifier function

Based on the criteria indicated previously, three functions were selected for the evaluation stage (see Table 9.2):

mod1: $\quad d_{t2} = \text{ipot} \times \dfrac{1}{1 + e^{a_0 + a_1 \text{RBM} + a_2\, 100/N + a_3 \text{cr}}} + d_{t1}$

subset 1_1: R^2_{adj} 0.909, RMS 0.762; subset 1_2: R^2_{adj} 0.917, RMS 0.689

mod2: $\quad d_{t2} = \text{ipot} \times e^{(a_0 + a_1 G)\frac{(\text{ddom}_{t1} - d_{t1})^b}{\text{ddom}_{t1}^c}} + d_{t1}$

subset 1_1: R^2_{adj} 0.906, RMS 0.792; subset 1_2: R^2_{adj} 0.921, RMS 0.659

mod3: $\quad d_{t2} = \text{ipot} \times e^{(a_1 G + a_2\, 100/N)\frac{(\text{ddom}_{t1} - d_{t1})^b}{\text{ddom}_{t1}^c}} + d_{t1}$

subset 2_1: R^2_{adj} 0.987, RMS 0.285; subset 2_2: R^2_{adj} 0.987, RMS 0.278

Distance-dependent modifier function

Assuming two different stages of stand development, the expectation was: (i) higher importance of two-sided competition indices (DR, PD) on the first stage of stand development; and (ii) higher importance of two-sided asymmetrical competition indices (APA) or one-sided competition indices (indices of dominance) in the second stage of stand development.

In both stages of stand development the modifier function selected was the same and included PDU and APA competition indices. However, the difference between PD in its traditional version and its unilateral version was very small. In the subset associated with the first stage of stand development, the available area for tree growth and the density around the tree seemed to be important for the definition of tree growth. Measures of dominance had no significance, possibly as a consequence of the fact that dominant trees were not considered in the fitting of the modifier functions.

Based on the criteria indicated previously, the following function was selected for the evaluation stage:

mod4: $d_{t2} = \text{ipot} \times e^{a_1 \text{PDU}} \times \left(1 - e^{b_0 + b_1 \text{APA}}\right) + d_{t1}$

subset 1_1: R^2_{adj} 0.904, RMS 0.802; subset 1_2: R^2_{adj} 0.917, RMS 0.691;

subset 2_1: R^2_{adj} 0.986, RMS 0.288; subset 2_2: R^2_{adj} 0.986, RMS 0.288

Distance-independent $\times$ distance-dependent modifier function

Several combinations of distance-independent and distance-dependent functions were tested. When applied to the data subset representing the second stage of stand development, all the combinations showed high collinearity. For that reason, none was selected. From the data subset representing the first stage of stand development and based on the criteria indicated previously, the following function was selected for the evaluation stage:

mod5: $d_{t2} = \text{ipot} \times \dfrac{1}{1 + e^{a_0 + a_1 \text{RBM} + a_2 100/N + a_3 cr}} \times e^{b_1 \text{PDU}}$

$$\times \left(1 - e^{c_0 + c_1 \text{APA}}\right) + d_{t1}$$

subset 1_1: R^2_{adj} 0.912, RMS 0.738; subset 1_2: R^2_{adj} 0.920, RMS 0.664

Model evaluation

Table 9.6 presents bias and precision measures associated with the selected functions. Based on those results as well as on the other procedures indicated previously:

- The first stage of stand development: mod2 was eliminated because the modifier function assumed values greater than 1; when considering both data subsets (1_1 and 1_2) the combined model was the most precise and the less biased; the dependent function was not superior to the independent function.
- The second stage of stand development: both independent and dependent functions showed similar behaviour; however, the distance-independent function was the less biased and the most precise, in spite of the fact that for the data subset 2_1 it presented a maximum value slightly superior to 1.

Table 9.6. Evaluation of the tree diameter increment equations.

Function	$\dfrac{y-\hat{y}}{n}$	$\dfrac{|y-\hat{y}|}{n}$	$RSS_p = \sum_{i=1}^{n}(y_i - \hat{y}_i)^2$	ME	Q99–Q1	Modifier functions (maximum value)
Stands where the asymmetric competition is not evident						
Evaluation with the subset 1_1 (n = 3155)						
mod1	0.362	0.713	2684.7	0.898	2.54−(−1.77) = 4.31	0.99
mod2	0.267	0.701	2629.8	0.900	2.46−(−1.91) = 4.37	1.40
mod4	0.374	0.726	2775.3	0.895	2.60−(−1.69) = 4.29	0.99
mod5	0.360	0.696	2580.0	0.902	2.49−(−1.74) = 4.23	0.99
Evaluation with the subset 1_2 (n = 3279)						
mod1	−0.135	0.671	2475.8	0.909	1.95−(−2.18) = 4.13	0.99
mod2	−0.212	0.667	2407.6	0.911	1.85−(−2.14) = 3.99	1.06
mod4	−0.217	0.683	2562.3	0.906	1.82−(−2.34) = 4.16	0.99
mod5	−0.118	0.655	2358.2	0.913	2.14−(−1.88) = 4.02	0.99
Stands where the asymmetric competition is evident						
Evaluation with the subset 2_1 (n = 6515)						
mod3	−0.055	0.409	1919.8	0.986	1.47−(−1.31) = 2.78	1.07
mod4	−0.048	0.411	1887.0	0.986	1.47−(−1.23) = 2.70	0.99
Evaluation with the subset 2_2 (n = 6355)						
mod3	0.083	0.400	1840.7	0.986	1.54−(−1.21) = 2.75	0.99
mod4	0.091	0.403	1846.6	0.986	1.57−(−1.15) = 2.72	1.00

Tree survival probability model

The two stages of stand development defined in this chapter were considered: (i) first stage: it is assumed that mortality is independent of the intraspecific competition; and (ii) second stage: it is assumed that mortality is dependent on intraspecific competition. In that case, observations were restricted to annual periods: pairs of observations (live tree, dead tree) = 489; pairs of observations (live tree, live tree) = 62,349.

Most of the one-variable models were significant in expressing the probability of tree survival; however, basal area and mean stand crown ratio, when analysed individually, were not significant.

Table 9.7 shows a summary of the results obtained with the stepwise logistic regression. The cut-off of the model was selected based on Fig. 9.2 and Table 9.8. The model shows difficulty in predicting 'dead trees', and that difficulty is increased for smaller cut-off values (Table 9.8).

Table 9.7. Analysis of maximum likelihood estimates for the selected tree survival probability model.

Model		χ^2	$P(>\chi^2)$
−2LogL (likelihood)	4270.6	1452.7	0.0001
	Analysis of maximum likelihood estimates		
Variable	Parameter estimate	Wald statistic	$P(>\chi^2)$
$b0$	2.2735	356.8	0.0001
G	−0.0469	68.4	0.0001
RBM	0.2841	163.6	0.0001
d	1.5340	35.5	0.0001

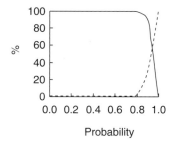

Fig. 9.2. Sensitivity (—) and specificity (----) for the tree survival probability model.

Table 9.8. Comparison of the tree state (live vs. dead) predicted by the model with the fitting data set, for annual growth periods, and cut-off values of 0.8, 0.85 and 0.9.

Cut-off	Observed value	Predicted value	
		Live tree (1)	Dead tree (0)
0.80	Live tree (1)	62,308	41
	Dead tree (0)	474	15
0.85	Live tree (1)	62,187	162
	Dead tree (0)	434	55
0.90	Live tree (1)	61,798	560
	Dead tree (0)	344	145

Conclusions

1. For the first stage of stand development defined by values of mean crown ratio >0.69, the use of the combined model is suggested:

$$d_{t2} = \text{ipot} \times \frac{1}{1 + e^{2.85858 - 1.11309 \times \text{RBM} - 31.21862 \times 100/N - 1.79650 \times cr}} \times e^{-0.05780 \times \text{PDU}} \times \left(1 - e^{-1.51267 - 0.35122 \times \text{APA}}\right) + d_{t1}$$

2. For the second stage of stand development defined by values of mean crown ratio ≤0.69, the results were indifferent as to the choice between using the distance-independent (A) or the distance-dependent (B) functions:

(A) $\quad d_{t2} = \text{ipot} \times e^{(-0.59196 \times G + 10.17083 \times 100/N)^{\frac{(\text{ddom}_{t1} - d_{t1})^{1.32943}}{\text{ddom}_{t1}^{1.62516}}}} + d_{t1}$

(B) $\quad d_{t2} = \text{ipot} \times e^{-0.05851 \times \text{PDU}} \times \left(1 - e^{-0.14559 - 0.30165 \times \text{APA}}\right) + d_{t1}$

The probability of a tree surviving, defined for an annual period, is given by:

$$\pi(x) = \frac{e^{2.2735 - 0.0469G + 1.5340\,\text{RBM} + 0.2841\,d}}{1 + e^{2.2735 - 0.0469G + 1.5340\,\text{RBM} + 0.2841\,d}}$$

where the symbols are as defined above.

A cut-off value of 0.85 was selected: trees with a probability of survival less or greater than 0.85 were considered as dead or living, respectively.

References

Amateis, R. and McDill, M.E. (1989) Developing growth and yield models using dimensional analysis. *Forest Science* 35, 329–337.

Ek, A.R. and Monserud, R.A. (1974) *FOREST: a Computer Model for Simulating the Growth and Reproduction of Mixed Species Forest Stands*. Research Report R2635, University of Wisconsin, Madison School of Natural Resources, Madison, Wisconsin, 13 pp.

Hosmer, D.W. and Lemeshow, S. (1989) *Applied Logistic Regression.* John Wiley & Sons, New York, 307 pp.

Korf, V. (1973) Vymezeni vyskoveho rustoveho oboru pro smrkove rustove tabulky. *Lesnictvi-Forestry* 19, 855–868.

McDill, M.E. and Amateis, R. (1992) Measuring forest site quality using the parameters of a dimensionally compatible height growth function. *Forest Science* 38, 409–429.

Monserud, R.A. and Sterba, H. (1999) Modeling individual tree mortality for Austrian forest species. *Forest Ecology and Management* 113, 109–123.

Richards, F.L. (1959) A flexible growth function for empirical use. *Journal of Experimental Botany* 10, 290–300.

Schumacher, F.X. (1939) A new growth curve and its application to timber-yield studies. *Journal of Forestry* 37, 819–820.

Soares, P. (1999) Modelação do Crescimento da Árvore em Eucaliptais em 1ª Rotação Localizados nas Regiões Norte e Centro Litoral. PhD thesis, Instituto Superior Agronomia, Lisbon, Portugal.

Soares, P. and Tomé, M. (1996) Changes in eucalyptus plantations structure, variability and relative growth rate pattern under different intraspecific competition gradients. In: Skovsgaard, J.P. and Johannsen, V.K. (eds) *Proceedings of the Conference on Modelling Regeneration Success and Early Growth of Forest Stands*. Danish Forest Landscape Research Institute, Forskningsserien 3, Copenhagen, pp. 255–269.

Soares, P. and Tomé, M. (1999a) Distance dependent competition measures for eucalyptus plantations in Portugal. *Annales Science Forestière* 56, 307–319.

Soares, P. and Tomé, M. (1999b) A distance dependent diameter growth model for first rotation eucalypt plantations in Portugal. In: Amaro, A. and Tomé, M. (eds) *Empirical and Process-based Models for Forest Tree and Stand Growth Simulation*. Ed. Salamandra, Lisbon, pp. 255–270.

Soares, P. and Tomé, M. (2001) A tree crown ratio prediction equation for eucalypt plantations. *Annals of Forest Science* 58(2), 193–202.

Soares, P. and Tomé, M. (2002) Height–diameter equation for first rotation eucalypt plantations in Portugal. *Forest Ecology and Management* 166(1–3), 99–109.

Stage, A.R. (1963) A mathematical approach to polymorphic site index curves for grand fir. *Forest Science* 9, 167–180.

Tomé, J. (1990) Estimação do Volume Total, de Volumes Mercantis e Modelação do Perfil do Tronco em *Eucalyptus globulus* Labill. MSc thesis, Instituto Superior Agronomia, Lisbon, Portugal.

Tomé, M. and Burkhart, H. (1989) Distance dependent competition measures for predicting growth of individual trees. *Forest Science* 35, 816–831.

Tomé, M., Ribeiro, F. and Soares, P. (2001) *O modelo GLOBULUS 2.1*. Relatórios Técnico Científicos, GIMREF, DEF/ISA, Lisbon, 1/2001.

Vanclay, J.K. (1991) Compatible deterministic and stochastic predictions by probabilistic modelling of individual trees. *Forest Science* 37, 1656–1663.

Vanclay, J.K. (1994) *Modelling Forest Growth and Yield. Applications to Mixed Tropical Forests*. CAB International, Wallingford, UK, 312 pp.

Vanclay, J.K. and Skovsgaard, J.P. (1997) Evaluating forest growth models. *Ecological Modelling* 98, 1–12.

Zhang, S., Amateis, R. and Burkhart, H. (1997) Constraining individual tree diameter increment and survival models for loblolly pine plantations. *Forest Science* 43, 414–423.

10 Modelling Dominant Height Growth: Effect of Stand Density

C. Meredieu,[1] S. Perret[2] and P. Dreyfus[3]

Abstract

In many timber species, height growth of dominant trees (100 largest trees per hectare) in even-aged stands is usually assumed to remain unchanged over a wide range of stand density. This assumption allows us to use the stand dominant height (mean height of the 100 largest trees per hectare) at a specified reference age as an index of site quality.

A 'tree–distance independent' growth model was developed for Corsican pine (*Pinus nigra* Arn. ssp. *laricio* (Poir.) Maire) in France (Meredieu, 1998) to describe individual growth and mortality according to tree size (diameter at breast height) but not to tree location within the stand, silviculture (modifying stand density and structure) and site quality. The five relationships were: stand dominant height growth, tree diameter growth ('potential × modifiers' form), mortality, stem profile and a static height–diameter function.

Data analyses evidenced the density-dependence of height growth even for dominant trees. Therefore, the dominant height growth relationship supports an original feature: a stand density effect is included in addition to age at breast height and site index effects.

This result was then evaluated using a wider range of experimental stands, namely 27 experimental plots of Corsican pines planted in the region 'Centre' (France) and managed with different thinning regimes. Annual height increments since planting were measured (non-destructive method) for each tree that had been dominant at least once since thinning (plots had been measured every 2 years). This independent data set allowed us to:

1. Determine more precisely the influence of stand density on the dominant tree population; and
2. Improve the density-related function for dominant height growth.

Lastly, we suggest the use of a 'potential dominant height increment' (i.e. corrected for stand density effect) as the 'potential' growth component in the diameter growth function.

[1] INRA Unité de Recherches Forestières, France
Correspondence to: Celine.Meredieu@pierroton.inra.fr
[2] Cemagref Division Ressources génétiques et Plants forestiers, France
[3] INRA Unité de Recherches Forestières Méditerranéennes, France

Introduction

Most often, top height or dominant height is defined as the average height of the hundred largest trees per hectare. In many timber species, height growth of dominant trees in even-aged stands is usually assumed to remain unchanged over a wide range of stand density. This assumption allows us to use stand dominant height at a specified reference age as an index of site quality. On the other hand, the relationship (dominant height = f(age)) is essential in forest growth and yield models (Arney, 1974).

Corsican pine (*Pinus nigra* Arn. ssp. *laricio* (Poir) Maire var *corsicana*) is a rather fast growing conifer with good quality wood and a straight stem. Corsican pine forests cover 150,017 ha (1.02% of the total forest area) in France and are widely used for afforestation (IFN, 2002).

One model has been developed for Corsican pine growth and development in France (Meredieu, 1998). This model contains a dominant height growth relationship, which supports an original feature: a stand density effect is included in addition to age at breast height and site index effects. Nevertheless, the data used to calibrate this relationship still need to be evaluated with a wider range of experimental data and with a strict definition of dominant tree. To demonstrate the effect of density on height growth, we need experiments in which it is possible to separate the effects of site productivity and density on height growth.

The aim of this study was to evaluate the influence of thinning on the dominant tree population and to state more precisely the influence of stand density on dominant height growth.

Materials and Methods

Experimental design

The permanent sample plots used in this analysis were installed and measured by the Cemagref (French National Centre for Agricultural and Environmental Engineering) and the IDF (French Institute for Forest Development). These plots were located in the region 'Centre' (France). They were established in pure even-aged stands to test the effects of early thinning on one site. The oldest plots were established in 1983. From this network, seven permanent thinning experiments were carried out. Each experiment included at least two plots of different densities: a control plot in which neither pre-commercial thinning, nor pruning, nor thinning was performed and one or more plots in which one or two thinnings were performed (Table 10.1). These plots were measured every 2 years. At each measurement, girth at breast height (to the nearest cm) was recorded for all trees and total height (to the nearest 10 cm) was recorded for a subsample of trees. This subsample was made up of the 100 largest trees per hectare, in the dominant storey, well distributed in the plot and without any forking below 4.5 m. In order to take this definition into account, the non-forked largest tree of a subplot of 0.1 ha was chosen; furthermore a distance of 5 m between the selected trees had to be maintained. Thereby, dominant tree selection was more restricted than the simple definition of dominant tree, which did not include any restriction on spatial tree distribution, tree form or tree height.

Table 10.1. Summarized statistics of each plot in each experiment.

Plot[a]	Age[b]	N[c]	Nb invent.[d]	RSI_i[e]	RSI_f[f]	Nb trees[g]
DR11 A	14	2100	9	37.11	15.50	18
DR11 C	14	800	9	60.59	31.86	13
DR15 A	16	2000	7	27.37	16.97	20
DR15 D	16	800	7	41.05	31.12	13
DR18 A	18	1100	6	45.46	25.53	14
DR18 B	18	2300	6	31.78	18.59	21
DR18 C	18	800	6	53.52	31.99	14
DR19 C	14	800	7	74.26	33.73	11
DR19 D	14	500	7	96.46	45.95	13
DR19 E	14	2200	7	44.93	21.85	17
DR19 F	14	1100	6	64.67	30.63	11
DR19 G	14	800	6	78.39	35.00	11
DR19 H	14	500	6	100.61	48.08	10
DR19 I	14	2200	6	43.33	21.64	17
DR19 J	14	1100	7	63.79	31.25	11
DR31 A	11	2300	7	52.77	19.09	19
DR31 B	11	800	7	72.43	33.03	15
ER40 H	21	1400	6	32.26	28.70	19
ER40 I	21	1100	6	34.86	40.31	14
ER40 J	21	800	6	41.65	28.36	18
ER40 L	21	1100	6	25.04	24.92	16
ER40 M	21	2200	6	25.20	18.04	21
ER40 R	21	1100	6	24.00	24.87	14
ER40 S	21	500	6	51.71	35.95	15
ER42 A	23	2400	5	20.51	16.49	21
ER42 B	23	800	5	35.71	26.97	16
ER42 D	23	1100	5	29.79	25.53	15
Total						417

[a] The first four letters and figures are the code of the experiment; the last letter is for the plot.
[b] Age since germination at the beginning of the growth period.
[c] Target number of trees/ha after first thinning. The plots over 2000/ha are the control plots without thinning. ER40 L and ER40 R were thinned after the third inventory.
[d] Number of inventories during the growth period.
[e] Relative site index after first thinning. The Hart–Becking index was calculated the first year of the growth period, after thinning with the heights measured by climbing.
[f] Relative site index at the end of the growth period.
[g] Number of trees climbed to retrace height growth.

Data

In order to obtain unbiased values of dominant height (H_{dom}) during the growth period, we retraced the past height growth of all the trees which had been among the dominant trees. A total of 417 trees were selected (Table 10.1). They were chosen on the basis of the inventories conducted every 2 years. Every tree which had been dominant during the growth period (according to the above-mentioned definition) was selected. In each plot, for each inventory, the trees were chosen considering only their girth at breast height among the 100 largest ones. When the experimental plot measured less than 0.5 ha, then $(n-1)$ trees were considered as dominant trees, with n being the surface area of the plot (in ha) multiplied by 100.

Corsican pine is a monocyclic species. Branch and stem elongation are annual, and each annual shoot constitutes one growth unit (GU). Lateral branching usually occurs 1 year after bud formation (proleptic branching). Branches are located only at the top of each annual growth unit, clustered in pseudo-whorls. There are no inter-whorl branches. Because the trees selected belonged to a permanent experimental plot, growth height reconstruction was performed by climbing the trees. The successive heights were measured from the top to the stump based on the annual pseudo-whorl positions.

Every 2 years during the growth period studied, the successive dominant heights of the plots were calculated as the average height of the selected dominant trees.

Stand density

The selected stand density indexes were: (i) basal area (BA); (ii) Reineke's stand density index (SDI) (Reineke, 1933); and (iii) Hart and Becking relative spacing index (RSI) (Pardé and Bouchon, 1988).

Although commonly used as a measure of stand density in growth and yield modelling (Arney, 1985), the Crown Competition Factor (Krajicek *et al.*, 1961) could not be evaluated in Corsican pine since no open-grown tree data were available.

The exponent parameter in SDI is not known for Corsican pine in the region studied. However, on the basis of values of the exponent parameter of SDI defined for other similar conifers, we fixed the value of the exponent to 1.7 (Meredieu, 1998):

$$\text{SDI} = N \times (D_g / 25)^{1.7}$$

where N is the number of stems/ha and D_g is the diameter at breast height equal to the quadratic mean diameter: $D_g = \sqrt{\dfrac{4}{\pi} \times \dfrac{\text{BA}}{N}}$.

The Hart–Becking index does not include any intrinsic information about the diameter of trees in the stand but refers to the dominant height of the stand. Therefore, its structure and content greatly differ from BA and SDI:

$$\text{RSI} = 100 \times \frac{\text{AS}}{H_{\text{dom}}} = \frac{10^4 \times \sqrt{\dfrac{2}{N \times \sqrt{3}}}}{H_{\text{dom}}}$$

where H_{dom} is the dominant height of the stand (m) and AS is the average spacing between trees (m) (for estimation of AS using N, trees are assumed to be positioned on a triangular grid). All these indices were calculated at the beginning of each 2-year increment period.

Results

Effects of thinning on the dominant tree population

Evolution of the dominant tree population during growth period

Generally, the composition of the dominant population did not remain constant over the years in the experimental plots. In the control plots, between 11 and 46% of dominant trees at the beginning of the growth period had lost their status after 6 years (mean value 26.67%) (Fig. 10.1). The number of superseded dominant trees increased with the duration of the growth period. The dominant trees at the beginning of the period were progressively replaced at the end of the period.

In thinned stands, up to 44% of dominant trees selected after thinning had lost

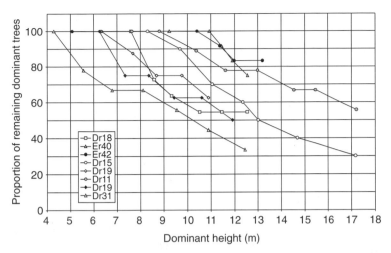

Fig. 10.1. Relationship between proportion of remaining dominant trees and dominant height in control plots. The level 100 corresponds to the dominant tree population at the beginning of the growth period.

their status after 6 years (mean value 24.45%). This mean level had reached 37.64% after 10 years.

Losing dominant status depended on the silvicultural management. It was low for both moderate thinning and severe thinning (where density after thinning was greater than 1000 stems/ha or below 500 stems/ha).

In both cases – control and thinned plot – important status modifications continuously occurred during the growth period.

Influence of thinning on dominant girth

In first thinning or pre-commercial thinning, between 33% (22.2% in systematic + 11.1% in selective) and 100% (62.5% in systematic + 37.5% in selective) of the dominant trees were removed (Table 10.2). These removals depended on the total number of trees cut (intensity of thinning), dominant height at the date of thinning and overall stem quality of the stand. Systematic thinning (line thinning) resulted in a number of dominant trees cut, ranging between 8.3 and 62.5% (from one line in six to one line in three), whereas after selective thinning between 0 and 87.5% of dominant trees were cut. Thus, the removal of dominant trees was a deliberate act. The criteria of form and branching, which are characteristics linked to vigour, were usually used in selective thinning and explained the sharp decrease in the dominant population.

Consequently, the decrease in dominant girth was important (Table 10.2). With very few exceptions, the decrease between dominant girth before and after thinning varied from 2 to 9 cm (4–24%). Thus, thinning had a very high technical effect on dominant girth.

Influence of thinning on dominant height

Eight control plots (i.e. without thinning) were included in the study. To compare the different thinning regimes, we simulated thinning in the control plots and simultaneously performed tree marking for real thinning in the other plots. Thus, in the control plots, we were able to calculate the real dominant height and the simulated dominant height with the simulated thinning.

C. Meredieu et al.

Table 10.2. Effect of first thinning on dominant breast height girth.

Plot	H_{dom}[a] (m)	i_{th}[b] (%)	kg[c]	Dom-sy[d] (%)	Dom-se[e] (%)	Gbh_{be}[f] (cm)	Gbh_{af}[g] (cm)	$Bias_{rel}$[h] (%)
DR31 B	3.95	65.0	0.89	22.2	11.1	22.6	22	−2.5
DR19 H	4.72	73.0	1.00	62.5	37.5	36.1	27.4	−24.2
DR19 G	4.82	62.2	0.95	37.5	0.0	33.8	31.4	−7.0
DR19 D	4.83	77.4	0.97	12.5	87.5	34	28.4	−16.5
DR19 F	5.01	54.4	0.93	25.0	37.5	33.5	30.1	−10.1
DR19 J	5.10	45.1	1.01	25.0	50.0	33.8	29.9	−11.5
DR19 C	5.16	63.0	0.95	37.5	37.5	34.4	31.8	−7.6
DR11 C	6.11	62.8	0.99	44.4	11.1	41	38.9	−5.1
DR18 C	7.12	60.8	0.99	22.2	55.6	46.2	42.2	−8.7
DR18 A	7.19	52.7	1.01	20.0	50.0	47.2	44.4	−5.9
ER40 H	8.92	37.7	1.04	25.0	33.3	68.5	63	−8.0
ER40 J	9.08	66.9	0.94	8.3	75.0	70.4	64.3	−8.8
ER40 S	9.20	75.4	0.93	8.3	66.7	71.5	65.8	−8.0
ER40 I	9.20	52.1	0.98	8.3	58.4	72.1	64.8	−10.2
ER42 B	10.63	67.0	0.94	16.7	50.0	64.1	60.8	−5.2
ER42 D	11.04	55.0	0.97	25.1	33.3	66.6	63.8	−4.1
ER40 L	11.05	49.3	0.89	8.3	25.0	76.8	74.7	−2.8
ER40 R	11.11	52.1	0.90	8.3	41.7	76.9	74.7	−2.9

[a] Dominant height.
[b] Thinning intensity expressed as the proportion of cut trees.
[c] Thinning nature expressed as the ratio between the mean cut tree basal area and the mean tree basal area before thinning.
[d] Proportion of dominant trees felled by systematic thinning.
[e] Proportion of dominant trees felled by selective thinning.
[f] Mean breast-height girth of dominant trees before thinning.
[g] Mean breast-height girth of dominant trees after thinning.
[h] Relative bias expressed as $100 \times (gbh_{af} - gbh_{be})/gbh_{be}$.

As we reconstituted the past height growth of all these trees, we were able to calculate all the possible dominant height values during the growth period (Table 10.3).

First thinning slightly reduced dominant height. The differences evolved from a few centimetres to 44 cm as a maximum value and were not significant (threshold 95%) with only one exception (plot DR31 ($P = 0.019$)). Thinning had no significant technical effect on the calculated dominant height.

Density effect on dominant height increment

Homogeneity of plot fertility in each stand

One of the interesting characteristics of the site index theory is that productivity can be estimated from height growth alone over a rather wide range of stand densities. The only way to check this assumption is to apply density treatments to the plots having similar topography, soil and past use, so that they can be assumed to have an equal productive potential.

Each experiment was conducted in a homogeneous stand with the same past silvicultural management. No thinning was performed before the beginning of the experiment; only weed control was carried out. Therefore, the only difference between the plots of each experiment at the beginning of the growth period was the site fertility. In order to check the uniformity of the site fertility, we used the calculation of dominant height as an indicator of site fertility.

The ANOVA showed no significant differences at the threshold 95% for the plots of the same experiment. Thus, we concluded that the plots had the same site fertility. Nevertheless, the analyses showed a significant difference at the 90% threshold for the plots of two trials: DR19 and DR31.

Relationship between dominant height increment and Hart–Becking index

Figure 10.2 shows dominant height increments in each plot by 2-year period vs. Hart–Becking index. Two trends were identified: dominant height increment was lower for both the lowest and the highest values of the Hart–Becking index.

Table 10.3. Technical effect of first thinning on dominant height (H_{dom}).

Plot	$H_{dom_{be}}$[a] (m)	$H_{dom_{af}}$[b] (m)	Bias[c] (m)	$Bias_{rel}$[d](%)
DR31 A	4.29	3.84	−0.44	−10.3
DR11 A	6.31	6.26	−0.06	−0.9
DR19 E	5.06	5.05	−0.01	−0.3
DR19 I	5.27	4.98	−0.29	−5.4
DR15 A	8.72	8.58	−0.15	−1.7
DR18 B	7.08	7.05	−0.02	−0.3
ER40 M	9.11	8.97	−0.14	−1.5
ER42 A	10.71	10.68	−0.04	−0.3

[a] Dominant height before thinning.
[b] Dominant height after thinning.
[c] Bias expressed as $H_{dom_{af}} - H_{dom_{be}}$.
[d] Relative bias expressed as $100 \times bias/H_{dom_{be}}$.

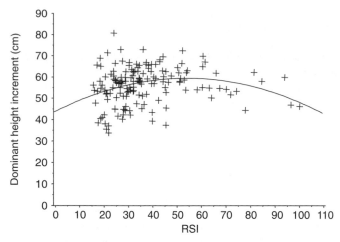

Fig. 10.2. Dominant height increments for each plot by 2-year period vs. Hart–Becking index (RSI).

Calculation of a relative mean dominant height increment

To avoid the variability due to site fertility between the different experiments, a relative mean dominant height was calculated for each combination plot 2-year period. The ratio between mean dominant height increment for each combination plot 2-year period and the maximum dominant height increment of the experiment for the same growth period provided a rate of growth for a plot vs. a potential increment.

Relationship between relative dominant height increment and stand density indices

Independent variables (i.e. the predictors) such as SDI, BA, RSI (calculated at the beginning of the 2-year period) and variation of the RSI over the 2-year period were tested against the relative dominant height increment.

The Hart–Becking index appeared to be an appropriate indicator for taking into account the variation of stand density in plots with a very low number of stems (Fig. 10.3).

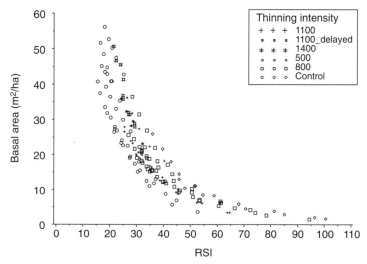

Fig. 10.3. Basal area vs. Hart–Becking index (RSI).

The ANOVA of the relative mean increment of dominant height vs. the Hart–Becking index confirmed the link between these two variables. Relative dominant height increment was lower with a high Hart–Becking index (Fig. 10.4). A similar trend was observed for the lower level of Hart–Becking index but this trend was not significant.

According to our observations, an optimal zone of dominant height increment can be defined: between 17 and 62%. Outside this zone, none of the plots reached the maximum relative dominant height increment.

Discussion

These new data analyses showed the density-dependence of height growth in dominant trees. This result was obtained using a wide range of experimental stands, i.e. 27 experimental plots of Corsican pine planted in the region 'Centre' (France) and managed with different thinning intensities. The choice of dominant trees was consistent with the usual definition (Pardé and Bouchon, 1988) and was not biased by other criteria such as the tree form or the spatial distribution in the stand. When such criteria are added, this dominant population could get close to the definition of crop tree population and the values of dominant height or dominant girth could decrease.

Lloyd and Jones (1982) have pointed out the influence of stand tree number on dominant height for loblolly pine and for slash pine: dominant height growth was reduced when stand density increased. In lodgepole pine stands, growth reduction can be dramatic in dense, stagnant stands (Cieszewski and Bella, 1993). Such growth reduction can also be noticed in very open stands (Pardé and Bouchon, 1988), or on less productive sites (Braathe, 1957). This phenomenon has been observed in different species and in particular in the *Pinus* species: *Pinus pinaster* (Illy and Lemoine,

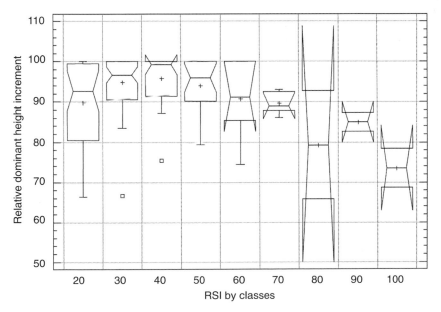

Fig. 10.4. Box and whisker plot: dominant height increments for each plot and every 2-year period vs. Hart–Becking index (RSI).

1970; Lemoine, 1980), *Pinus contorta* (Ottorini, 1978; Cieszewski and Bella, 1993) and *Pinus sylvestris* (Braathe, 1957).

In our study, we showed that there is an optimal zone of dominant height growth which is a function of the Hart–Becking index (Fig. 10.5). For a Hart–Becking index between 17 and 62%, the observed plots had a maximum relative dominant height increment (value = 100). Nevertheless, density-dependence could change during the life of the stand (Cieszewski and Bella, 1993). Our hypothesis is that the influence of stand density increases in magnitude in young and intermediate-age stands as crowding intensifies, until it reaches a certain stability in mature stands. Figure 10.5 illustrates this assumption: with the increase in age and with moderate thinning regimes, stand density reaches a mean value around 15–20% for a Hart–Becking index which is in the optimal zone. Thus, in mature stands where the range of stand density is narrowed, the effect of density on height could probably be not significant.

Our data did not show any effect of site fertility on the relationship between stand density and dominant height growth such as reported by Braathe (1957) or Ottorini (1978). Furthermore, all experimental plots were measured during the same growth period, so climatic factors might have interacted with the stand density effect.

Future Directions

We confirm the existence of a density-dominant height effect in the early stages of development (below 30 years old) and for a Hart–Becking index below 17% and above 62%. Consequently, a model to determine dominant height growth needs to combine site fertility, age and Hart–Becking index such as:

$$H_{dom} = f(H_{dom30}, \text{Age, RSI})$$

where H_{dom30} is the dominant height at age 30 (breast height age).

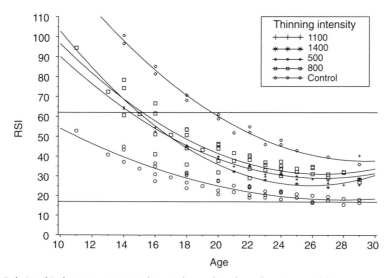

Fig. 10.5. Relationship between Hart–Becking index and total age for a range of thinning intensities. The horizontal lines show the optimal height growth zone (between 17 and 62%).

Another solution would be to build site index curves for a fixed density and, thereafter, to correct it with a function of the Hart–Becking index.

In the growth model SPS, Arney (1985) used a negative effect of stand density on dominant height growth for the highest stand density.

In the 'tree–distance independent' growth model built for Corsican pine in France (Meredieu, 1998), the dominant height growth relationship included a stand density effect in addition to age at breast height and site index effects.

Diameter growth has been described and fitted as the product of potential growth (POT), and reduction factors, or modifiers, to quantify global competition within the stand (RED1) and the status of the tree in the stand (RED2).

DIAMETER GROWTH = POT × RED1 × RED2

The potential term can be related to site fertility through dominant height increment. In order to take the influence of density into account, instead of using real dominant increment, we could use the corrected term of dominant height increment with the optimum density index.

Thus, the potential term of diameter growth could account for site fertility, age and period of growth through the dominant height increment.

Further studies are needed to improve the relationship between density index and dominant height growth and to connect this relationship with the potential term of the diameter growth relation.

References

Arney, J.D. (1974) An individual tree growth model for stand simulation in Douglas-fir. In: Fries, J. (ed.) *Growth Models for Tree and Stand Simulation*. Research Notes, Royal College of Forestry, Stockholm 30, 38–46.

Arney, J.D. (1985) A modelling strategy for the growth projection of managed stands. *Canadian Journal of Forest Research* 15, 511–518.

Braathe, P. (1957) *Thinning in Even-aged Stands: a Summary of European Literature*. Faculty of Forestry, University of New Brunswick, 92 pp.

Cieszewski, C.J. and Bella, I.E. (1993) Modeling density-related lodgepole pine height growth, using Czarnowski's stand dynamic theory. *Canadian Journal of Forest Research* 23, 2499–2506.

Illy, G. and Lemoine, B. (1970) Densité de peuplement, concurrence et coopération chez le pin maritime. I. Premiers résultats d'une plantation à espacement variable. *Annales des Sciences Forestières* 27, 127–155.

Inventaire Forestier National (2002) Available at: www.ifn.fr

Krajicek, J.E., Brinkman, K.A. and Gingrich, S.F. (1961) Crown competition: a measure of density. *Forest Science* 7, 35–42.

Lemoine, B. (1980) Densité de peuplement, concurrence et coopération chez le pin maritime. II. Résultats à 5 et 10 ans d'une plantation à espacement variable. *Annales des Sciences Forestières* 37, 217–237.

Lloyd, F.T. and Jones, E.P. (1982) Density effects on height growth and its implications for site prediction and growth projection. In: *2nd Biennial Southern Silvicultural Research Conference, USDA GTR SE-24*. 4–5 November, Atlanta, Georgia, pp. 329–333.

Meredieu, C. (1998) Croissance et branchaison du Pin laricio (*Pinus nigra* Arn. ssp. *laricio* (Poir.) Maire): élaboration et évaluation d'un système de modèles pour la prévision de caractéristiques des arbres et du bois. MSc thesis, Université Cl. Bernard Lyon I, Lyon, France.

Ottorini, J.M. (1978) Aspects de la notion de densité et croissance des arbres en peuplement. *Annales des Sciences Forestières* 35, 299–320.

Pardé, J. and Bouchon, J. (1988) *Dendrométrie*, 2nd edn. In: ENGREF (ed.), Nancy, France, 328 pp.

Reineke, L.H. (1933) Perfecting a stand-density index for even-aged forests. *Journal of Agricultural Research* 46, 627–638.

11 Testing for Temporal Dependence of Pollen Cone Production in Jack Pine (*Pinus banksiana* Lamb.)

S. Magnussen,[1] F.A. Quintana,[2] V. Nealis[1] and A.A. Hopkin[1]

Abstract

Nine-year records of the presence or absence of pollen cones (microsporangiate strobili) in 1299 jack pine trees growing in 38 locations (plots) across northern Ontario were tested for temporal dependence. The null hypothesis of a zero-order Markov chain was tested with three statistics at the tree level and at the location level. Tree-level test results were generally in agreement. Only about 5% of the trees had records indicating a significant departure from the null hypothesis. About one-third of the trees carried pollen cones every year or not at all. After adjustment for multiple comparisons, no trees violated the null hypothesis. Location-level tests relied on Monte-Carlo reference distributions. Depending on the test statistic, we found 2–15 locations with significant departures from the null hypothesis. Conditional (on location) temporal independence is accepted as a reasonable working model. A β binomial model described reasonably well the overall frequency distribution of the number of years (out of nine) with pollen cones. The intra-location correlation coefficient of pollen cone production was strong (0.40).

Introduction

Jack pine (*Pinus banksiana* Lamb.) growing in Canada's boreal and sub-boreal forests and in the Lake States of the USA is subject to periodic defoliation by the jack pine budworm (*Choristoneura pinus pinus* Freeman) causing significant mortality and growth losses (Gross, 1992; Conway *et al.*, 1999; Hall *et al.*, 1999; Volney, 1999; McCullough, 2000). Pollen cones (microsporangiate strobili) constitute a food source critical for the survival and vitality of jack pine budworm larvae when they emerge in the early spring (Rose, 1973; Nealis and Lomic, 1997). In the absence of defoliation, pollen cones are produced in great abundance in most years (Benzie, 1977; Rudolph and Yeatman, 1982), yet climate, soil properties, and tree crown characteristic modify the periodicity and frequency of production. Trees defoliated by the budworm often respond by a marked decline in the production of pollen cones the following

[1] Natural Resources Canada, Canadian Forest Service, Canada
Correspondence to: smagnuss@PFC.Forestry.CA
[2] Department of Probability and Statistics, School of Mathematics, Pontificia Catholic University of Chile, Chile

spring (Nealis, 1990; Nealis and Lomic, 1997; Nealis *et al.*, 1998). Quantification of the decline in pollen cone production induced by defoliation of the jack pine budworm requires an understanding of the temporal dynamics of pollen cone production in the absence of defoliation. Is the presence of pollen cones on an undisturbed tree in year t independent of the presence/absence of pollen cones the previous year $(t − 1)$? Many coniferous forest tree species show periodicity of seed set and pollen cone production (Smith, 1997) which, everything else being equal, points to a lack of temporal independence. A lack of temporal independence carries with it the potential of confounding 'carry-over effects' (Kunert, 1987; Matthews, 1990; Yang and Tsiatis, 2002) in simple estimates of temporal contrasts of pollen cone production.

This study tests the hypothesis of temporal independence of the production (presence/absence) of jack pine pollen cones during a period of 9 years in undisturbed stands in northern Ontario, Canada. Hypotheses of temporal independence are tested at both the tree and at the location (plot) level.

Materials and Methods

Data from 1299 jack pine trees located in 38 plots across Ontario (Canada) are used in this study. The plots are part of a network of 180 Forest Insect and Disease Survey plots established in 1992. Each plot contained 50 dominant/codominant mature jack pine trees (age 30–110 years). Tree size, crown characteristics and health status were recorded for all trees in 1992. Defoliation by the jack pine budworm (%) and the presence/absence of pollen cones were recorded for all trees from 1992 to 2001. Counts of egg masses and surviving larvae were completed (1992–1998) on branch samples taken from a subset of ten trees from each plot. The 1299 trees used in this study are the set of trees in the data with no defoliation during the 9 years of observation. Defoliation prior to the first observation is considered unlikely because the plots were established in a quiescent period between two outbreaks of the budworm. Data records consist of tree and location identifiers and nine binary scores indicating annual presence (1) or absence (0) of pollen cones between 1992 and 2001.

Binary Markov chains of pollen cone scores

We assume that each individual series of binary scores forms a stationary Markov chain of some order (Kedem, 1980). A zero-order chain implies temporal independence of scores and indicates that the presence and absence of pollen cones depend only on the long-term average (stationary mean) and not on past patterns of pollen cone production. In a first-order chain, the conditional distribution of the state in year t given the whole sequence of observed past states depends only on the state in year $t − 1$. For second-order chains this dependency is extended to $t − 2$, etc.

Let x_t for $t = 0,1,...,n$ form a binary Markov chain of ith order ($x_t = 1$ if pollen cones are present, and 0 if absent) with expectation (over time) $E_t (x_t) = p$. For a binary sequence $x_0,x_1,x_2,...,x_t$ the probability of the transition from $x_{t−1}$ to x_t is:

$$P_{x_{t-1}x_t} = (p_{11}^{x_t} \times (1 - p_{11})^{1-x_t})^{x_{t-1}} \times \left(p_{10}^{x_t} \times (1 - p_{10})^{1-x_t}\right)^{1-x_{t-1}}$$ (1)

where p_{11} and p_{10} are the one-step transition probability that a score of 1 is followed by a score of 1 and 0, respectively. By extension, we obtain the probability of three successive scores as:

$$P_{x_{t-2}x_{t-1}x_t} = \left\{ \left(p_{111}^{x_t} \times p_{011}^{1-x_t}\right)^{x_{t-1}} \times \left(p_{101}^{x_t} \times p_{001}^{1-x_t}\right)^{1-x_{t-1}} \right\}^{x_{t-2}}$$
$$\times \left\{ \left(p_{110}^{x_t} \times p_{010}^{1-x_t}\right)^{x_{t-1}} \times \left(p_{100}^{x_t} \times p_{000}^{1-x_t}\right)^{1-x_{t-1}} \right\}^{1-x_{t-2}} \tag{2}$$

After further expansion, the joint distribution of a time series $x_1,...,x_n$ becomes:

$$P_{x_1,...,x_n} = p^{x_1}(1-p)^{1-x_1} \times \prod_{t=2}^{n} p_{i_1 i_2,...,i_t}^{I_1 I_2,...,I_t} \tag{3}$$

where the second product is over all 2^t 0–1 t-tuples $(i_1,i_2,...,i_t)$ and

$$I_t = \begin{cases} x_t \text{ if } i_t = 1 \\ 1-x_t \text{ if } i_t = 0 \end{cases} \tag{4}$$

Hence the product $I_1 \times I_2 \times ... \times I_t$ is a sufficient statistic for the chain. Given a binary time series $x_1,...,x_n$ the number of observed transitions $i_1,i_2,...,i_k$ of order k is defined as $T_k = \{t_{i_1 i_2,...,ik+1}\}$. It follows that T_k is a sufficient statistic for the binary time series. In other words, counting transition patterns suffice for the statistical inference of a binary time series.

A subsequence in a binary time series consisting of identical scores (0 or 1) is called a run (streak). The number of runs in a series of length n is denoted by R_n. Although not a sufficient statistic, R_n, due to ease of interpretation and known zero-order distribution (Swed and Eisenhart, 1943), is often used for testing the temporal randomness of a binary sequence.

Tests of temporal independence of pollen cone scores

Values of R_n and T_k ($k = 1$) were obtained for each tree and compared with the probability of obtaining a similar or more extreme value under the null hypothesis of a zero-order Markov chain (temporal independence of pollen cone production).

Under the conditional (on n, u and v) null hypothesis of temporal independence, the probability of obtaining $\hat{R}_n$ runs in a sequence of length n composed of u ones and v zeroes (we assume without loss of generality that $u \leq v$) is (Swed and Eisenhart, 1943):

$$P(\hat{R}_n \mid n,u,v) = \binom{u+v}{u}^{-1} \times \begin{cases} 2 \times \binom{u-1}{\hat{R}_n/2-1} \times \binom{v-1}{\hat{R}_n/2-1} \text{ for } \hat{R}_n \text{ even} \\ \binom{u-1}{(\hat{R}_n-1)/2} \times \binom{v-1}{(\hat{R}_n-3)/2} + \binom{u-1}{(\hat{R}_n-3)/2} \times \binom{v-1}{(\hat{R}_n-1)/2} \text{ otherwise} \end{cases} \tag{5}$$

where the binomial coefficient $\binom{a}{b}$ is the number of different arrangements of b elements of one kind and a-b elements of another kind (Jeffreys, 1961). The conditional null distribution of $\hat{R}_n$ is obtained by computing the probability of each possible number of runs given the observed values of u, v and the constraint $u + v = 9$.

When the null hypothesis of complete randomness (no temporal dependency) of the individual binary scores is true, we would expect, conditional on x_1 (the first binary score), the number of '00' sequences to be similar to the number of '10' sequences. Hence, the test statistic $U_{0,1}(\mathbf{X})$ defined in Equation 6 (Quintana and Newton, 1998) as a function of the vector $\mathbf{X}$ of observed binary scores tends to zero for $n \rightarrow \infty$ when the null hypothesis is true.

$$\hat{U}_{0,1}(\mathbf{X} \mid x_1,u,v,n = 9) = \left(\frac{t_{00}}{t_{01}+t_{00}} - \frac{t_{10}}{t_{10}+t_{11}}\right)^2 \tag{6}$$

The significance of $\hat{U}_{0,1}(\mathbf{X} \mid x_1, u, v, n = 9)$ is computed as the probability of obtaining a more extreme value when the null hypothesis is true. The distribution of this test statistic under the null hypothesis was obtained by a complete enumeration of the sequences '00', '01', '10' and '11' and subsequent computation of $U_{0,1}$ for all possible permutations of nine-digit binary sequences with fixed values of x_1, u and v.

The likelihood of obtaining the counts t_{00}, t_{01}, t_{10} and t_{11} is (see Equation 1):

$$L_1(p_{00}, p_{10} \mid n, t_{00}, t_{10}, t_{01}, t_{11}) = p_{00}^{t_{00}}(1 - p_{00})^{t_{01}} p_{10}^{t_{10}}(1 - p_{10})^{t_{11}} \tag{7}$$

Under the null hypothesis the likelihood is (Katz, 1981):

$$L_0 = \sup_{p_{00} = p_{10}} L_1(p_{00}, p_{10} \mid n, t_{00}, t_{10}, t_{01}, t_{11}) \tag{8}$$

where $p_{00} = t_{00}/t_0$ and $p_{10} = t_{10}/t_1$. Asymptotically ($n \to \infty$), the likelihood ratio test statistic of a zero-order Markov chain is $lr_{0,1} = -2(\log(L_0) - \log(L_1))$. Under the null hypothesis, $lr_{0,1}$ is distributed as a χ^2 distribution with 1 degree of freedom (Billingsley, 1961).

Location level tests of zero-order Markov chains of binary pollen scores

Conditional on the location (a random effect), the tree-specific sequences are assumed to be exchangeable (independent). To test the hypothesis that the binary pollen score sequences of all n_i sample trees in a given location (i) are consistent with a zero-order Markov chain, we computed the location-specific sums of the tree-level test statistics R_n, $U_{0,1}(\mathbf{X})$ and $lr_{0,1}$. The probabilities of obtaining a more extreme value of the test statistics $\sum_{i=1}^{n_i} R_n(i)$ and $\sum_{i=1}^{n_i} U_{0,1}(\mathbf{X}_i)$ under the null hypothesis were computed as the observed frequency of more extreme values in a zero-order Monte-Carlo reference distribution (Quintana and Newton, 1998). In particular, 5000 random realizations of $\sum_{i=1}^{n_i} R_n(i)$ and $\sum_{i=1}^{n_i} U_{0,1}(\mathbf{X}_i)$ under the null hypothesis constituted the location-specific distribution under the null hypothesis of a zero-order Markov chain. Individual random tree-level realizations of the binary sequences of pollen scores under the null hypothesis were generated as outlined above. Location-level likelihood ratio test statistics of the null hypothesis were computed as $\sum_{i=1}^{n_i} lr_{0,1}(i)$ and its probability under the null hypothesis was computed from the cumulative distribution function of a χ^2 random variable with n_i degrees of freedom (Quintana and Newton, 1998). Quintana and Newton (1998) found that the Monte-Carlo distribution was a very good approximation to the exact (asymptotic) distribution.

Results

A majority of trees (77%) produced pollen cones in an average year. Annual pollen cone frequencies varied from a low of 56% in 1995 to a high of 92% in 2000. Location-specific frequencies varied considerably, from a low of 39% to a high of 100%. Figure 11.1 shows the empirical cumulative distribution function (*cdf*) of location-level pollen cone frequencies (p_{loc}) and the *cdf* of a β distribution with parameters 3.55 and 1.21 estimated by methods of maximum likelihood (Johnson and Kotz, 1970).

A total of 373 trees (29%) produced pollen cones every year, while 28 trees (2%)

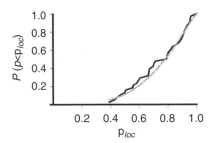

Fig. 11.1. An empirical (full black line) and a fitted β (dashed grey line) cumulative distribution function (*cdf*) of the location-specific mean propensity p_{loc} to produce pollen cones in any given year.

did not produce pollen cones in any year. Thus temporal independence of the binary sequence of pollen cone scores could only be tested for the remaining 898 trees (69%) since the order of a constant sequence is undefined (unknown). The results below refer to the 898 trees with intermittent pollen cone production.

For all trees with no pollen cones in a particular year, the conditional probability of carrying no pollen cones the next year (p_{00}) was 0.50 and equal to p_{01}, the transition probability of going from no pollen cones in year t to presence of pollen cones in year $t + 1$. Conversely, for trees with pollen cones in year t, the probability p_{10} was 0.12 while p_{11} was 0.88. Application of the fundamental matrix equations for Markov chains (Kemeny and Snell, 1976) gives a mean waiting time of 2.0 years (SE 0.5) for trees with no pollen cones in year t to produce pollen cones again. Trees carrying pollen cones in year t will, on average, stay in this mode for 8.5 years (SE 2.7).

Years with successive pollen cones tend to cluster more than expected under the assumption of a temporal independence of pollen scores. Given the individual number of years with pollen cones, the expected overall number of runs of identical pollen cone scores ($\hat{R}_n$) would be 3.0 under the null hypothesis. The observed average number of runs was 2.7. Forty-seven trees (5%) from 17 locations (median two trees per location) had significantly ($P \leq 0.05$) fewer runs than expected under the null hypothesis. Under the null hypothesis the frequency distribution of expected number of runs was:

Runs	1	2	3	4	5	6	7	8
Observed	401	209	327	203	96	51	11	1
Expected under H_0	401	0	373	210	315	0	0	0

About 5% (45) of the trees with intermittent pollen cone production produced a $\hat{U}_{0,1}(\mathbf{X})$ test statistic with a probability of 0.05 or less under the null hypothesis of a zero-order Markov chain. An additional 2% (18) were significant at the 0.05–0.10 level of significance. Significant (0.05) departures were found in 17 locations (of 38) with a median of two significant departures per location. The most common pollen score sequence amongst the significant departures had the first 5 years with no pollen cones followed by 4 years of pollen cone production.

The asymptotic likelihood ratio tests were more liberal, with 55 (6%) rejections of the null hypothesis ($P \leq 0.05$) representing 19 locations with a median of two trees per location.

The three test statistics R_n, $U_{0,1}(\mathbf{X})$ and $lr_{0,1}$ were generally complementary. All trees with a significant deficit of runs also exceeded the 0.05 significance level of the likelihood ratio test. A common subset of 37 trees achieved significance in each test.

Note that a Bonferroni-type adjustment of the significance level, to control the overall rate of Type I errors in multiple comparisons (Miller, 1980) would result in no significant departures from the null hypothesis of temporal independence of pollen cone scores.

Location-level test of significance

Location-specific sums of runs of pollen scores were in 22 cases (58%) significantly ($P \leq 0.05$) smaller than expected under the null hypothesis of temporal independence of all pollen scores in a location. The number of significant departures dropped to 15 after a Bonferroni-type adjustment of the significance levels. In contrast, a comparison of location-specific values of $\sum_{i=1}^{n_i} \hat{U}_{0,1}(\mathbf{X}_i)$ to values expected under the null hypothesis identified only nine significant ($P \leq 0.05$) departures, of which only four qualified after the Bonferroni adjustment of the significance levels. Location-level likelihood ratio tests largely confirmed the $\sum_{i=1}^{n_i} \hat{U}_{0,1}(\mathbf{X}_i)$ test statistics with six significant departures at the 0.05 level of significance, of which just two were significant after adjusting for multiple comparisons.

A β binomial model of pollen scores

The above results make it reasonable to accept the null hypothesis at the tree level. Conditional on the location 'effect', which introduces a strong intra-location correlation of pollen scores, it seems reasonable to assume temporal independence of pollen scores. A variation in the probability of pollen cones among locations and among trees within locations coupled with a binomial within-tree process gives rise to a β binomial distribution (Collett, 1991) of the number of years (out of 9) that a tree carries pollen cones. Figure 11.2 shows the observed tree-level frequency distribution of the number of years a tree carried pollen cones between 1992 and 2001. A fitted β binomial frequency distribution is shown for comparison. Maximum likelihood estimates of the parameters of the β binomial distribution were obtained as outlined by Griffiths, (1973). Fitted values were $\hat{\alpha} = 2.13 \pm 0.12$ and $\hat{\beta} = 0.66 \pm 0.04$. The corresponding intra-location correlation coefficient was a highly significant $\hat{\delta} = 0.40 \pm 0.02$. The overall fit between

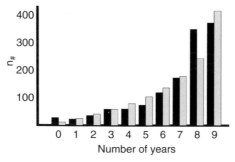

Fig. 11.2. Observed and estimated frequency ($n_\#$) distribution of years out of 9 with pollen cones. Estimated frequencies are obtained from a maximum likelihood estimation of a β binomial model.

observed and expected frequencies is quite reasonable, yet a Kolmogorov–Smirnov test (Conover, 1980) indicated statistical significance of their differences (maximum absolute difference $\hat{D}_{max} = 0.05$, $P < 0.01$). The observed 'surplus' of trees with pollen cones in 8 of the years accounted for most of the discrepancies.

Discussion and Conclusions

Statistical tests for the order of a finite binary Markov chain are most powerful when the frequencies of zeroes and ones are in the intermediate range between 0.3 and 0.6 (Billingsley, 1961). When jack pine carries pollen cones, on average of three out of every 4 years, the number of non-estimable or poorly estimated transition probabilities becomes a non-trivial issue. Also, the 9-year records used here are probably at the lower limit for statistical inference.

Our results do not clearly reject the null hypothesis of temporal independence of pollen cone production, at least not at the individual tree level. Conditional on the location effect, it appears reasonable to accept, as a working model, the notion of temporal independence. A first-order model, however, would, from an ecological viewpoint, be more realistic. Annual variations in growing conditions and disturbances of the physiological conditions can be expected to influence the tree beyond the year in which they occur. A precocious pollen cone production amounts to a significant carbon sink for the tree. A feedback control of pollen cone production governed by environmental cues would seem a reasonable hypothesis. Periodicity of seed set is the norm (Greene and Johnson, 1994); production of pollen cones is expected to show a similar pattern.

Although we settled for a β binomial model to explain the observed tree level frequencies of years with pollen cones, we recognize that a hierarchical model, with a first-order Markov chain at the first level, a Dirichlet distribution describing the mixing distribution of the transition probabilities within a location (p_{00}, p_{01}, p_{10}, p_{11}) at the second level, and at the third level a multivariate γ distribution of the Dirichlet distribution parameters to capture the effects of locations, might provide a more satisfactory fit overall. A practical reason for assuming order 1 is that order 0 becomes a special case, which could be tested under a hierarchical model, in much the same way as done here. We are currently exploring hierarchical models.

References

Benzie, J.W. (1977) *Manager's Handbook for Jack Pine in the Northern Central States.* NC-32, USDA Forest Service.

Billingsley, P. (1961) *Statistical Inference for Markov Processes.* University of Chicago Press, Chicago, Illinois, 194 pp.

Collett, D. (1991) *Modelling Binary Data.* Chapman and Hall, London, 369 pp.

Conover, W.J. (1980) *Practical Nonparametric Statistics.* Wiley, New York, 493 pp.

Conway, B.E., Leefers, L.A. and McCullough, D.G. (1999) Yield and financial losses associated with a jack pine budworm outbreak in Michigan and the implications for management. *Canadian Journal of Forest Research* 29, 382–392.

Greene, D.F. and Johnson, A.E. (1994) Estimating the mean annual seed production of trees. *Ecology* 77, 642–647.

Griffiths, D.A. (1973) Maximum likelihood estimation for the beta-binomial distribution and an application to the household distribution of the total number of cases of a disease. *Biometrics* 29, 637–648.

Gross, H.L. (1992) The distribution and estimation of jack pine budworm defoliation. *Canadian Journal of Forest Research* 22, 1079–1088.

Hall, R.J., Volney, W.J.A. and Wang, Y. (1999) Yield and financial losses associated with a jack pine budworm outbreak in Michigan and the implications for management. *Canadian Journal of Forest Research* 29, 382–392.

Jeffreys, H. (1961) *Theory of Probability.* Clarendon Press, Oxford, 241 pp.

Johnson, N.L. and Kotz, S. (1970) *Continuous Univariate Distributions.* Houghton Mifflin, Boston, Massachusetts, 719 pp.

Katz, R.W. (1981) On some criteria for estimating the order of a Markov chain. *Technometrics* 23, 243–249.

Kedem, B. (1980) *Binary Time Series.* Marcel Dekker, New York, 140 pp.

Kemeny, J.G. and Snell, J.L. (1976) *Finite Markov Chains.* Springer, New York, 210 pp.

Kunert, J. (1987) On variance estimation in crossover designs. *Biometrics* 43, 833–845.

Matthews, J.S. (1990) The analysis of data from crossover designs: the efficiency of ordinary least squares. *Biometrics* 46, 689–696.

McCullough, D. (2000) A review of factors affecting the population dynamics of jack pine budworm (*Choristoneura pinus pinus* Freeman). *Population Ecology* 42, 243–256.

Miller, R.G. Jr (1980) *Simultaneous Statistical Inference,* 2nd edn. Springer, New York, 293 pp.

Nealis, V.G. (1990) Jack pine budworm populations and staminate flowers. *Canadian Journal of Forest Research* 20, 1253–1255.

Nealis, V.G. and Lomic, P.V. (1997) Host-plant influence on the population ecology of the jack pine budworm. *Ecological Entomology* 19, 367–373.

Nealis, V.G., Lomic, P.V. and Meating, J.H. (1998) Forecasting defoliation by the jack pine budworm. *Canadian Journal of Forest Research* 28, 228–233.

Quintana, F. and Newton, M.A. (1998) Assessing the order of dependence for partially exchangeable binary data. *Journal of the American Statistical Association* 93, 194–202.

Rose, D.W. (1973) Simulation of Jack pine budworm attacks. *Journal of Environmental Management* 1, 259–276.

Rudolph, T.D. and Yeatman, C.W. (1982) *Genetics of Jack Pine.* WO-38, USDA Forest Service, 112 pp.

Smith, D.M. (1997) *Silviculture: Applied Forest Ecology.* Wiley, New York, 537 pp.

Swed, F.S. and Eisenhart, C. (1943) Tables for testing randomness of grouping in a sequence of alternatives. *Annals of Mathematical Statistics* 14, 83–86.

Volney, W.J.A. (1999) Long-term effects of jack pine budworm outbreaks on the growth of jack pine trees in Michigan. *Canadian Journal of Forest Research* 29, 1510–1517.

Yang, L. and Tsiatis, A.A. (2002) Efficiency study of estimators for a treatment effect in a pretest–posttest trial. *American Statistician* 55, 314–321.

12 Spatial Stochastic Modelling of Cone Production from Stone Pine (*Pinus pinea* L.) Stands in the Spanish Northern Plateau

Nikos Nanos,[1] Rafael Calama,[2] Nieves Cañadas,[3] Carlos García[4] and Gregorio Montero[2]

Abstract

The spatial structure of the mean cone production from even-aged stone pine stands of the Northern Plateau of Spain has been studied. Available data consisted of 123 five-tree plots, where cone crop was collected during a 5-year period (1996–2000). The experimental variogram for the mean cone crop has shown that cone production is a variable with a high percentage of spatially structured variance and a range of spatial correlation of approximately 2000 m.

A kriging map for the mean stand production (over the 5-year period) has been built. Additionally we used conditional simulation to study the spatial variability and to quantify the uncertainty of predictions. This method permits the construction of uncertainty and probability maps. Geostatistical prediction of cone production at a regional scale has been shown to be a useful tool for modelling a low-value natural resource such as stone pine cones, without needing additional covariate measurements.

Introduction

Stone pine (*Pinus pinea* L.) is a typical Mediterranean species occupying in Spain an area of about 400,000 ha (more than 50% of the species' total area worldwide). Stone pine stands play an important role as soil protection elements in the semiarid zones of the Mediterranean basin (Ximénez de Embún, 1960) due to the ability of the species to occupy sandy areas, in both coastal and continental dune systems. Stone pine stands have been used and transformed by human activities from time immemorial. Among the products obtained from these stands, firewood, bark, wood and pinyon (the stone pine edible nut) are the most important. Stone pine pinyon is a highly prized food, widely used in pastry making. Cones are collected in natural or artificial stands, as well as in seed orchards and grafted plantations from genetically

[1] Unidad de Anatomía, Fisiología y Genética Forestal, ETS de Ingenieros de Montes, Spain
[2] Centro de Investigación Forestal CIFOR. Instituto Nacional de Investigación y Tecnología Agraria y Alimentaria, Spain
Correspondence to: rcalama@inia.es
[3] Dirección General del Medio Natural, Consejería de Medio Ambiente, Spain
[4] Servicio Territorial de Medio Ambiente. Junta Castilla y León

selected individuals. Despite the high cost of harvesting (cones are usually harvested manually), the actual increasing value of pinyon justifies its commercial utilization.

Forest management of these stands has been influenced by the main production objective (either wood or pinyon). The guidance of forest management to one or other product has been conditioned by the price and the demand for the corresponding product. The silvicultural practice had to be flexible enough in order to change the objective of the management as a response to market oscillations. Forest managers need useful, simple and flexible tools that can help them to decide on the principal forest product, depending on the market prices and the predicted stand productivity. Forest growth and yield models are nowadays among the most powerful available tools for forest management.

Cone Production Modelling

Forest edible fruit production has seldom been modelled, due to its great complexity and variability. Among the numerous factors that control cone production we can mention genetic variability, site factors (climatic, edaphic, etc.), forest stand factors (stand density, basal area), single tree (breast height diameter, crown size, etc.) and other exogenous factors (pests, rodents, robbery, etc.).

Stone pine flowering and fruiting is a 3-year process. Interannual variability in the production is very pronounced. Despite the interest shown in stone pine stands as fruit producers, most of the existing research work on cone production is merely descriptive. These works generally associate a silvilcultural intervention with the mean cone production for a given area. Prediction models have been carried out only in few studies, especially in Italy (Cappelli, 1958; Pozzera, 1959; Castellani, 1989) and Spain.

In Spain, the Department of Silviculture of CIFOR-INIA has been carrying out a research programme for the development of sustainable management schedules for stone pine stands. One of the main objectives of this research is the modelling of cone crop production. Within this project more than 400 experimental plots have been installed in the four most important stone pine regions of Spain. In these plots annual cone crop is collected and measured. As a result of this programme, several research projects, including cone production analysis, have been developed.

García Güemes (1999), in his silviculture simulation model for *P. pinea* in Valladolid, includes an analysis of the mean cone production per hectare. Cone production was found to be positively correlated with both mean square diameter and basal area and negatively correlated with stand density. Maximum cone production is found at ages ranging between 80 and 100 years. Montero *et al.* (2000) used data from stone pine stands of Western Andalucia to study the influence of age and density in cone production. In this case, significant differences in production were not found among different stand density classes. Cañadas (2000) developed a single-tree growth and yield model for even-aged stands of stone pine growing in the Central Range region. Individual tree cone crop was positively correlated to squared breast height diameter of the tree and dominant height of the stand, while stand density showed a negative correlation.

Justification for a geostatistical prediction for cone production

Current prediction models for stone pine cone production are deterministic. Actual knowledge of the factors that control cone production is very limited. Generally, single-tree models are considered to be better at predicting cone production (compared with whole-stand models). The inclusion in single-tree models of stand variables, such as stand density, as well as environmental factors – edaphic, climatic and topographical – improves the prediction ability of the model. However, single-tree models are not very useful in practice, since numerous covariates need to be measured.

In any case, the prediction ability of cone production models is generally very low. Attempts at predicting annual crop from these models are useless, since crop cycle variability is not included in the models. The low value of the stone pine cones does not justify a large monetary investment in predicting the production for a given area.

The use of geostatistical models to predict cone production shows several advantages over the existing regression models. Geostatistical estimations are generally much cheaper to obtain, since no additional covariates need to be measured. This property makes geostatistics an interesting tool for modelling a low-value natural resource such as cones. Kriging and simulation techniques give an unbiased estimation for the mean stand production as well as the spatial uncertainty of these estimates. Models for predicting the stand production at a regional scale could provide a helpful framework for regional forest management and planning, since these models allow the identification of the most productive areas. Finally, the extension of geostatistics to include time as an additional dimension could provide a useful approach for predicting the stand production at both the spatial and temporal scales (Stein *et al.*, 1998), giving the possibility for modelling inter-annual variability and crop cycles.

Several variables have been successfully predicted with geostatistics: site index (Hock *et al.*, 1993), total tree height (Samra *et al.*, 1989), timber volume (Holmgren and Thuresson, 1997) and stand density (Mandallaz, 2000). Yields of non-timber forest products, such as resin, have also been modelled with the use of geostatistical tools (Nanos *et al.*, 2001).

The aim of the present work is the spatial analysis of the average cone production per hectare (mean production of the years 1996–2000) as well as the construction, following kriging and simulation techniques, of maps for the average production and the associated spatial uncertainty.

Materials and Methods

Data

Data consisted of 123 plots established in even-aged stands and distributed all over the stone pine public forests of southern Valladolid (Fig. 12.1). Plots were circular, of variable radius, and included five trees. Among other variables, we measured the diameter at breast height, crown projection, total height and the height to the crown base. Cones were collected each autumn during 5 consecutive years (1996–2000), a period that is considered as a productive cycle (Ximénez de Embún, 1960).

Cones were classified into two groups according to amount of damage observed (as a consequence of insect attack: *Pissodes validirostris* or *Dioryctria mendacella*). For each group, cones were counted and weighted. In this work, the studied variable was the mean weight of undamaged cones per hectare computed as the average value over a 5-year period by plot.

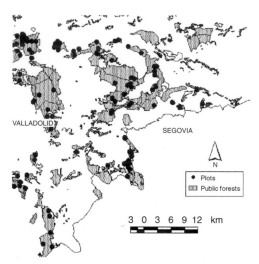

Fig. 12.1. Location of study area and plots.

Geostatistical methods

A classical geostatistical analysis can be divided into two parts: *structural analysis* and *spatial prediction*. The aim of the first is to quantify the spatial correlation for a given attribute. When spatial correlation among observations is detected, one usually proceeds to the second step. Spatial prediction allows the study and estimation of the value of the attribute at unsampled locations within the study area.

Structural analysis

Let $Z(x)$ denote a random function defined over the domain D of R^2, and sampled over a set of $i = 1,2,...n$ points. In the geostatistical framework, the set of observations is considered as a single realization of the random function $Z(x)$. The spatial structure of a single realization is usually described by the semivariance $\hat{\gamma}(h)$ for the lag $\mathbf{h}$:

$$\hat{\gamma}(h) = \frac{1}{2N(h)} \sum_{a=1}^{N(h)} \left[z(u_a) - z(u_a + \mathbf{h}) \right]^2 \qquad (1)$$

where $N(h)$ is the number of pairs of data locations a vector $\mathbf{h}$ apart, while $z(u_a)$ and $z(u_a + \mathbf{h})$ are measurements at locations $\mathbf{u}_a$ and $\mathbf{u}_a + \mathbf{h}$, respectively. A plot of the semivariance versus the distance $\mathbf{h}$ is called a (experimental) variogram and it describes the spatial behaviour of the function.

Typically, the semivariance exhibits an ascending behaviour near the origin ($\mathbf{h} = 0$), while at larger separation distances it levels off at a maximum value called the *sill* of the variogram. The distance at which the sill is reached is called the *range* of the variogram, while the term *nugget* is used for the semivariance value at a distance $\mathbf{h} = 0$.

Detecting anisotropy

Anisotropy directions were detected by variogram maps (Isaaks and Srivastava, 1989), which were constructed from 18 directional variograms, computed with a 40°

angular tolerance and 4000 m bandwidth. A gridding algorithm was then applied to the resulting variogram values and the directions of maximum and minimum continuity were judged visually.

Variogram modelling

Semivariogram modelling is needed for the posterior spatial prediction and for the conditional simulation. A basic semivariogram model should be a positive definite function, in order to guarantee the existence and singularity of the solution for the kriging system.

Anisotropy was corrected by rotating clockwise the coordinate system so as to identify the main axes of anisotropy and linearly transforming the rotated coordinates according to the anisotropic variogram model (Isaaks and Srivastava, 1989; Goovaerts, 1997). Kriging was finally performed in the transformed coordinate system using a global search neighbourhood.

Spatial prediction

The existence of a model of spatial dependence allows us to tackle the problem of estimating attribute values at unsampled locations. To do this, we used the data obtained from the sampled plots, and the semivariogram model constructed previously. Prediction was made following the kriging method, so called in recognition of the pioneering work of D. Krige.

Kriging estimators are among the BLUEs (best linear unbiased estimators), and are obtained in a similar way to linear regression estimators. Their two main properties are:

- prediction from these estimators is unbiased; and
- the error variance is minimized (the precision is going to be maximized).

Ordinary kriging estimators for the mean value of the attribute $z^*(\mathbf{u})$ at an unsampled location $\mathbf{u}$ is written as a linear combination of the values of the same attribute in the n sampled points $\{z(\mathbf{u_a}), a=1,2,\dots,n\}$:

$$z^*(\mathbf{u}) = \sum_{a=1}^{n(\mathbf{u})} \lambda_a(\mathbf{u}) z(\mathbf{u_a}) \tag{2}$$

where $\lambda_a(\mathbf{u})$ is the assigned weight for data $z(\mathbf{u_a})$.

From this formulation, to find the best estimator we have to determine the optimum value for the weights $\lambda_a(\mathbf{u})$. In order to get non-bias conditions for the estimator $z^*(\mathbf{u})$, we have to impose the constraint that sum of the weights $\lambda_a(\mathbf{u})$ should be equal to 1 (universality condition):

$$\sum_{a=1}^{n(\mathbf{u})} \lambda_a(\mathbf{u}) = 1 \tag{3}$$

In order to verify the property of minimum error variance, prediction error is defined as $R(\mathbf{u})=Z^*(\mathbf{u})-Z(\mathbf{u})$, and its variance equals zero:

$$\text{Var}\{R(\mathbf{u})\} \rightarrow 0 \tag{4}$$

Variance error defined in Equation 4 is only related to spatial covariance, which can be estimated from the variogram. In order to minimize Equation 4 we use the Lagrange technique, under the universality condition imposed by Equation 3. We get the following equation system, whose solution gives us the optimum weights:

$$\sum_{b=1}^{n(u)} \lambda_a \gamma \left(u_a - u_b \right) + \mu = \gamma \left(u_a - u \right) \qquad a = 1,...,n(u)$$

$$\sum_{a=1}^{n(u)} \lambda_a(u) = 1 \tag{5}$$

where $\gamma(u_a - u_b)$ is the semivariance among points u_a and u_b, where the attribute has been measured; $\gamma(u_a - u)$ is the semivariance among u_a and u, unsampled location, and μ is the value for the Lagrange multiplier.

Conditional simulation

Every prediction has an associated estimation error that should be quantified. In the geostatistical framework the main tool for quantifying the kriging estimation errors has been the kriging variance. However, it has been shown that this measure of spatial uncertainty is not reliable, since its value depends only on the spatial arrangement of the sampled plots and not on the actual values of the studied attribute.

Simulation methods are preferred, then, as primary tools for quantifying the spatial uncertainty of the estimation. Several methods have been proposed and used in practice. In this chapter we have used sequential Gaussian simulation (SGS). The steps that have been followed during the simulation can be summarized as follows.

1. The original data should be transformed to normal-score data. This transformation is usually achieved through the normal-score transformation.
2. A grid of the desired resolution is defined over the studied area.
3. Each node of the grid is visited (only once) in a random sequence. At each grid-node the parameters of the random variable (mean and standard deviation) are determined by simple kriging.
4. Draw randomly a simulated datum from the distribution specified in the previous step and add this datum to the data set.
5. Proceed to the next grid-node (chosen randomly) and repeat this process until all the nodes are simulated.
6. Repeat steps 3 to 5 until the desired number of simulations is achieved. In this study we realized 50 simulations.
7. The final step of the simulation is the back-transformation of the normal data to their original values, as described in Goovaerts (1997).

Simulation post-processing

There are several ways to summarize the results of the conditional simulations, the most important being the measurement of uncertainty. In the present study we used the standard deviation of the simulated data as an overall measurement of uncertainty. Other approximations include the interquartile range or the entropy of the simulated distribution (Goovaerts, 1997).

We also constructed a map showing the probability for the mean cone production to be larger than 200 kg/ha. This production can be considered as the minimum stand production for cone collection to be profitable.

Results and Discussion

Structural analysis

In Fig. 12.2 we present the histogram of the original values over the 123 sampled plots. The directional experimental variograms of the normal-transformed values are presented in Fig. 12.3. The variogram shows the existence of spatial correlation for the mean cone production per hectare, up to a distance of about 2000 m (range of spatial correlation).

The practical absence of nugget variance in the experimental variogram is very important for the present study. The absence of nugget variance implies that cone production is very continuous at a larger scale of variation and that factors controlling the production have a large range of spatial correlation.

Detection for anisotropic spatial continuity in the production showed that the main axes of anisotropy laid on the 0° and 90° direction (the former is the direction of maximum spatial continuity). Since the directional experimental variogram exhibits significant differences, it is preferable to model spatial variance using directional variogram models. These models are the following:

$$\gamma(h_0) = 1.028 \times SPHER_{2000}(h_0) + 0.756 \times EXP_{7000}(h_0) \tag{6}$$

$$\gamma(h_{90}) = 1.028 \times SPHER_{1000}(h_{90}) \tag{7}$$

where SPHER and EXP are the spherical and the exponential variogram models (the subscript indicates the range of the variogram model), while h_0 and h_{90} are the distances following 0° and 90° direction.

Prediction and simulation of the mean cone production

A kriged map for the mean cone production can be seen in Fig. 12.4. There are four nuclei with high production surrounded by large areas where the production seems to be very low. The kriged map for the mean cone production is in general accordance with the practical experience of forest rangers. Besides being a useful tool for forest management, the kriged map represents a certain scale of variation of the phenomenon under study and its use should be restricted to that specific scale of variation. The cone production from stone pine seems not to be influenced

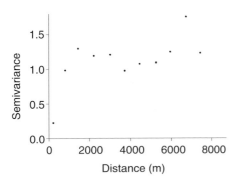

Fig. 12.2. Experimental variogram of mean cone production.

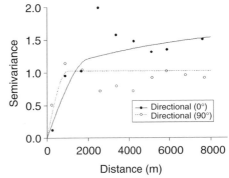

Fig. 12.3. Directional experimental variograms and models.

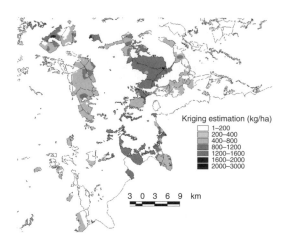

Fig. 12.4. Kriged map of mean cone production.

by factors with a small range of spatial variation (less than 1000–1500 m) since the nugget effect of the experimental variogram has been shown to be null.

In Figs 12.5 and 12.6 we present the results of the simulations. The standard deviation of the simulated data, shown in Fig. 12.5, may be used as an indication of uncertainty, while the probability map, Fig. 12.6, should be used in order to guide the future campaigns for collection of cones. This map shows the probability of a stand producing more than 200 kg/ha of cones.

Conclusions

The mean cone production from stone pine stands has been shown to be a variable exhibiting spatial correlation up to a distance of 2000 m. Furthermore, it has been shown that the production is very continuous at a large scale of variation (absence of nugget variance). This result may be used for the optimization of the sampling design in future sampling campaigns. On the other hand, this result has some direct theoretical implications: cone production is not influenced by micro-environmental conditions. Intuitively,

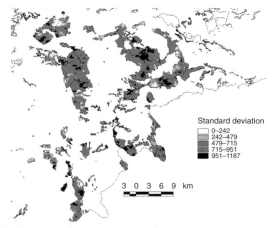

Fig. 12.5. Map of standard deviation for mean cone production.

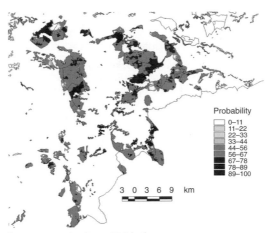

Fig. 12.6. Probability map for cone production >200 kg/ha.

the mean stand production is influenced by small-scale climatic factors such as precipitation and temperature.

Regarding the spatial prediction of the mean stand production, we believe that, at this stage of the modelling process, the geostatistical approach is satisfactory, since other predictive models require expensive measurements to be realized (single-tree models) or they have a very low predictive power (whole-stand models).

We compared the results obtained with those from the work by Gordo Alonso *et al.* (2000). The authors used data from post-crop visual estimation made by forest rangers every year in all the public forestlands. Data used are the mean value from a series of 34 years. Despite the roughness of the data, the work is a good approach to a geographical classification of cone productivity for stone pine stands in Valladolid (Fig. 12.7).

In comparing both studies, we find the main differences in the total amount of the cone crop per hectare, ranging in Gordo's work from 0 to 600 kg/ha and in our work

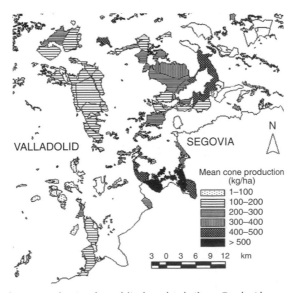

Fig. 12.7. Mean annual cone production for public forestlands (from Gordo Alonso *et al.*, 2000).

from 0 to 2500 kg. The reason for this difference is that kriging estimation is more accurate, detecting differences at a small spatial scale, while mean production has been calculated for a whole forest. Apart from this, in our work, the complete crop was collected in fully stocked stands, while in real harvesting work, there are many areas in the forestlands where cones are not collected (young stand areas, small productive areas, isolated trees). Despite differences in total crop amount, we find similarities in locating the most productive areas in the region.

The probability map (Fig. 12.6), showing the probability of a crop larger than 200 kg/ha, could be a very useful tool for designing cone collection campaigns.

Stone pine area in Valladolid province is about 30,000 ha. Valladolid is one of the provinces in Spain where cone collection and the pinyon market have been a traditional activity. Since ancient times, cone production has been included as a main objective in forestland management (Romero y Gilsanz, 1884; Olazabal, 1917). Nowadays, the pinyon and transformed products industry invoices an annual amount of 12 million euros, generating nearly 2000 jobs, which gives an idea of the importance of this industry in the province (Gordo Alonso, 1998). On account of this, any attempt to improve the ability for predicting cone production should be considered positive.

However, we believe that the geostatistical model presented in this study should be considered as a temporary prediction model, until other more sophisticated modelling tools, including site, stand and single-tree variables, as well as temporal variability, become available.

Acknowledgements

The authors wish to thank Javier Gordo Alonso, from the Servicio Territorial de Medio Ambiente, Valladolid, for technical support and data supply. We are also grateful to Angel Bachiller, for his work in installing the plots, and to Sven Mutke, for his valuable comments on the manuscript. The research was partially supported by a grant to R. Calama from the Consejería de Educación, Comunidad de Madrid, in the context of the INIA project SC-99-017.

References

Cañadas, M.N. (2000) *Pinus pinea* L. en el Sistema Central (valles del Tiétar y del Alberche): desarrollo de un modelo de crecimiento y producción de piña. PhD thesis, ETSI Montes, Universidad Politécnica de Madrid, Spain.

Cappelli, M. (1958) Note preliminari sulla produzione individuale de stroboli in *Pinus pinea* L. *L'Italia Forestale e Montana* 13(5), 181–203.

Castellani, C. (1989) La produzione legnosa e del fruto e la durata economico delle pinete coetanee di pino domestico (*Pinus pinea* L.) in un complesso assestato a prevalente funzione produttiva in Italia. *Annali ISAFA* 12, 161–221.

García Güemes, C. (1999) Modelo de simulación selvícola para *Pinus pinea* L. en la provincia de Valladolid. PhD thesis, ETSI Montes, Universidad Politécnica de Madrid, Spain.

Goovaerts, P. (1997) *Geostatistics for Natural Resources Evaluation*. Oxford University Press, New York.

Gordo Alonso, F.J. (1998) Programa de Mejora Genética de *Pinus pinea* L. en Castilla y León. *Montes* 52, 71–84.

Gordo Alonso, F.J., Mutke, S. and Gil, L. (2000) La producción de piña de *Pinus pinea* L. en los montes públicos de Valladolid. In: de Castilla y León, J. (ed.) *Primer Simposio sobre el pino piñonero* (Pinus pinea *L.*), Valladolid, pp. 269–277.

Hock, B.K., Payn, T.W. and Shirley, J.W. (1993) Using a geographical information system and geostatistics to estimate site index of *Pinus radiata* for Kaingaroa Forest, New Zealand. *New Zealand Journal of Forestry Science* 23, 264–277.

Holmgren, P. and Thuresson, T. (1997) Applying objectively estimated and spatially continuous forest parameters in tactical planning to obtain dynamic treatment units. *Forest Science* 43, 317–326.

Isaaks, E.H. and Srivastava, R.M. (1989). *An Introduction to Applied Geostatistics*. Oxford University Press, New York.

Mandallaz, D. (2000) Estimation of the spatial covariance in universal kriging: application to forest inventory. *Environmental and Ecological Statistics* 7, 263–284.

Montero, G., Candela, J.A., Ruiz-Peinado, R., Gutiérrez, M.F., Pavón, J., Bachiller, A., Ortega, C. and Cañellas, I. (2000) Density influence in cone and wood production in *Pinus pinea* L. forests in the south of Huelva province. In: CIFOR-INIA (ed.) *Mediterranean Silviculture with Emphasis in* Quercus suber, Pinus pinea *and* Eucalyptus *sp.* IUFRO Working Group 1.05.14, Seville, pp. 129–142.

Nanos, N., Tadesse, W., Montero, G., Gil, L. and Alía, R. (2001) Spatial stochastic modelling of resin yield from pine stands. *Canadian Journal of Forest Research* 31, 1140–1147.

Olazabal, D. (1917) Datos sobre el problema silvopastoral en España. *Revista de Montes* 41.

Pozzera, G. (1959) Raporti fra produzione di stroboli in *Pinus pinea* L. de andamento stagionale. *L'Italia Forestale e Montana* 14(5), 196–206.

Romero y Gilsanz, F. (1884) De los pinares en las arenas sueltas de la provincia de Valladolid. *Revista de Montes* 9, 220–228.

Samra, J.S., Gill, H.S. and Bhatia V.K. (1989) Spatial stochastic modelling of growth and forest resource evaluation. *Forest Science* 35, 663–676.

Stein, A., Van Groeningen, J.W., Jeger, M.J. and Hoosbeek, M.R. (1998) Space–time statistics for environmental and agricultural related phenomena. *Environmental and Ecological Statistics* 5, 155–172.

Ximénez de Embún, J. (1960) *Arenas movedizas y su fijación (Dunas continentales)*. Ministerio de Agricultura, Madrid, Spain.

13 Modelling the Carbon Sequestration of a Mixed, Uneven-aged, Managed Forest Using the Process Model SECRETS

G. Deckmyn,[1] R. Ceulemans,[1] D. Rasse,[2]
D.A. Sampson,[3] J. Garcia[4] and B. Muys[4]

Abstract

Currently there is a high demand for knowledge on the actual and attainable carbon sequestration in existing forests, and the influence of forest management thereon. Although process models have proved their worth in simulating and forecasting growth and yield of even-aged, single-species, regularly spaced forests, their applicability to uneven-aged, mixed-species, patchy forests is less well documented. By describing a complex forest as a combination of multiple simple patches, it is possible to simulate the total ecosystem with a relatively small number of parameters.

In this case study, the process model SECRETS was adapted and parameterized to simulate C sequestration in the different compartments (both above- and below-ground) of Meerdaalwoud, a mixed deciduous–coniferous forest in central Belgium. The current management consists of an increase in untouched forest reserve area (to 10% of the total area) and a gradual replacement of exotic species (several pine species) with native species managed in a low-impact silvicultural system.

The results indicate that SECRETS is able both to simulate the current yield and to predict the future effect of current changes in management. The results indicate that a gradual change from the current situation to a more natural one will increase C content of the ecosystem by 22.9 t/ha under current climatic conditions or 46.5 t/ha under a global climate change scenario over the next 150 years.

Although forest productivity will decline slightly (from 5.9 to 5.3 t/ha/year), the sequestration in wood products will increase slightly. This is, however, not due to a larger proportion of long-lived wood products from oak and beech forest, but to a change in age-class structure. Under global climate change conditions, carbon stocks in soil, biomass and wood products are predicted to increase. The use of small-sized timber as a fuel, substituting for fossil fuels, can significantly increase the total carbon sequestration in the forest.

[1] University of Antwerp (UIA), Belgium
Correspondence to: gaby.deckmyn@ua.ac.be
[2] University of Liege (UCL), Belgium
[3] VPI and USDA Forest Service, USA
[4] Catholic University of Leuven, Belgium

Introduction

Forests play an important role in the global carbon (C) cycle (Schimel, 1995). The importance of forests to atmospheric CO_2 levels was acknowledged in the commitments for greenhouse gas emission reductions negotiated in 1997 in the Kyoto Protocol (UNFCCC, 1998; Schlamadinger and Marland, 2000) but accurate estimations of C stocks and fluxes are still lacking. The need to estimate and predict C sequestration in forests has stimulated the development of forest models.

The purpose of the present study was firstly to simulate the current growth of an old, mixed forest, using the plot-scale process model SECRETS (Sampson *et al.*, 2001) on multiple forest patches. As a case study, the Meerdaalwoud, a mixed deciduous–coniferous forest in central Belgium, was used. Secondly, the model was used to predict how the current changes in management options would affect growth and yield in the coming 150 years, under both current and global change scenarios. Finally, the obtained growth and yield data were used to calculate the total C balance of the system, including the sequestration in forest products and the substitution for fossil fuels using GORCAM (Graz Oak Ridge Carbon Accounting Model; Schlamadinger *et al.*, 1997).

State of the Art

A multitude of forest growth models have been developed, ranging from very simple, deterministic to sophisticated process-based models (McMurtrie and Landsberg, 1992; Lüdeke *et al.*, 1994; Tiktak and van Grinsven, 1995). There are models operating at different scales, from the single tree (Jarvis, 1993) to the regional scale (Cao and Woodward, 1998). At present it is possible to simulate and predict the growth of a single tree in a forest, in detail, with good reliability (Wang and Jarvis, 1990; Jarvis, 1993). The models to do this, however, require an enormous input of parameters and they are generally not easily scaled-up to a total forest. On the other hand, forest-scale models are well able to simulate so-called normal forest. They consist of monospecific, uniform, even-aged stands on identical sites with stands of each age class equally represented. Such modelling results are very relevant to the understanding of forest growth, for large-scale yield predictions and to aid management decisions. However, many multiple-use woodlands in urbanized Western Europe have a distinctive structure. Due to continuous changes in silvicultural treatment, and the choice of tree species over the years, and due to differences in site characteristics within the forest, a mosaic of small patches containing different species and age classes has developed. It is unclear whether current process models are able to produce reliable results for forests with such complex structure. Furthermore, carbon sequestration includes wood products as sinks and replacement issues such as substitution of fossil fuel by fuelwood (Schlamadinger *et al.*, 1997) or substitution of wood for more energy-intensive materials such as steel or concrete. Also, the impact of yield quality and durability of the wood products on total carbon sequestration is often ignored although it can be of great importance (Marland and Schlamadinger, 1995).

Besides an adequate simulation of the present status and productivity of a forest, a main goal of many modelling efforts is to predict the effects of changes in climate (global climate change) and management over time. More recently, forest management models seek to maximize more than just yield or carbon sequestration. Other functions, such as biological conservation, recreation and sustainability in general, have become important management objectives in many forests, especially in densely populated European countries such as Belgium (Englin and Richard, 1991). In view of these diverse goals, more emphasis is being placed on the use of native species, minimization of human impact in some reserve areas, and development of uneven-aged stands (where all age categories are present in more or less equal proportions). For the managers to optimize between these sometimes conflicting functions, models should be made more relevant in order to provide decision support concerning the impact of management options on yield and C sequestration. Some attention must be paid to the scale factor, because effects on a small spatial scale, or short time scale, can differ from the large scale (Harmon, 2001).

Methodology

Forest description

Meerdaalwoud (4°40'–4°45' E, 50°45'–50°50' N, 100 m above sea level) covers 1195 ha in central Belgium, south of Leuven. In area it has been 80% woodland for at least the last 300 years (Bossuyt *et al.*, 1999). However, it has been intensively used and managed since early historical times and very little is left of the original tree-species composition. The forest consists of many patches varying in species composition, soil and age class structure (Table 13.1).

Although many different tree species are found, the deciduous patches are dominated by beech (*Fagus sylvatica*, 31% of total area) and oak (*Quercus robur* and

Table 13.1. Soil type (A, loam; L, sandy loam; S, sandy), species and age-class distribution of the different patches in Meerdaalwoud.

Species	Age class (years)	Soil	Area (ha) Uniform	Mixed	Reserve
Deciduous	0–50	A	11.99	0.03	5.8
		L	7.24	0.00	6.4
		S	3.70	0.00	9.0
	50–100	A	20.60	9.32	98.4
		L	15.18	6.86	2.2
		S	10.45	3.62	4.2
	100–150	A	32.12	0.42	3.9
		L	24.01	0.42	17.6
		S	19.21	0.63	1.7
	Uneven aged	A	357.61	0.54	1.7
		L	152.43	0.35	2.9
		S	107.21	0.24	3.5
					0.0
Coniferous	0–20	A	0.78	0.00	0.3
		L	1.39	0.03	2.6
		S	0.51	0.02	0.7
	20–60	A	24.91	2.96	2.9
		L	39.40	3.44	0.7
		S	24.35	2.38	2.4
	60–100	A	18.09	21.17	3.4
		L	29.82	22.49	5.1
		S	17.41	10.69	0.2
	Uneven aged	A	0.01	3.85	1.1
		L	0.00	6.85	0.8
		S	0.00	1.52	1.1

Quercus petraea, 25% of total area). The coniferous species include Corsican black pine and are dominated by Scots pine (*Pinus sylvestris*, 24% of total area). For this study we used four forest types (i.e. deciduous (<20% coniferous), mixed deciduous (20–50% coniferous), mixed coniferous (20–50% deciduous) and coniferous (<20% deciduous)) and three soil types (i.e. A, loam; L, sandy loam; S, sand). All soils are well drained and no groundwater table is within reach of the roots. Since many patches are more or less even aged we used seven age classes: six even-aged classes ranging from 0 to 150 years and one class of uneven-aged patches. Each patch was modelled as a mixed patch reflecting the overall composition of the mentioned tree species. The simulations were based on the available stand inventory data (density, standing biomass, yield and age distribution).

Management

The current management of Meerdaalwoud is not homogeneous over the entire forest. Current silvicultural practice is mainly related to the species (deciduous vs. coniferous), but also to patch productivity. Furthermore, 178.6 ha (covering different patches of species and age composition) are set aside as strict forest reserves (no longer under active management, referred to as unmanaged in the Discussion). In the model simulations, differences in management due to patch productivity were ignored. Differences in modelled management consisted of the differences between deciduous patches (plant density 5000 trees/ha, thinning from age 40 years onward every 10 years, and clearfelling at year 150) and coniferous patches (plant density 5000 trees/ha, thinning from age 20 years onward every 10 years and felling at year 100), and the difference between managed (1016 ha) and unmanaged (178.6 ha) patches.

Model description

SECRETS

The model, termed SECRETS (Stand to Ecosystem CaRbon and EvapoTranspiration Simulator) is a modular, process-based model (Sampson and Ceulemans, 2000). SECRETS incorporates different modules and submodels based on several well-known models:

- light penetration: sun–shade model (De Pury and Farquhar, 1997);
- photosynthesis: (Farquhar *et al.*, 1980);
- surface and soil litter, C and nitrogen (N) pools: grassland dynamics (Thornley, 1998);
- carbon storage and partitioning: Frankfurt Biosphere Model (Lüdeke *et al.*, 1994) and adaptations (Sampson and Ceulemans, 2000); and
- maintenance respiration: biomass (McMurtrie and Landsberg, 1992).

Unlike the individual submodels above, SECRETS can model multiple patches of forest with different species composition and vegetation in over-, sub- and under-storey. SECRETS has been used and parameterized to simulate a mixed coniferous–deciduous forest (pine, oak, beech) in the Campine region, about 70 km from Meerdaalwoud (Sampson *et al.*, 2001).

A set of weather data including hourly values of irradiation, rainfall and humidity, day- and night-time temperatures was created using a weather generator and based on a data set from a meteorological tower in the coniferous–deciduous forest in the Campine region (Rasse *et al.*, 2001).

Although different sets of weather data may give different results in tree growth, with the large time and spatial scale used in this study, differences between different sets of weather data were very small (<2%) and will be ignored.

To simulate global climatic change conditions, the data set from the European project ECOCRAFT was used (1999) which gives predicted monthly changes in temperature, humidity and rainfall in 2100 for different latitudes (Table 13.2). A CO_2 level of 700 ppm was used in these simulations. The simulations were run for the future climate, not for a gradual transition from current to future climate.

Model simulations were run for 150 years, starting from the current situation in age distribution and standing stock (from stand inventory data). Although the model yields hourly values of photosynthesis and respiration, as well as daily values for forest growth, only yearly (annual) values were used for this application.

GORCAM

The spreadsheet model GORCAM (Schlamadinger *et al.*, 1997) calculates the total carbon balance of different land use scenarios. The model accounts for the carbon in the ecosystem (soil and living biomass) and in the products (from short-lived pools in paper up to long-lived pools in durable wood for furniture or window frames), the fossil fuel displacement when the biomass yield is used for producing bioenergy and also the fossil fuels used for stand tending and harvesting operations.

A harvested pine tree delivers relatively more stemwood than either oak or beech but only part of it is useful for construction timber. Oak and beech trees contain relatively less stemwood but can be almost entirely used by the furniture industry. As a consequence, a final cut of all three species leads to approximately the same allocation of products: 30% in long-lived products, 30% in medium-lived products, 18% in short-lived products and 22% staying in the forest. For this study, long-lived wood products (furniture and construction materials) were assumed to last for 40 years (half-life at 20 years) in the case of broadleaved trees and 30 years in the case of pine; medium-lived wood products (particle board) for 10 years; while for short-lived products (paper and fuelwood from thinnings and part of the final harvest) a life span of 1 year (half-life of 6 months) was assumed. The fuelwood was assumed to be used in home heating systems, where it replaced fossil fuels. Finally, the fossil fuel consumption caused by forest management, harvesting and log transportation was subtracted from the final sequestration balance. Instead of using existing yield tables as an input, we fitted the output of the SECRETS model as input for the sequestration calculations. The advantage of this serial connection of models was that it allows climate change scenarios to be taken into account in GORCAM.

Table 13.2. Predicted changes in minimum and maximum air temperature (°C) and in precipitation (%) and humidity (%) under global climate change conditions for Belgium. Data from the European ECOCRAFT project (final report 1999).

	Jan	Feb	Mar	April	May	June	July	Aug	Sept	Oct	Nov	Dec
T_{max}	1.3	1.1	0.9	0.9	1	1.2	1.5	1.9	2.1	1.6	1.3	1.3
T_{min}	1.4	1.4	1.2	1	1.1	1.3	1.4	1.6	1.66	1.4	1.3	1.3
Precipitation	7	8	5	3	2	0	−7	−8	−3	5	8	7
Humidity	12	11	9	8	8	8	7	6	7	9	11	11

Data

Current and future C content

The simulated biomass of the mixed forest (on an area basis) as a function of time, and for the different tree compartments (roots and above-ground woody parts) is presented in Fig. 13.1 (time zero in Fig. 13.1 corresponds to the present situation of the forest). About 20% of the living biomass is found in the roots of the trees, which is similar to values found in most studies on deciduous European forests (Cannell and Dewar, 1994). The simulated overall biomass in the forest is 244.4 t/ha (of which half is carbon), which is comparable to values from the literature (Liski *et al.*, 2000). The predicted biomass in 150 years time is only slightly higher (249.7 t/ha) than the current biomass, although pine is being replaced by beech and oak. This might seem contradictory to many reports that suggest that using species with a longer rotation period will increase standing stock (Thornley and Cannell, 2000). Indeed, in our simulations of the individual species, average standing stock of pine is lower than for oak or beech, as expected.

Replacing pine with hardwoods increases standing biomass, and thus carbon stock in the long run (150 years). However, this effect is partly neutralized by the fact that, in the current age-class distribution, old stands are over-represented, which means that the future will bring more young patches with lower standing. Also, oak and beech have a lower growth rate during the early years, so replacing pine results in a decrease in standing stock over the first 80 years, followed by an increase.

The current soil carbon content is 175.4 t/ha and the simulation runs indicate an increase to 193 t/ha in the managed area and 237 t/ha in the forest reserves. Although Meerdaalwoud is an ancient woodland, many patches have been disturbed in the past 200 years. It is therefore not surprising that the soil carbon content

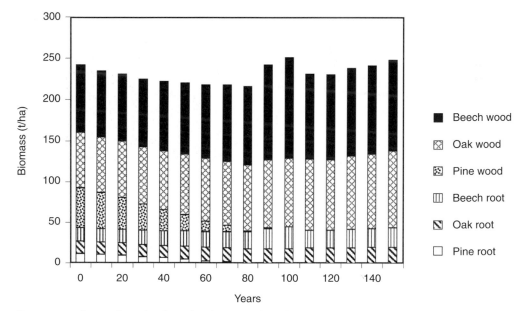

Fig. 13.1. Evolution of simulated standing biomass (t/ha) of Scots pine, oak and beech in the mixed coniferous–deciduous Meerdaalwoud (Belgium) over the next 150 years (present = year 0). Biomasses of both roots and above-ground woody parts (stems + branches) are presented.

has not reached steady state. Furthermore, due to atmospheric deposition, there is an important influx of N in the system (Neirynck *et al.*, 2001) which can increase the C-sequestering capacity of the soil (McMurtrie *et al.*, 2001). Therefore, Meerdaalwoud is predicted to be a moderate sink for carbon (0.040 t/ha/year) for the future 150 years even when the C sequestration in harvested wood and fossil fuel substitution is ignored, and assuming no climatic changes.

Harvest yield

The current simulated yield of the forest is 5.9 t/ha/year dry biomass (or 2.9 t C/ha/year, see Table 13.3) of which most is deciduous (4.36 t/ha/year or 5.8 m^3/ha/year), and the rest is coniferous (1.54 t/ha/year or 3.1 m^3/ha/year, see Fig. 13.2); this compares well with the actual harvested wood of previous years (i.e. 7.7–8.9 m^3/ha/year).

Pine yield is slightly higher than beech or oak yield over the total rotation period (150 years). Since pine thinning starts at a younger age, the economic constraints seem to favour pine. However, beech and oak yield is not much lower in the long run (Fig. 13.3), and the yield is of a higher quality and commands much higher prices per unit volume. Although the wood of oak is much more durable than that of pine or beech, this does not necessarily lead to higher sequestration in the products because of the lower stemwood allocation. By replacing pine with beech and oak, overall yield of Meerdaalwoud will be reduced by 10% (Table 13.3). In terms of wood volume the decrease is much larger (because pine has a lower density; Jansen *et al.*, 1996). C seques tration in products, however, increases slightly, not because of the more durable charac- ter of oak or beech timber, but because of changes in the age-class structure. Total sequestration increases significantly, especially when substitution of fuelwood for fossil fuels is taken into account. These results confirm previous studies indicating the impor- tance of including wood products in total C sequestration (Schlamadinger *et al.*, 1997).

C sequestration

Defining C sequestration as the total C in the ecosystem, and additional C sequestra- tion as an increase in sequestration (Ingram and Fernandes, 2001), additional

Table 13.3. Current and future (150 years) carbon content of soil, living trees, wood products (with long and short life time), C in fossil fuels displaced (by use of fuelwood for energy) and used (for management and harvesting) and additional sequestration in a mixed forest (Meerdaalwoud). Simulated data (SECRETS and GORCAM) for the managed area and the forest reserve under unchanging conditions and for the managed area under global climate change conditions are given.

| | Soil C (t C/ha) | Trees (t C/ha) | Yield (t C/ha) | Wood production | | Fossil fuels | | Additional sequestration (t C/ha/year) |
				Short (t C/ha)	Long (t C/ha)	Displ. (t C/ha)	Used (t C/ha)	
Current								
Managed	175.4	122.2	2.9	24.5	8.7	0	0	
Reserve	223	166.6	0	0	0	0	0	
Future								
Managed	193	124.9	2.6	26.5	12.2	26.8	−3.98	0.330
Reserve	237	192.5	0	0	0	0	0	0.266
Global change	212	136.5	2.8	29.8	16.1	56.2	−7.62	0.478

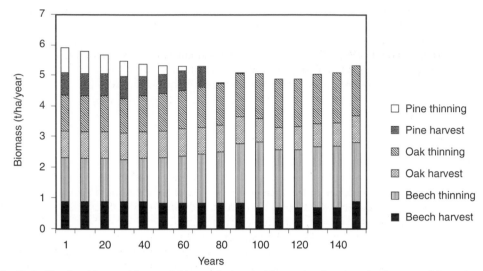

Fig. 13.2. Simulated harvest biomass (t biomass/ha/year) of Scots pine, beech and oak removed through thinning or harvest from Meerdaalwoud over the next 150 years (present = year 0).

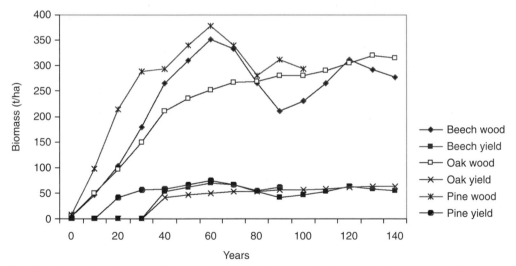

Fig. 13.3. Evolution of simulated biomass of Scots pine, beech and oak stands in Meerdaalwoud (Belgium) over time after planting (= year 0) on loamy soil. Both standing woody stock (stems + branches) and harvested yearly yield are presented.

C sequestration of a mature, existing forest is always quite low because only changes in C pools are used and, from this positive contribution to C sequestration, the C used for managing and harvesting the ecosystem is subtracted. Although the Meerdaalwoud has quite a high yield, and some of this yield is immobilized in wood products, eventually all wood products decay. Assuming that the yield is relatively constant then the amount of C in wood products will not increase either, and additional C sequestration remains low. For the managed area of Meerdaalwoud, the additional sequestration over the coming 150 years is predicted to be 0.330 t/ha/year and roughly half of this amount is due to fossil fuel displacement (fossil fuels that have not been used because energy was won from burning wood; Table 13.3). These results confirm previous studies, indicating the importance of including wood products and fossil fuel displacement in total C sequestration (Schlamadinger *et al.*, 1997). Furthermore, the use of even a small part of the yield as fuelwood has an important effect on C sequestration.

Of course, eventually, continuously removing wood from forests should reduce soil fertility, but with the current high nitrogen deposition this is not an issue in Belgium.

Patch effects

The differences between the different forest patches are quite large. In the patches with a more sandy soil, tree growth is less for all species (Fig. 13.4). However, pine is less affected than beech, and the difference in yield between beech and pine on sandy soils is considerably enhanced. Pine had been planted on the sandy patches because it grows better on sandy soils. When comparing the species, it is obvious that pine growth is rapid in the early years but soon stabilizes. Growth of beech is slow in young stands and higher in older stands (Fig. 13.3). Oak growth is intermediate in young stands but never reaches high values. The higher growth rate of beech compared with oak is not related to higher photosynthetic rates (model inputs for maximum rate of electron transport (J_{max}) and maximal carboxylation velocity (Vc_{max}) of beech are lower (Medlyn *et al.*, 1999)), but to a higher light absorption by the canopy due to the higher allocation of biomass to leaves resulting in higher leaf area index (compared with oak). The simulated growth curves compared well with the available growth tables of the three forest species (results not shown; see Jansen *et al.*, 1996). However, because SECRETS uses varying weather inputs, the simulated growth curves are generally less smooth than average growth curves, and important reductions in growth can occur (Fig. 13.3).

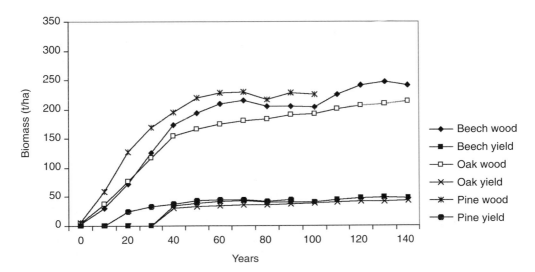

Fig. 13.4. Evolution of simulated biomass of Scots pine, beech and oak stands in Meerdaalwoud (Belgium) over time since the present (= year 0) on sandy soil. Both standing woody stock (stems + branches) and harvested yield are presented.

Reserves

The forest reserves are mainly important for conservational and biodiversity goals. Above-ground and total biomass are considerably higher than in the managed forest patches (total biomass is 258.7 vs. 332.2 t/ha) and are predicted to increase further (to 384.9 t/ha) over the next 150 years. Soil C content is predicted to increase slightly from 223 to 237 t/ha. Forest reserve areas will continue to be moderate C sinks over the coming years (Table 13.3). However, since there is no harvested yield, additional C sequestration is lower than in the managed forest patches (0.266 t/ha/year). Productivity is very low, since reserves evolve to an equilibrium where no carbon is gained or lost. The present model results suggest, however, that this equilibrium has not yet been reached. Several studies have shown European forests to be C sinks (Liski *et al.*, 2000), which is attributed to increasing standing stock, due to land-use and management changes, but also to global climate change effects and nitrogen deposition. The increase in winter temperature has been shown to prolong the growing phase, while increased CO_2 has enhanced growth (see below). Possibly even old forests can function as carbon sinks (total carbon storage in the ecosystem increases; see Carey *et al.*, 2001). However, other studies show contradictory results and proclaim boreal forests currently to be carbon sources (because total carbon in the ecosystem is decreasing; see Wang and Polglase, 1995).

Global change effects

Many studies concerning the effects of global climate change on trees and forests have been performed (Norby *et al.*, 1999). Although most experiments clearly show a growth stimulation on young trees (Gielen and Ceulemans, 2001), it is often suggested that the effects will be less pronounced on older trees (Luo *et al.*, 1999). In some cases, acclimation of photosynthesis reduces the effect of global climate change (Medlyn *et al.*, 1999). In other cases, forest growth is limited by N or phosphorus (P) (Oren *et al.*, 2001). The effect of global climate change on a forest can only be predicted by extrapolating results from small and short-term experiments to a larger scale and longer term. The ability of models for such large-scale predictions is dependent on the level of sophistication and scaling-up potential of the model, which can vary widely (e.g. is acclimation included? is there a maximal tree biomass? is respiration affected? is bud-burst simulated? etc.). This explains some of the contradictory results from using different models (Wang and Polglase, 1995; Rasse *et al.*, 2001).

SECRETS simulates temperature effects in detail, i.e. bud-burst, soil respiration, water use, etc. are all included and influenced by temperature. The effect of CO_2 is limited to the effect on stomatal resistance and on the stimulation of photosynthesis. The result on the growth curves of the individual species (data not shown) compares well with the existing literature (Gielen and Ceulemans, 2001) in that growth is stimulated primarily during the early years. In older trees, the effect becomes marginal due to various mechanisms (acclimation is not specifically modelled, but plant competition increases as trees grow, and growth becomes light, nutrient and/or water limited). The overall effect of global climatic changes on Meerdaalwoud shows an 8% increase in biomass yield (from 5.3 to 5.7 t/ha/year; Table 13.3) and in standing stock (from 249 to 273 t/ha). Under global climate change conditions, the simulations predict that Meerdaalwoud will continue to be an important sink for C (0.478 t/ha/year; Table 13.2). There is no indication of N or P limitation on forest growth in Meerdaalwoud (N input is high and exceeds demands; see Neirynck *et al.*, 2001). As yet, too little is known about soil processes to accurately predict the interaction of global change with soil processes, although this is probably of key importance in the global climate change issue (McMurtrie *et al.*, 2001).

Conclusions

Forests (including soils) contain, worldwide, 1150 Pg C, roughly 1.5 times the total amount of carbon in the atmosphere (Liski *et al.*, 2000). Therefore, changes in C sequestration in forests due to human land use or global climate change can be of vital importance in the total carbon cycle. In this study we showed that changes in species (oak and beech versus pine) increased the standing stock and thus C sequestration, although C in durable wood did not increase. Results from the GORCAM model indicated that using part of the harvest as fuelwood to substitute for fossil fuels could double additional C sequestration. The model simulations further suggest that conversion of managed forest to forest reserve increases standing stock, but decreases total C sequestration (other goals such as biodiversity are positively affected). Due to global warming and increasing CO_2, the reserves will remain an important sink for carbon over the coming 150 years. For the managed forest, global change increases total C sequestration from 0.330 to 0.478 t/ha/year. This study clearly illustrates the usefulness of process-based growth models in combination with accounting models for calculating C sequestration of heterogeneous and multi-species forest ecosystems.

Acknowledgements

This study was financially supported by the Flemish Minister for Scientific and Technological Innovation (Brussels, Belgium) as project no. PBO98/41-12/16. The authors especially thank Bernhard Schlamadinger from Joanneum Research in Graz, Austria, for offering to work with GORCAM. The authors also acknowledge Ellen Moons (K.U. Leuven Centre for Economic Studies) for useful discussions, and the Forest Administration of the Flemish Region for providing inventory data of Meerdaalwoud.

References

Bossuyt, B., Deckers, J. and Hermy, M. (1999) A field methodology for assessing man-made disturbance in forest soils developed in loess. *Soil Use and Management* 15, 14–20.

Cannell, M.G.R. and Dewar, R.C. (1994) Carbon allocation in trees: a review of concepts for modelling. *Advances in Ecological Research* 25, 59–104.

Cao, M. and Woodward, F.I. (1998) Net primary and ecosystem production and carbon stocks of terrestrial ecosystems and their response to climate change. *Global Change Biology* 4, 185–198.

Carey, E.V., Sala, A., Keane, R. and Callaway, R.M. (2001) Are old forests underestimated as global carbon sinks? *Global Change Biology* 21, 339–344.

De Pury, D.G.G. and Farquhar, G.D. (1997) Simple scaling of photosynthesis from leaves to canopies without the errors of big-leaf models. *Plant, Cell and Environment* 20, 537–557.

Englin, J. and Richard, M. (1991) A hedonic travel cost analysis for evaluation of multiple components of site quality: the recreation value of forest management. *Journal of Environmental Economics and Management* 21, 275–290.

Farquhar, G.D., von Caemmerer, S. and Berry, J.A. (1980) A biochemical model of photosynthetic CO_2 assimilation in leaves of C_3 species. *Planta* 149, 78–90.

Gielen, B. and Ceulemans, R. (2001) The likely impact of rising atmospheric CO_2 on natural and managed *Populus:* a literature review. *Environmental Pollution* 115, 335–358.

Harmon, M.E. (2001) Carbon sequestration in forests: addressing the scale question. *Journal of Forestry* April, 24–29.

Ingram, J.S.I. and Fernandes, E.C.M. (2001) Managing carbon sequestration in soils: concepts and terminology. *Agriculture, Ecosystems and Environment* 87, 111–117.

Jansen, J.J., Sevenster, J. and Faber, P.J. (1996) *Opbrengsttabellen voor belangrijke boomsoorten in Nederland.* IBN Report 221, IBN-DLO, Wageningen, 202 pp. [in Dutch].

Jarvis, P.G. (1993) MAESTRO: a model of CO_2 and water vapour exchange by forests in a globally changing environment. In: Schulze, E.D. and Mooney, H.A. (eds) *Design and Execution of Experiments on CO_2 Enrichment.* CEC Ecosystems Research Report 6, 107–116.

Jarvis, P.G. (ed.) (1993) ECOCRAFT: predicted impacts of rising carbon dioxide and temperature on forests in Europe at stand scale. EC project ENV4CT95-0077, IC20-CT96-0028, final report, 346 pp.

Liski, J., Karjalainen, T., Pussinen, A., Nabuurs, G.-J. and Kauppi, P. (2000) Trees as carbon sinks and sources in the European Union. *Environmental Science and Policy* 3, 91–97.

Lüdeke, M.K.B., Badeck, F.-W., Otto, R.D., Häger, C., Dönges, S., Kindermann, J., Würth, G., Lang, T., Jäkel, U., Klaudius, A., Ramge, P., Habermehl, S. and Kohlmeier, G.H. (1994) The Frankfurt Biosphere Model: a global process-oriented model of seasonal and long-term CO_2 exchange between terrestrial ecosystems and the atmosphere. I. Model description and illustrative results for cold deciduous and boreal forests. *Climate Research* 4, 143–166.

Luo, Y., Reynolds, J., Wang, Y. and Wofe, D. (1999) A search for the predictive understanding of plant responses to elevated CO_2. *Global Change Biology* 5, 143–156.

Marland, G. and Schlamadinger, B. (1995) Biomass fuels and forest-management strategies: how do we calculate the greenhouse-gas emission benefits? *Energy* 20, 1131–1140.

McMurtrie, R.E. and Landsberg, J.J. (1992) Using a simulation model to evaluate the effects of water and nutrients on the growth and carbon partitioning of *Pinus radiata. Forest Ecology and Management* 52, 243–260.

McMurtrie, R.E., Medlyn, B.E. and Dewar, R.C. (2001) Increased understanding of nutrient immobilization in soil organic matter is critical for predicting the carbon sink strength of forest ecosystems over the next 100 years. *Tree Physiology* 21, 831–839.

Medlyn, B.E., Badeck, F.-W., De Pury, D.G.G., Barton, C.V.M., Broadmeadow, M.S.J., Ceulemans, R., Forstreuter, M., DeAngelis, P., Jach, M.E., Kellomäki, S., Laitat, E., Marek, M., Portier, B., Rey, A., Strassemeyer, J., Laitinen, K., Liozon, R., Robertnz, P., Wang, K. and Jarvis, P.G. (1999) Effects of elevated CO_2 on photosynthesis in European forest species: a meta-analysis of model parameters. *Plant, Cell and Environment* 22, 1475–1495.

Neirynck, J., Van Ranst, E., Roskams, P. and Lust, N. (2001) Impact of decreasing throughfall depositions on soil solution chemistry at coniferous monitoring sites in northern Belgium. *Forest Ecology and Management* 160, 127–142.

Norby, R.J., Wullschleger, S.D., Gunderson, D.A., Johnson, D.W. and Ceulemans, R. (1999) Tree responses to rising CO_2 in field experiments: implications for the future forest. *Plant, Cell and Environment* 22, 683–714.

Oren, R., Ellsworth, D.S., Johnson, K.H., Philips, N., Ewers, B.E., Maier, C., Schäfer, K.V.R., McCarthy, H., Hendrey, G., McNulty, S.G. and Katul, G.G. (2001) Soil fertility limits carbon sequestration by forest ecosystems in a CO_2-enriched atmosphere. *Nature* 441, 469–471.

Rasse, D.P., Nemry, B. and Ceulemans, R. (2001) Stand-thinning effects on C fluxes in 20th and 21st-century Scots pine forests: a sensitivity analysis with ASPECTS. In: Carnus, J.-M., Dewar, R., Loustau, D., Tomé, M. and Orazio, C. (eds) *Models for the Sustainable Management of Temperate Plantation Forests.* Proceedings 41, European Forest Institute, Joensuu, Finland, pp. 18–30.

Sampson, D.A. and Ceulemans, R. (2000) SECRETS: simulated carbon fluxes from a mixed coniferous/deciduous Belgian forest. In: Ceulemans, R., Veroustraete, F., Gond, V. and Van Rensbergen, J.B.H.F. (eds) *Forest Ecosystem Modelling, Upscaling and Remote Sensing.* SPB Academic Publishing, The Hague, pp. 95–108.

Sampson, D.A., Janssens, I.A. and Ceulemans, R. (2001) Simulated CO_2 efflux and net ecosystem exchange in a 70-year-old Belgian Scots pine stand using the process model SECRETS. *Annals of Forest Science* 58, 31–46.

Schimel, D.S. (1995) Terrestrial ecosystems and the carbon cycle. *Global Change Biology* 1, 77–91.

Schlamadinger, B. and Marland, G. (2000) *Land Use and Global Climate Change: Forests, Land Management and the Kyoto Protocol.* Pew Center on Global Climate Change, Arlington, Virginia, 54 pp.

Schlamadinger, B., Canella, L., Marland, G. and Spitzer, J. (1997) Bioenergy strategies and the global carbon cycle. *Sciences Géologiques* 50, 157–182.

Thornley, J.H.M. (1998) *Grassland Dynamics: an Ecosystem Simulation Model.* CAB International, Wallingford, UK, 241 pp.

Thornley, J.H.M. and Cannell, M.G.R. (2000) Managing forests for wood yield and carbon storage: a theoretical study. *Tree Physiology* 20, 477–484.

Tiktak, A. and van Grinsven, H.J.M. (1995) Review of sixteen forest–soil–atmosphere models. *Ecological Modelling* 81, 35–53.

UNFCCC (1998) *The Kyoto Protocol to the Convention on Climate Change.* Climate Change Secretariat, Bonn, 34 pp.

Wang, Y.P. and Jarvis, P.G. (1990) Description and validation of an array model – MAESTRO. *Tree Physiology* 7, 297–316.

Wang, Y.P. and Polglase, P.J. (1995) Carbon balance in the tundra: scaling up from leaf physiology and soil carbon dynamics. *Plant, Cell and Environment* 18, 1226–1244.

14 An Allometric-Weibull Model for Interpreting and Predicting the Dynamics of Foliage Biomass on Scots Pine Branches

Richard A. Fleming[1] and Tim R. Burns[1]

Abstract

The foliar biomass dynamics on the branches of young, open-grown, Scots pine (*Pinus sylvestris* L.) are modelled as functions of branch length and age. These dynamics are rooted in a biological foundation by assuming that foliage production depends allometrically on branch length and that foliage survival on a branch follows an age-dependent Weibull distribution. Like previously constructed descriptive models, the more process-based model developed here fitted the data well. In contrast to these purely descriptive models, however, this more process-based model demonstrated some predictive ability when extrapolated, and produced biologically meaningful parameter estimates.

Introduction

There are a number of motivations for modelling the foliage biomass dynamics of trees and of even individual branches. Most motivations derive from an interest in ecophysiology, growth and yield, and the recognition that foliage constitutes the photosynthetic engine driving plant growth.

Studies of the impact of different treatments (e.g. partial defoliation, fertilization, increased atmospheric CO_2, spacing) on the foliage complement are typically conducted by comparing a 'control' or untreated group of trees with corresponding treated groups (e.g. Ericsson *et al.*, 1985; Långström *et al.*, 1990; Piene and Fleming, 1996; Skre and Nes, 1996; Lyytikäinen-Saarenmaa, 1999). It is often a logistical challenge, however, to acquire sample sizes large enough to provide such studies with much statistical power. As big organisms with complex rooting and branching systems, each individual sampling unit (of one or more trees) requires ample space.

A possible solution is suggested by the observation that trees have a modular structure where branches may function as more or less independent units. This offers the possibility of using branches as experimental units instead of whole trees (Långström *et al.*, 1990; Sprugel *et al.*, 1991, and references therein). By applying

[1] Canadian Forest Service, Great Lakes Forest Research Centre, Canada
Correspondence to: rfleming@nrcan.gc.ca

treatments to individual branches, the impact on the foliage development of individual branches could be studied with less effort than studying the impact on the foliage development of whole trees. By comparing the development of treated (e.g. artificially defoliated) branches on trees with only a few selected branches receiving treatment with the development of whole trees receiving a whole-tree treatment, the suitability of individual branches as substitutes for completely treated trees could be evaluated (Långström *et al.*, 1998).

Measuring treatment effects on the foliage development of a branch requires accurate estimates of pre-treatment foliage biomass. This poses a problem because accurate estimates of foliage biomass typically require that foliage be removed from its branch and oven-dried. This precludes measuring pre-treatment (and consequently the change in) foliage dry weight directly, because foliage removal affects the sample branches and trees and thus confounds the treatment. In addition, foliage drying is typically very inconvenient and time-consuming.

State of the Art

One approach to getting accurate estimates of pre-treatment foliage biomass is to use other variables which can be measured non-destructively. For instance, Piene (1983) estimated the total foliage biomass per branch of 25- to 30-year-old balsam fir (*Abies balsamea* (L.) Mill.) using an algorithm that required as input the number of shoots on the branch, and for each age class of shoot it required the average shoot and needle lengths and the percentage litter fall. Although much more efficient than previous methods, Piene's procedure still relied on detailed information about each foliage age class on each branch and he found that the predicted foliage biomass consistently underestimated the actual foliage biomass.

While working with Scots pine (*Pinus sylvestris* L.), Långström *et al.* (1998) tried to further streamline the estimation of foliar biomass per branch by using only variables relevant to the branch as a whole and excluding variables associated with individual age classes of foliage. (This use of whole-branch characteristics may be reliable within the range of the data for conifer species such as balsam fir and Scots pine, which have regular, orderly branching patterns; it is probably less accurate on species such as jack pine (*Pinus banksiana* Lamb.) which branch irregularly.) Långström *et al.* (1998) began by assuming an allometric relationship between foliar biomass and the length of the branch. Recognizing that branch age was also likely to be a factor, since few Scots pine needles remain longer than 3 years on their branch, these authors expanded the allometric relationship by incorporating branch age along with additional powers of branch length to produce a polynomial relationship with interactions. Accordingly, they expressed the mean foliar biomass, measured as total needle dry weight (g), on a branch A growing seasons old and of length L (cm) as

$$Y(A,L) = \exp\left\{b_0 + b_L L + b_A A + b_{L2} L^2 + b_{AL} AL + b_{A2} A^2\right\} \tag{1}$$

This equation was logarithmically transformed to produce a second-order polynomial model which was then fitted by linear regression (Fig. 14.1). The data were from the undefoliated or 'control' trees in their (Långström *et al.*, 1998) study of the effects of artificial defoliation on the foliage dynamics of Scots pine (see section on Data below for more detail). Parameter b_A was eliminated because it was not significantly ($\alpha = 0.05$) different from zero. The resulting five-parameter (transformed) model fitted the data well: $r^2 = 92.5\%$ and standard error of estimate (SEE) = 0.341.

When the model was extrapolated, however, to estimate the foliage biomass of unusually large branches, which are important because they often bear much

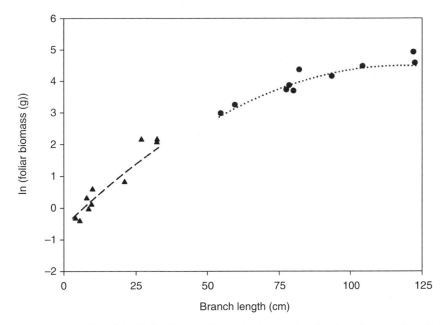

Fig. 14.1. Natural logarithm (ln) of foliar biomass (dry weight) plotted against branch length. The dashed and dotted curves show the fit of Equation 1 to cross-sections of all ($n = 97$) data for branches in their first (triangles, $A = 0$) and fifth (solid circles, $A = 4$) growing seasons, respectively.

foliage, predictions contradicted biologically based expectations. For instance, the dashed line in Fig. 14.2 shows the predictions of the second-order polynomial expansion of the allometric model (Equation 1) when fitted to only the 77 shortest branches. These predictions deviate from the biological expectation of continued allometric growth for longer branches (see Fig. 14.1). That this dashed line also represents the corresponding third-order expansion shows that higher order models are not necessarily better. This long-recognized difficulty with polynomial models is important whenever predictions are needed near the edges of the data or beyond, and Scots pine branches can exceed 150 cm in length. In what follows, modelling the foliage biomass of Scots pine branches is pursued with a more process-oriented approach than used earlier, in the hope of fitting the data well and improving on the potential for extrapolation.

Methodology

This chapter refines previous exploratory work (Fleming, 2001) in developing another approach to modelling relationships between a branch's length, age and foliar biomass for conifers with regular branching patterns. Ideally such a model will be based on biological relationships so that its mathematical structure will produce both good fits to the data and realistic extrapolations beyond the range of the data set. This approach is applied to the same data used to fit Equation 1.

The model is developed by thinking of changes in foliage biomass as being the net result of different demographic processes (Harper and Bell, 1979). Assuming for the moment that changes in the biomass of individual needles make relatively minor contributions to changes in foliar biomass, changes in the latter represent the difference between the biomass of new needles produced each year by the new

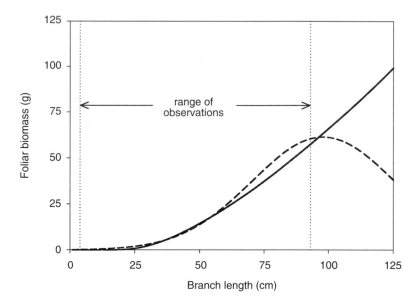

Fig. 14.2. Predictions of the models when fitted to the 77 shortest branches. Branch age is fixed at its mean, $A = 4.12$ years. The dashed curve represents the second- (Equation 1) and third-order polynomial expansions of the allometric model. The solid curve illustrates the corresponding basic allometric growth.

shoots on the branch and the mortality of needles as they age. Hence, following Långström *et al.* (1998), it is assumed that needle production is an allometric function of branch length, and following Fleming and Piene (1992a,b), it is assumed that an age-dependent population survival model can describe needle survival on the branch. Putting these two assumptions into mathematical form and adopting the simplifying assumption that the processes of needle production and survival are independent, the foliar biomass (g) on a branch of age A growing seasons and length L cm can be written:

$$Y(A,L) = P(L)S(A) \qquad (2)$$

The factor $P(L)$ represents the dry weight of needles produced over the years on the branch. $P(L)$ is assumed to be an allometric function of L, the length (cm) of the branch, so that

$$P(L) = wL^{c+1} \qquad (3)$$

where $w = P(1) > 0$ ($P(1) = P(L)$ when $L = 1$) and c represents a constant. This equation satisfies the definition for allometric growth (Pienaar and Turnbull, 1973) in that the specific rate of change in $P(L)$ bears a constant proportional relationship to the specific rate of change of length: $dP(L)/P(L) = (c+1)\, dL/L$ where d represents the differential operator. (The '+1' in the exponent is convenient for parameter estimation: if c is not significantly different from 0 then the corresponding exponent can be considered equal to 1 and dropped.)

When comparing the foliage productivity of different branches of the same length, branch age, A, becomes an additional consideration. Younger branches have less time than older branches to build networks of side branches. Side branches provide extra surface area for growing foliage, so young branches of a given length tend to have less branch surface area on which to grow foliage than older branches of similar length (Fig. 14.3).

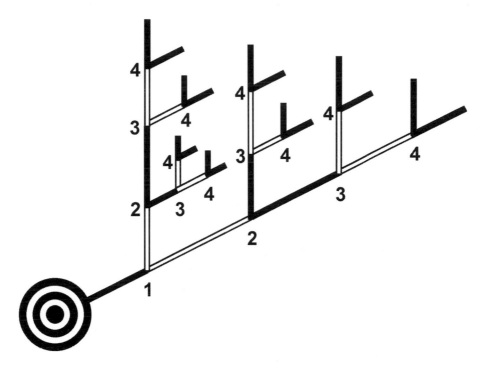

Fig. 14.3. Schematic of a 5-year-old Scots pine branch (after Flower-Ellis *et al.*, 1976; Långström, 1980). Numbers 1–4 at forks indicate cumulative growth along the main and side axes over each of the branch's first four growing seasons. The final increments beyond the last (*A* = 4) forks were added in the fifth growing season and represent the youngest shoots.

In contrast, although the very oldest branches may have extensive networks of side branches, they occur at the base of the crown (Fig. 14.4) where light is more limited. Under limited light, a decline in foliage production rates (per unit branch surface area) is conceivable. For these reasons, the expression for foliage productivity was expanded beyond its basic allometric form to

$$P(A,L) = w(A)L^{c(A)+1} \qquad (4)$$

where $w(A) = P(A,1) > 0$ ($P(A,1) = P(A, L)$ when $L=1$) and $c(A)$ now represents unspecified functions of branch age. The biology of foliage development on growing branches restricts the range of the allometric function, $c(A)$. Branches tend to grow side branches as they age (Fig. 14.3), so long branches tend to have disproportionately more available surface area than short branches for growing foliage. Accordingly, $c(A) > 0$.

In Equation 2, $S(A)$ represents the proportion of the foliage surviving on branches of age A. The Weibull model was adopted to approximate $S(A)$ because: (i) it is relatively simple and flexible when compared with other survivorship models (Kalbfleisch and Prentice, 1980), and (ii) it has been successfully applied to foliage survivorship in other conifers (Fleming and Piene, 1992a,b; Piene and Fleming, 1996). Therefore, we model the proportion of the original foliage biomass surviving on a branch of age A as

$$S(A) = \exp\left\{-(A/u)^{v+1}\right\} \qquad (5)$$

where $u > 0$ and $v > -1$ represent constants.

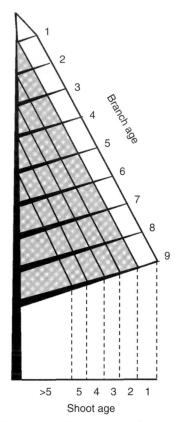

Fig. 14.4. Schematic of a 10-year-old Scots pine (after Ericsson *et al.*, 1980). Branch ages in completed growing seasons are shown vertically on the right. Ages of shoots on the main branch axis are shown along the bottom.

Branch length has been considered a factor when comparing foliage survival on branches produced in the same year. Fleming (2001) noted that foliage on the longest branches, especially on the outer extremities of those branches, is exposed to foliar mortality agents such as freezing rain and high winds. However, for open-grown trees, branch height is thought to be more important in this regard than branch length, with the uppermost branches being particularly vulnerable. Therefore, since a branch's height (crown position) is closely related to its age (Fig. 14.4), it is reasonable to initially and tentatively assume that foliage survival can be largely explained by branch age alone. (An added benefit of this assumption is that it simplifies model fitting.) Consequently, Equations 4 and 5 were substituted into Equation 2 to produce an allometric-Weibull model to approximate the total dry weight of foliage on a branch of age A and length L:

$$Y(A,L) = w(A)L^{c(A)+1} \exp\left\{-(A/u)^{v+1}\right\} \tag{6}$$

For simplicity, we assume that $w(A)$ and $c(A)$ are linear. (More complex relationships can be entertained later if this simple approach does not prove satisfactory.) After explicitly writing in the linear functions, Equation 6 was log-transformed to produce homogeneity of variance in the dependent variable and symmetry of the residuals. Non-linear regression was used to fit the resulting log-transformed equation. This process involved recursive model building and the application of a pseudo Gauss–Newton algorithm for non-linear least squares estimation (Ralston, 1992) of each model constructed. At each stage in this recursive process, parameters with estimates not significantly different from zero ($\alpha = 0.05$) according to the partial F-test (Draper and Smith, 1981) were removed and the reduced model refitted. This procedure was continued until only parameters with statistically significant estimates remained. Residual distributions and residuals plotted against predictions and predictor variables were examined to verify reasonable compliance with the assumptions of regression.

Predicted values resulting from fitting the log-transformation of Equation 6 are not directly useful because they represent geometric means and it is arithmetic means that are needed. To get arithmetic predicted means, the fitted equation must be back-transformed and the right side multiplied by the correction factor (Sprugel, 1983), $\exp\{\text{SEE}^2/2\}$. Then the (arithmetic) mean foliar biomass (g dry weight) for a branch L cm long and A years of age can be predicted by substituting the parameter estimates and SEE (standard error of estimate) from fitting the log-transformation of Equation 6 into the general expression,

$$Y(A,L) = \exp\left\{\ln\left(w_0 + w_A A\right) + \left(c_0 + 1 + c_A A\right)\ln(L) - \left(A / u\right)^{v+1} + \text{SEE}^2 / 2\right\} \tag{7}$$

Test of extrapolation reliability

It was suggested above, in the context of Fig. 14.2, that the two fitted polynomial expansions of allometric growth appeared less biologically plausible than the more process-oriented, allometric-Weibull model when extrapolated to large branches. To test this notion more thoroughly, these three models were each fitted to the observations from the 77 shortest ($L < 93$ cm) branches as described above. By inputting appropriate combinations of A and L, predictions were obtained from each model for the 20 longest branches (93 cm $< L <$ 125 cm) in the data set. This amounted to extrapolating each model up to 34.4% beyond the range of the data used for fitting.

The extrapolated predictions were compared with the observed foliar biomass of the 20 longest branches to estimate the relative reliability of each model for extrapolation. Three options for estimating bias were considered (Zar, 1996). The paired t-test was applied if the deviations (observations − corresponding predictions from extrapolation) had approximately normal distributions. Otherwise, if the variance of the extrapolated predictions did not differ substantially from the variance of the observations, the Mann–Whitney test (Mann and Whitney, 1947) was applied. Failing that, a two-sample t-test without pooled variances was applied.

Data

The data were produced in an artificial defoliation study (Långström *et al.*, 1998) conducted at Ivantjärnsheden, Jädraås, in central Sweden, about 200 km north-west of Stockholm (61° N, 16° E, 185 m above sea level). The site was in a heath Scots pine forest, of dry dwarf shrub type, with a podzolic soil profile. (Axelsson and Bråkenhielm (1980) provide a detailed description of the area.) The site had been clearcut in 1971, and planted in 1972 with Scots pine seedlings of local provenance at 2000 seedlings per hectare. Naturally regenerated pine seedlings had developed in gaps following seedling mortality. The stand was cleaned in 1985 to about 1500 stems per hectare, and by spring 1991, was a pure pine stand with some open patches.

The 97 observations used in this chapter were made on open-grown, untreated (i.e. 'control') Scots pines in August 1991, at the end of the growing season that year. The data consist of branch lengths and corresponding foliage biomasses for branches which had experienced 1–10 full growing seasons (i.e. ages A = 1–10). These data were supplemented by observations taken in June 1991, before the 1991 growing season, on branches initiated in 1990. These supplementary observations provide information on new branches (age A = 0) but their use requires the assumption that foliage losses from these new branches were minimal in the winter of 1990/91.

Results

Fitting the allometric-Weibull model, Equation 6, after log-transformation, to all 97 observations resulted in SEE = 0.343 and a pseudo r^2 (Ralston, 1992) of 92.4%. The estimates of the statistically significant parameters, their SEs (standard errors) and their P values were w_0 = 0.0799 (SE = 0.0246, P = 0.0016), c_0 = 0.279 (SE = 0.1060, P = 0.01), c_A = 0.237 (SE = 0.0327, P < 0.0001), u = 1.57 (SE = 0.226, P < 0.0001) and v = 0.312 (SE = 0.0638, P < 0.0001). According to these results, foliage biomass is predicted to increase allometrically with branch length on branches of all ages and the rate of this allometric increase depends on branch age (Fleming, 2001; Fig. 14.3). It is slow on old branches, presumably because of high foliage losses. It is also slow on very young branches, presumably because such branches have had little time to develop extensive networks of lower order side branches which provide much additional surface for growing foliage (Långström *et al.*, 1998). The youngest branches also occupy some of the most exposed parts of the crown where their foliage may be especially vulnerable to mortality from freezing rain or high winds. The low foliar biomass on very short branches may be because these branches are very young, are shaded by overhanging branches or were broken off before the data were collected, and much foliage was lost when this happened.

To gauge the adequacy of Equation 7, a predicted arithmetic mean was calculated using the values of A and L corresponding to each observation. The r^2 value (i.e. the square of the Pearson correlation coefficient between the observations) and these predictions was 85.4%. The 97 paired differences were not normally distributed (P < 0.0005) and the variances of the distributions of observations and predictions differed little (P = 0.909), so the Mann–Whitney test was applied to reveal bias. The high P value (0.810) of this test is consistent with the practical absence of statistically significant bias. The solid curve in Fig. 14.2 shows the allometric growth predicted by fixing A at its average, 4.12 growing seasons, and substituting the estimated parameter values into Equation 7.

Extrapolation reliability

In fitting both the second- (Equation 1) and third-order polynomial expansions of allometric growth to the 77 shortest branches, six parameters were statistically significant, and hence were retained. Only parameters w_0, c_0, u and v were retained in fitting the allometric-Weibull model. Characteristics (SEE, r^2) of fitting log-transformations of these models were: second-order polynomial (Equation 1) (0.349 g, 91.8%), third-order polynomial (0.340 g, 92.2%) and allometric-Weibull model (0.374 g, 90.3%). Table 14.1 gives the results of comparing extrapolations of these fits to the 20 longest branches.

Table 14.1. Performance of model fits to the 77 shortest ($L \leq 93$ cm) branches when extrapolated to the 20 longest (93 cm $\leq L \leq$ 125 cm) branches.

	Deviations (observation − extrapolated prediction)			Extrapolations		
	Model mean	Standard error	Normal?[a]	Equal variances?[b]	Two-sample *t*-test[c]	Pearson correlation
2nd-order polynomial expansion	13.4 g	6.43 g	$P = 0.012$	$P = 0.062$	$P = 0.041$	$r = -0.102$ $P = 0.670$
3rd-order polynomial expansion	13.2 g	6.86 g	$P = 0.055$	$P = 0.053$	$P = 0.066$	$r = -0.004$ $P = 0.986$
Allometric-Weibull model	2.96 g	4.87 g	$P = 0.045$	$P = 0.084$	$P = 0.639$	$r = -0.533$ $P = 0.015$

[a] Anderson–Darling test (Anderson and Darling, 1954).
[b] Levene's test (Levene, 1960).
[c] Without pooled variances.

Discussion and Conclusions

Table 14.1 shows that the allometric-Weibull model provides a more reliable basis for extrapolating to large branches than using polynomial expansions of allometric relationships such as Equation 1. Together, the mean deviations and the results of the two-sample *t*-tests show tendencies for the polynomial expansions to underestimate the foliage biomass of long branches. In contrast, for the allometric-Weibull model, the relatively high P value of its two-sample *t*-test and its relatively low mean deviation provide no indication of bias. The correlation coefficients and their associated P values show that: (i) the extrapolated polynomial expansions provide little information about trends in foliage biomass for long branches, while (ii) the allometric-Weibull still explains a statistically significant part of the variation in foliage biomass when extrapolated to the 20 longest branches. Clearly, the allometric-Weibull model is an improvement on the polynomial approach, but these results also indicate the dangers of extrapolating: the model's r^2 is 84.2% when fitted to the 77 shortest branches but falls to 28.4% when extrapolated to the 20 longest.

One advantage of using well-studied models for constructing subcomponents of larger models is that it can simplify the interpretation of results. In developing the model presented in Equation 6, the Weibull distribution model was used to describe the process of foliage mortality/survival on a branch (Equation 5). Because changes to v and u, respectively, do and do not affect the distribution's shape, they have become known as the 'shape' and 'scale' parameters.

In our case, $v > 0$ ($P < 0.0001$), so the age-specific mortality rate increases monotonically and asymptotically with age towards its maximum. This situation occurs in many ageing populations in which individuals suffer 'wear-out' or 'old age' effects. Since it was branch age, not foliage age, that was recorded, inferences about the distribution of foliage longevities must be made in terms of branch age. For instance, $v > 0$ indicates that rates of branch age-specific foliage survivorship decreased with branch age.

Finally, that c_A was significantly greater than zero suggests that foliar production increased with branch age. This implies that the tendency for very young branches to have relatively restricted networks of lower order side branches (and thus relatively little potential branch surface on which to grow foliage per unit length) outweighs the effect of reduced light availability limiting foliage production in the lower crown. (Stand thinning probably weakened this latter effect.)

Acknowledgements

We thank Bo Långström for organizing accommodation at Jädraås and Stockholm through the Swedish University of Agricultural Sciences, and for arranging financial support through the Swedish Royal Academy of Agriculture and Forestry (KSLA). We thank STORA Skog for permission to work at their premises.

References

Anderson, T.W. and Darling, D.A. (1954) A test for goodness of fit. *Journal of the American Statistical Association* 49, 765–769.

Axelsson, B. and Bråkenhielm, S. (1980) Investigation sites of the Swedish Coniferous Forest Project – biological and physiographical features. In: Persson, T. (ed.) *Structure and Function of Northern Coniferous forests: an Ecosystem Study. Ecological Bulletins* 32, 25–64.

Draper, N.R. and Smith, H. (1981) *Applied Regression Analysis.* Wiley, New York, 709 pp.

Ericsson, A., Larsson, S. and Tenow, O. (1980) Effects of early and late season defoliation on growth and carbohydrate dynamics in Scots pine. *Journal of Applied Ecology* 17, 747–769.

Ericsson, A., Hellqvist, C., Långström, B., Larsson, S. and Tenow, O. (1985) Effects on growth of simulated and induced shoot pruning by *Tomicus piniperda* as related to carbohydrate and nitrogen dynamics in Scots pine. *Journal of Applied Ecology* 22, 105–124.

Fleming, R.A. (2001) The Weibull model and an ecological application: describing the dynamics of foliage biomass on Scots pine. *Ecological Modelling* 138(1–3), 309–320.

Fleming, R.A. and Piene, H. (1992a) Spruce budworm defoliation and growth loss in young balsam fir: period models of needle survivorship for spaced trees. *Forest Science* 38(2), 287–304.

Fleming, R.A. and Piene, H. (1992b) Spruce budworm defoliation and growth loss in young balsam fir: cohort models of needlefall schedules for spaced trees. *Forest Science* 38(3), 678–694.

Flower-Ellis, J., Albrektsson, A. and Olsson, L. (1976) *Structure and Growth of Some Young Scots Pine Stands. 1. Dimensional and Numerical Relationships.* Technical Report 3, Swedish Coniferous Forest Project, Uppsala, Sweden.

Harper, J.L. and Bell, A.D. (1979) The population dynamics of growth and form in organisms with modular construction. In: Anderson, R.M., Turner, B.D. and Taylor, L.R. (eds) *Population Dynamics.* Blackwell, Oxford, pp. 29–52.

Kalbfleisch, J.D. and Prentice, R.L. (1980) *The Statistical Analysis of Failure Time Data.* Wiley, New York, 321 pp.

Långström, B. (1980) Distribution of pine shoot beetle attacks within the crown of Scots pine. *Studia Forestalia Suecica* 154, 1–25.

Långström, B., Tenow, O., Ericsson, A., Hellqvist, C. and Larsson, S. (1990) Effects of shoot pruning on stem growth, needle biomass, and dynamics of carbohydrates and nitrogen in Scots pine as related to season and tree age. *Canadian Journal of Forest Research* 20, 514–523.

Långström, B., Piene, H., Fleming, R. and Hellqvist, C. (1998) Shoot and needle losses in Scots pine: experimental design and techniques for estimating needle biomass of undamaged and damaged branches. In: McManus, M.L. and Liebhold, A.M. (eds) *Proceedings: Population Dynamics, Impacts, and Integrated Management of Forest Defoliating Insects.* General Techical Report NE-247. USDA Forest Service, Northeastern Research Station, Radnor, Pennsylvania, pp. 230–246.

Levene, H. (1960) Robust tests for equality of variances. In: Ingram, O., Hoeffding, W., Ghurye, S.G., Madow, W.G. and Mann, H.B. (eds) *Contributions to Probability and Statistics.* Stanford University Press, Stanford, California, pp. 278–292.

Lyytikäinen-Saarenmaa, P. (1999) Growth responses of Scots pine (Pinaceae) to artificial and sawfly (Hymenoptera: Diprionidae) defoliation. *Canadian Entomologist* 131, 455–463.

Mann, H.B. and Whitney, D.R. (1947) On a test of whether one of two random variables is stochastically larger than the other. *Annals of Mathematical Statistics* 18, 50–60.

Pienaar, L.V. and Turnbull, K.J. (1973) The Chapman–Richards generalization of von Bertalanffy's growth model for basal area growth and yield in even-aged stands. *Forest Science* 19, 2–22.

Piene, H. (1983) Nondestructive estimation of foliar biomass in balsam fir. *Canadian Journal of Forest Research* 13, 672–677.

Piene, H. and Fleming, R.A. (1996) Spruce budworm defoliation and growth loss in young balsam fir: spacing effects on needlefall in protected trees. *Forest Science* 42, 282–289.

Ralston, M. (1992) Derivative-free nonlinear regression. In: Dixon, W.J. (ed.) *BMDP Statistical Software Manual*. University of California Press, Berkeley, California, pp. 427–455.

Skre, O. and Nes, K. (1996) Combined effects of elevated winter temperatures and CO_2 on Norway spruce seedlings. *Silva Fennica* 30, 55–63.

Sprugel, D.G. (1983) Correcting for bias in log-transformed allometric equations. *Ecology* 64, 209–210.

Sprugel, D.G., Hinckley, T.M. and Schaap, W. (1991) The theory and practice of branch autonomy. *Annual Review of Ecology and Systematics* 22, 309–334.

Zar, J.H. (1996) *Biostatistical Analysis*. Prentice-Hall, Upper Saddle River, New Jersey, 918 pp.

15 Diameter Distribution Models and Height–Diameter Equations for Estonian Forests

A. Kiviste,[1] A. Nilson,[1] M. Hordo[1] and M. Merenäkk[1]

Abstract

To study the structure and growth of Estonian forest stands, a network of 501 permanent sample plots was established in 1995–2001. The plots were placed randomly using the grid of the European forest monitoring program ICP FOREST. The network of permanent plots covers the main forest types and actual age interval of commercial forests. The plots are to be re-measured at 5-year intervals.

The plots are circles holding a minimum of 100 trees each. On the plots, the polar coordinates and breast height diameters of all trees are measured. Additionally, the total height and crown length of selected sample trees are also measured. To date, the measurement data for 90,209 trees have been recorded in the database.

The aim of the permanent plot network is to create and recalibrate Estonian forest-growth models. However, to achieve this, long-term measurement series on the permanent plots are still required. Nevertheless, on the basis of the existing material, some forest structure models have been tested and developed for Estonia, such as: (i) diameter distribution models and (ii) height–diameter models.

The Johnson's SB distribution was flexible enough to describe the diameter distributions of Estonian forests. Regression methods of parameter estimation represented a better fit than percentile methods. A diameter distribution model, the parameters of which were predicted by stand variables, has been developed for Estonian forests.

A standardized height–diameter equation has been created for Estonian forests. Depending on the number of height–diameter measurements, the equation can be used as a one-parameter or two-parameter model. Model parameters can be estimated by solving a system of linear equations.

Introduction

Models of forest stand growth and structure are the basis for planning forest management activities such as cutting, thinning and regeneration. At the Estonian State Forest Management Centre, a computer-based decision-support system is being

[1] Estonian Agricultural University, Estonia
Correspondence to: akiviste@eau.ee

developed. Both adequate forest stand descriptions and stand growth and structure models are needed for the effective use of the system.

The current Estonian forest structure and growth models are poor (Nilson, 1999). To make the computer-based decision-support system work, many temporary models based on old tables generalized for large regions (e.g. the Orlov site index tables, the Tretyakov standard tables) or on limited empirical data have been used. It is expected that more accurate models will soon be produced by Estonian forest researchers to replace the incomplete models in the system.

The stand growth models based on the distance-independent individual tree growth models used in the Finnish MELA (Hökka, 1997; Hynynen *et al.*, 2002) and Swedish HUGIN (Hägglund, 1981) would be acceptable types of growth model in the computer-based decision-support system. However, much effort is required to establish and maintain long-term observations on permanent forest growth plots. An all-Estonian network of permanent forest growth plots has been designed after a similar Finnish system (Gustavsen *et al.*, 1988) to provide empirical data for the models. At present, the network consists of 501 plots, 193 of which have already been re-measured.

According to Finnish results (Hynynen *et al.*, 2002), for reliable forest growth modelling, data from at least 3–4 plot re-measurements (over 15–20 years) are required. Nevertheless, the collected plot data can be used for the modelling of Estonian forest structure. This chapter presents our first attempts to use the collected material for modelling of diameter distribution and height–diameter relationships for Estonian forests.

Materials and Methods

Data collection

Thousands of plots have been established in Estonian forests over many decades. Most of them are subjectively located temporary plots. However, construction of forest growth models using data from subjectively located temporary plots is problematical. In 1999, the National Forest Inventory (NFI) was started in Estonia. The NFI permanent plots have been optimized to estimate Estonian forest area and volume. Having a radius of only 10 m, they are, however, too small to reliably characterize stand regularities for research purposes. A new method of establishing a network of permanent forest growth plots has been developed by the Institute of Forest Management of the Estonian Agricultural University. The following principles were used for designing the network of permanent forest growth plots.

1. A long series of re-measured permanent plots would provide the best data for the modelling of forest stand growth.
2. All basic forest types and age and density classes should be represented throughout Estonia.
3. Plots should be located randomly.
4. In addition to measurements of trees, tree coordinates on plots should be determined in order to find the same trees at the next re-measuring and to enable the construction of both distance-independent and distance-dependent models.
5. Forest growth plots should be large enough to clearly characterize stand regularities.

The method of establishing permanent forest growth plots is mainly based on the experience of the Finnish Forest Research Institute (Gustavsen *et al.*, 1988). Some recommendations from Curtis (1983) were taken into consideration. The codes of species, site types, faults and measurement units were taken from the Estonian NFI guidelines (Estonian Forest Survey, 1999). The following procedures characterize the method.

1. The compartments for plots were selected before the beginning of fieldwork according to the desired plot distribution. For the placement of plot regions, the grid of the European forest-monitoring program ICP FOREST 1 was used. Also, the grid of the Estonian National Forest Inventory was used to place plot centres. To save on transport costs, two or three survey pairs worked together in the same plot region.
2. All trees were measured on circular plots with a radius of 15, 20, 25 or 30 m to get at least 100 first-storey trees. Smaller trees (second-storey and understorey trees) were measured in an inner circle with a radius of 8 m (the surrounding circle 15 m) or 10 m (the surrounding circle more than 15 m).
3. Measurements on plots were carried out by pairs of researchers. The recorder at the plot centre measured the azimuth and the measurer measured the distance from the plot centre, breast height diameter in two directions, and the faults of each tree. The measured trees were marked with a coloured spot at a height of 1.3 m. Every fifth tree, dominant trees, trees of rare species and trees in the inner circle were measured as sample trees. For sample trees, the total height, the crown height and the height of dead branches were also measured. Crown height was defined as height of the lowest live contiguous whorl. In addition to living trees, the coordinates and diameters of standing dead trees and fresh stumps were measured.
4. The main instruments for tree measurements were a compass, a calliper and a 'Forestor Vertex' hypsometer.
5. The age of stand components was determined by counting tree rings from cores extracted from sample trees. The thickness of the soil organic layer was measured at several places on the plot.
6. A metal stick was used to mark the plot centre. In addition, a couple of trees near the plot centre were marked with a coloured circle for easier location in 5 years time. The geographical coordinates of plot centres were determined using a GPS device. The plots and their neighbouring objects (roads, ditches, rocks, etc.) were mapped.

In Estonia, the first permanent forest growth plots, with tree coordinates, were established by Urmas Peterson in the Pikknurme and Aakre forest districts (Central Estonia) in 1995/96. In 1997/98, plots of the same type were established in the Sagadi forest district (North Estonia) and in several areas of South Estonia. During those years, various instruments and methods were tried and, as a result, an optimal fieldwork technique was developed. In 1999, a network of permanent forest growth plots to cover the whole of Estonia was designed. Figure 15.1 shows that most of Estonia is already covered by plots. In the summer of 2002, more than 100 forest growth plots were established in north-west Estonia and on some Estonian islands. With that, the first round of measuring was finished. At present, a total of 501 forest growth plots have been established in the network, 193 of which have already been re-measured. Figure 15.2 shows the distribution of plots by dominant species and age classes. Our database includes a total of 90,209 tree measurements, 70,002 of which pertain to the first layer. The number of model trees is 28,441. Most of them (22,993) are first-layer trees.

Analysis

Diameter distribution

Various density functions have been used for the modelling of diameter distributions. Most often, normal, log-normal, Weibull, Gram–Charlier, γ, β and Johnson's distributions have been fitted. In this study, we selected Johnson's SB distribution as

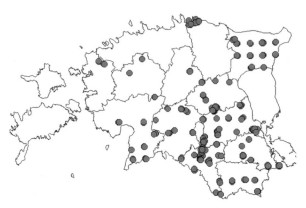

Fig. 15.1. Network of permanent forest growth plots in Estonia established in 1995–2001. Each circle on the map marks a group of plots.

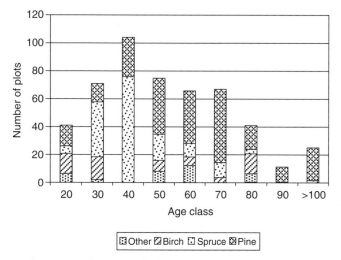

Fig. 15.2. Distribution of permanent forest growth plots by dominant species and age classes.

one of the most flexible among the classical distribution functions (Tarve, 1986; Kamziah *et al.*, 1999).

The density function of Johnson's SB distribution (1949) is

$$f(d) = \left(\frac{\delta}{\sqrt{2\pi}}\right) \cdot \left(\frac{\lambda}{(x-\varepsilon)\cdot(\varepsilon+\lambda-d)}\right) \cdot \exp\left[-\frac{1}{2}\cdot z^2\right]$$

where $z = \gamma + \delta \cdot \ln\left[\dfrac{d-\varepsilon}{\varepsilon+\lambda-d}\right]$

γ and δ are shape parameters, ε and λ are location and scale parameters, d is tree breast height diameter (cm).

In this study, six methods were tested for the estimation of Johnson's SB distribution parameters.

1. Four percentile (4PR) method (Slifker and Shapiro, 1980). According to this method, parameter estimates are calculated from four symmetrical points ($-3z$, $-z$, z, $3z$).

On the basis of 436 Estonian forest plot data points, the optimal value $z = 0.972$ was used.

2. Knoebel and Burkhart's (1991) (KB) method. According to this method, the lower limit, range, median and 95th percentile of the observed diameter distributions were used for parameter estimation.

3. Moments (MOM) method (Bowman and Shenton, 1983). According to this method, the lower limit, range, arithmetic mean and standard deviation of the observed diameter distributions were used for parameter estimation.

4. Maximum likelihood (MLE) method (Zhou and McTague, 1996) with predetermined ε and λ using the Knoebel and Burkhart (1991) method.

5. Linear regression (LIN) method (Zhou and McTague, 1996). According to this method, the parameters ε and λ were predetermined using Knoebel and Burkhart's (1991) method, and the shape parameters γ and δ were calculated from 10%, 20%, …,90% percentiles using linear regression.

6. Non-linear regression (NLIN) method (Kamziah *et al.*, 1999). According to this method, the parameters ε and λ were predetermined using Knoebel and Burkhart's (1991) method, and the shape parameters γ and δ were calculated from all points using Marquardt non-linear regression procedure NLIN (SAS, 1989).

Kolmogorov–Smirnov (KS) statistics were calculated and goodness-of-fit tests (Zar, 1999) at the significance level $\alpha = 0.05$ were performed for each distribution and parameter estimation method.

Height–diameter relationships

In this study we wanted to develop a standard equation for use in the procedure of approximation of plot height–diameter data in the Estonian forest-management decision-support system. A considerable number of different equations have been used for the approximation of height–diameter relationships (e.g. Huang *et al.*, 2000). The equations having parameters that can be estimated using the linear regression technique are preferred for practical use. Root mean square errors (RMSEs) of 27 different height–diameter equations used on the Estonian forest plot data (Mängel, 2000) were compared. Several equations had almost the same (minimum) residual standard error. Finally we settled on the model developed by Professor A. Nilson:

$$h = \frac{H}{1 - b \cdot \left[1 - \left(\dfrac{D}{d} \right)^c \right]}$$

where h is the tree height in m, d is the tree breast height diameter in cm, D is the quadratic mean diameter in cm, and c is a constant (1.31 for pine, 1.47 for spruce and 1.38 for other species).

Equation parameters H and b can be calculated from the following system of linear equations:

$$\begin{cases} H \cdot n + b \cdot \sum_{i=1}^{n} z_i \cdot h_i = \sum_{i=1}^{n} h_i \\ H \cdot \sum_{i=1}^{n} z_i + b \cdot \sum_{i=1}^{n} z_i^2 \cdot h_i = \sum_{i=1}^{n} z_i \cdot h_i \end{cases}$$

where $z_i = 1 - \left(\dfrac{D}{d_i} \right)^c$, d_i and h_i are the diameter and height of the ith sample tree, and D is the quadratic mean diameter (cm).

Results

Diameter distribution

For the modelling of diameter distributions, first-storey trees of the dominant species were selected from each plot. The plots holding less than 56 first-storey trees of the dominant species were excluded from the analysis of distributions. As a result, 436 diameter distributions, with 52,608 trees, were selected. Table 15.1 presents summary statistics of the 436 diameter distributions used in this study.

Kolmogorov–Smirnov (KS) statistics were calculated and goodness of fit tests were performed for each diameter distribution and parameter estimation method. Table 15.2 presents the means and standard deviations of parameter estimates and the KS statistics with the numbers of passes of goodness-of-fit tests. Figure 15.3 shows box-and-whiskers plots of KS statistics by different parameter estimation methods

A comparison of methods applied to Estonian forest plot data showed that the regression methods (LIN and NLIN) for parameter estimation represented the best fit. The methods based on certain percentiles (KB and 4PR) were unstable. For both methods, about 10% of plots did not pass the Kolmogorov–Smirnov goodness-of-fit test, and for the four percentile (4PR) method the parameter estimation failed in four cases. The moments (MOM) method and the maximum likelihood (MLE) methods represented a slightly worse fit than the regression methods; however, they are simpler to apply.

Table 15.1. Summary statistics of 436 diameter distributions used in this study.

Characteristic	Mean	Minimum	Maximum
Number of trees	121	56	378
Average diameter	17.1	4.6	37.1
Standard deviation	4.7	1.6	9.4
Coefficient of variation	29.9	13.8	72.1
Skewedness	0.51	−0.59	4.1
Kurtosis	0.23	−1.13	23.5

Table 15.2. Arithmetic means and standard deviations (SD) of Johnson's SB distribution parameters and Kolmogorov–Smirnov (KS) statistics on the basis of data from 436 plots. A plot passes the test when the null hypothesis (H_0) fails to be rejected.

		Parameters				KS test ($\alpha = 0.05$)		
Method		ε	λ	γ	δ	KS statistic	Number of passes (H_0)	%
4PR	Mean	7.39	28.67	0.87	1.18	0.079	393	90.0
	SD	4.50	21.65	0.86	0.40	0.029		
KB	Mean	6.89	26.31	0.64	1.20	0.084	399	91.5
	SD	4.47	7.20	0.41	0.20	0.032		
MOM	Mean	5.73	30.70	0.82	1.35	0.064	434	99.5
	SD	3.12	9.97	0.35	0.25	0.019		
MLE	Mean	5.73	30.70	0.81	1.36	0.067	436	100
	SD	3.12	9.97	0.35	0.24	0.020		
LIN	Mean	5.73	30.70	0.85	1.41	0.060	436	100
	SD	3.12	9.97	0.41	0.28	0.017		
NLIN	Mean	5.73	30.70	0.87	1.41	0.056	436	100
	SD	3.12	9.97	0.42	0.29	0.015		

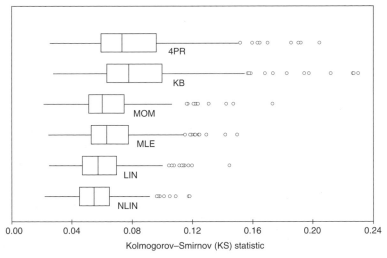

Fig. 15.3. The box and whiskers plot of Kolmogorov–Smirnov (KS) statistics for different methods.

Using the general linear method procedure (GLM) of the SAS software (SAS, 1989), equations to predict Johnson's SB distribution parameters from stand variables were developed. Initially, all main stand variables (site type, site index, stand age, dominant species, average height, quadratic mean diameter, basal area, relative density, number of trees per hectare and indicator of thinning) were analysed as independent variables in the model. Finally, only significant variables at a significance level of $\alpha = 0.05$ were included. Table 15.3 presents the variables and parameter estimates of linear models for the prediction of the minimum and the maximum diameters of first-layer trees in stands.

The standard deviation of diameter STD can be predicted using the following equation:

$$STD = 0.006 + 0.294 \cdot MAX - 0.233 \cdot D$$

where MAX is the maximum diameter (cm), and D is the quadratic mean diameter (cm).

Table 15.3. Parameter estimates of linear models for the prediction of the minimum (MIN) and the maximum (MAX) diameters of first-layer trees from stand variables.

Stand variable	MIN	MAX
Intercept according to dominant species		
Pine	−3.88	7.78
Spruce	−4.02	8.59
Birch	−3.43	7.68
Aspen	−3.21	6.71
Common alder	−1.87	5.26
Grey alder	−2.66	6.10
Site index H_{100} (m), base age 100 years	0.0730	−0.1085
Quadratic mean diameter D (cm)	0.6553	1.2499
Number of trees per hectare (N)	0.000285	−0.000577
Relative density of stand T (by Tretyakov)	−1.67	3.05
Indicator of thinning (thinned 1, otherwise 0)	0.31	0
Coefficient of determination (R^2)	0.89	0.92
RMSE	1.56	2.60

The coefficient of determination $R^2 = 0.66$ and the RMSE = 0.71 cm.

Using the moments method (Scolforo and Thierschi, 1998), the parameters of Johnson's SB distribution can be predicted as follows:

$$\varepsilon = 0.7 \, \text{MIN}$$
$$\lambda = \text{MAX}$$

$$\delta = \frac{\mu \cdot (1-\mu)}{s} + \frac{s}{4} \cdot \left[\frac{1}{\mu \cdot (1-\mu)} - 8 \right]$$

$$\gamma = \delta \cdot \ln\left(\frac{1-\mu}{\mu}\right) + \left(\frac{0.5-\mu}{\delta}\right)$$

where

$$\mu = \frac{\bar{d} - \varepsilon}{\lambda}$$

$$s = \frac{\text{STD}}{\text{MAX}}$$

$\bar{d}$ = arithmetic mean diameter

Height–diameter relationships

To date, a total of 28,441 sample trees with height–diameter measurements have been recorded in the database of the Estonian permanent plot network. Records were classified into groups by storeys and species, and only the groups containing at least 25 trees were used for the modelling of height–diameter relationships. As a result, a total of 14,821 height–diameter measurements from 292 stands were used in the analysis of height–diameter models.

On the basis of sample tree data, parameters H and b of the Nilsons' height–diameter equation were calculated for each stand using the system of linear equations. The residual standard error of the model was 1.36 m.

The scale parameter H of the model expresses the mean height of the stand corresponding to the quadratic mean diameter. The shape parameter b characterizes the curvature of the height–diameter relationship (Fig. 15.4). Analysis of the estimates of

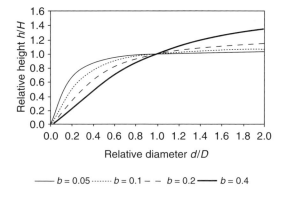

Fig. 15.4. Standardized height–diameter curves by Nilson's (1999) model for different values of shape parameter b.

the shape parameter b using Estonian growth plot data revealed a linear relationship to the quadratic mean diameter D and the tree species (Fig. 15.5). By replacing the parameter b with the linear relationship $b = a - 0.0056 \cdot D$, we get a one-parameter height–diameter model:

$$h = \frac{H}{1 - (a - 0.0056 \cdot D) \cdot \left[1 - \left(\dfrac{D}{d} \right)^c \right]}$$

where h is the tree height (m), d is the tree breast height diameter (cm), a is a constant (0.369 for pine, 0.394 for spruce and 0.359 for other species), c is a constant (1.31 for pine, 1.47 for spruce and 1.38 for other species), and H is the mean height parameter (corresponding to the quadratic mean diameter).

The residual standard error of the model was 1.50 m.

Discussion and Conclusions

The creation of a network of permanent forest growth plots has been started in Estonia. The network is designed for long-term observations of trees to be used for building forest growth models for Estonia.

Numbering of trees with metallic tags or colour is a common practice in forest permanent sample plots. We preferred the measurement of tree coordinates instead. Re-measuring of almost 200 plots convinced us that the precision of the measurements of azimuth and distance from plot centre was sufficient to ensure quick location of the same trees 5 years later. The additional advantage of measuring of tree

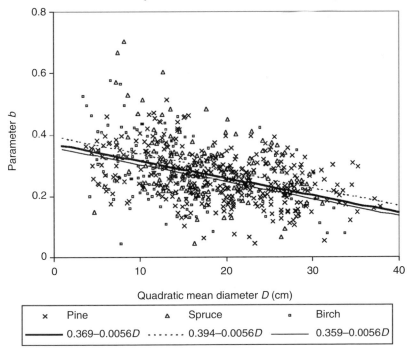

Fig. 15.5. Relationship between parameter b and quadratic mean diameter D on the basis of first-storey tree data.

coordinates is the availability of data for the construction of distance-dependent tree growth models.

The establishment and maintenance of the network is an ongoing process. Nevertheless, a considerable amount of data that can be used for modelling the forest structure has already been collected.

The Johnson's SB distribution was flexible enough to describe diameter distributions of Estonian forests. Regression methods of parameter estimation represented a better fit than percentile methods. A diameter distribution model, the parameters of which were predicted by stand variables, has been developed for Estonian forests.

Using the diameter distribution model, the parameters of which were predicted by stand variables, 9% of the plots did not pass the Kolmogorov–Smirnov goodness-of-fit test. When distribution parameters were estimated using empirical diameter distribution (using the moments method), then all the plots passed the Kolmogorov–Smirnov goodness-of-fit test. Thus, we did not manage to predict the distribution parameters perfectly. It is likely that regular correlations between distribution parameters and stand variables can be found in unmanaged forests; however, since thinning of stands is common practice in Estonia, it is unlikely that perfect predictions of diameter distributions can be made.

A standardized height–diameter equation has been created for Estonian forests. Depending on the number of height–diameter measurements, the equation can be used as a one-parameter or two-parameter model. Model parameters can be estimated by solving a system of linear equations.

For the height–diameter models, the values of parameter c were estimated according to main species. For this a data set was generated using Kuliešis' (1993) model for Lithuanian forests.

Residuals of the models were studied, and no significant effect of tree diameter, quadratic mean diameter or tree species was found either for the two-parameter or for the one-parameter model. However, a dependency of residuals on relative diameter d/D was found in the Estonian forest plot data. Therefore, further correction of the parameter c estimates is needed. Also, we cannot rule out additional dependence of the values of parameters b and c on stand age, stocking grade, management history and other stand variables.

Acknowledgement

This study was supported by the Estonian Science Foundation, Grant No. 4813.

References

Bowman, K.O. and Shenton, L.R. (1983) Johnson's system of distributions. In: *Encyclopedia of Statistical Sciences*, Vol. 4. John Wiley & Sons, New York, pp. 303–314.

Curtis, R.O. (1983) *Procedures for Establishing and Maintaining Permanent Plots for Silvicultural and Yield Research*. USDA General Technical Report PNW-155, 56 pp.

Gustavsen, H.G., Roiko-Jokela, P. and Varmola, M. (1988) *Kivennäismaiden talousmetsien pysyvät (INKA ja TINKA) kokeet. Suunnitelmat, mittausmenetelmät ja aineistojen rakenteet.* Metsäntutkimuslaitoksen tiedonantoja 292, 212 pp. [in Finnish].

Hägglund, B. (1981) *Forecasting Growth and Yield in Established Forests: an Outline and Analysis of Outcome of a Subprogram Within Hugin Project*. Swedish Univerity of Agricultural Sciences, Department of Forest Survey, Report 31, 145 pp.

Hökka, H. (1997) *Models for Predicting Growth and Yield in Drained Peatland Stands in Finland*. Finnish Forest Research Institute, Research Paper 651, 45 pp.

Huang, S., Price, D. and Titus, S.J. (2000) Development of ecoregion-based height–diameter models for white spruce in boreal forests. *Forest Ecology and Management* 129, 125–141.

Hynynen, J., Ojansuu, R., Hökka, H., Siipilehto, J., Salminen, H. and Haapala, P. (2002) *Models for Predicting Stand Development in MELA System*. Finnish Forest Research Institute, Research Paper 835, 116 pp.

Johnson, N.L. (1949) Systems of frequency curves generated by methods of translation. *Biometrika* 36, 149–176.

Kamziah, A.K., Ahmad, M.I. and Lapongan, J. (1999) Nonlinear regression approach to estimating Johnson SB parameters for diameter data. *Canadian Journal of Forest Research* 29, 310–314.

Knoebel, B.R. and Burkhart, H.E. (1991) A bivariate approach to modelling forest diameter distributions at two points in time. *Biometrics* 47, 241–253.

Kuliešis, A. (1993) *Lietuvos medynu prieaugio ir jo panaudojimo normatyvai*. Kaunas, 383 pp. [in Lithuanian].

Mängel, M. (2000) *Puu kõrguse ja diameetri vahelise seose modelleerimine proovitükkide andmeil*. EPMÜ metsakorralduse instituut. Lõputöö, 40 pp. [in Estonian].

Nilson, A. (1999) Pidev metsakorraldus – mis see on. In: *Pidev Metsakorraldus*. EPMÜ Metsandusteaduskonna toimetised 32, 4–13 [in Estonian].

SAS Institute (1989) *SAS/STAT User's Guide*, Version 6, 4th edn, Vol. 2, SAS Institute, Cary, North Carolina, 846 pp.

Scolforo, J.R.S. and Thierschi, A. (1998) Estimativas e testes da distribuição de frequência diâmétrica para *Eucalyptus camaldulensis*, através da distribuição Sb, por diferentes métodos de ajuste. *Scientia Forestalis* 54, 93–106.

Slifker, J.F. and Shapiro, S.S. (1980) The Johnson System: selection and parameter estimation. *Technometrics* 22, 239–246.

Estonian Forest Survey (1999) *Centre Instruction of National Forest Inventory Fieldworks*. 55 pp.

Tarve, T. (1986) *Puude rinnasdiameetri jaotus Rakvere metsamajandi puistutes*. EPA metsakorralduse kateeder, 36 pp. [in Estonian].

Zar, J.H. (1999) *Biostatistical Analysis*. Prentice-Hall, Upper Saddle River, New Jersey, 663 pp.

Zhou, B. and McTague, J.P. (1996) Comparsion and evaluation of five methods of the Johnson System parameters. *Canadian Journal of Forest Research* 26, 928–935.

16 Modelling the Diameter at Breast Height Growth of *Populus* × *euramericana* Plantations in Spain

Francisco Rodríguez,[1] José Antonio De la Rosa[1] and Álvaro Aunós[1]

Abstract

Preliminary results are presented in order to model the diameter at breast height growth of three clones of *Populus* × *euramericana* (I-214, Luisa Avanzo and MC) in the Cinca valley (Huesca, Spain) using data from re-measured permanent and measured temporary plots. The model employed was an integral form of the Chapman–Richards equation with a reference age established at 7 years. We built an anamorphic system of curves with different productivity indexes (PIs), elaborating an empirical height–diameter relationship to estimate average height based on a multiplicative model adapted by Cunia (1979).

Introduction

The hybrid poplar (*Populus* × *euramericana*) is a tree species of great economic importance in Spain, forming the basis of the peeler wood industry. According to the Spanish Agriculture Department (Padró, 1992), poplar plantations occupy about 100,000 ha of forested area in Spain. Recently, these plantations have undergone a remarkable expansion in areas such as the Cinca valley (Huesca province, northern Spain), in which the hybrid clones I-214, Luisa Avanzo and MC are extensively planted. This increase in forest land can be explained by two factors: (i) its rapid growth and, (ii) its suitability for vegetative reproduction (Rueda *et al.*, 1997). Nowadays, poplars can be used as an alternative to less profitable agricultural crops, partly due to the new orientations of the EU agrarian policies favouring the reforestation of former agricultural land. However, despite the many hectares currently dedicated to poplars, more than 750,000 m³ of wood is imported annually to Spain.

One of the major concerns of forest managers in poplar plantations is related to the possibility of predicting the production of peeler wood in the future. This is also critical in evaluating the sustainability of forest management practices within the framework of the national forest policies. In this context, development of growth models for the species represents a useful tool.

[1] University of Lleida, UdL-ETSEA, Spain
Correspondence to: Francisco.Rodriguez@pvcf.udl.es

Growth models are often a prerequisite for the prediction of stand-level volume growth, where height of dominant and codominant trees in even-aged stands is commonly used as a measure of forest productivity (Monserud, 1984). Traditionally, these models, whether empirical or mechanistic, have been obtained by fitting observed data to mathematical functions. Indeed, tree growth models usually employ a set of equations to describe stand development over time.

The concept of a productivity curve, as defined for poplar plantations (Padró, 1982), describes the relationship between the age of the stand and its mean diameter at breast height (DBH). In fact, the productivity index (i.e. DBH at a reference age, PI) is seldom determined through direct observations but estimated by productivity curves using the mean DBH and age of trees in the target stand. Such an estimation assumes that the DBH growth pattern is associated with PI for a given tree species. However, DBH–age curves for a given PI can differ significantly before and after the reference age due to factors such as climate and site characteristics (Wang and Huang, 2000).

For poplar plantations, a parallel is observed between productivity curves and quality curves (i.e. the relationship between age of the stand and its dominant height). This is because thinning is not performed in the area and pruning practices have little influence on the diameter growth (Rodríguez *et al.*, 2001). Under these circumstances, DBH is a good indicator of yield. The application of productivity curves to any given poplar plantation allows for an assignment of a productivity class to the stand and, knowing its mean DBH to any age, it also describes stand development with time and allows a choice of rotation period based on the stand PI.

The objectives of this study were: (i) to develop a stand-level DBH growth model specific for each clone (I-214, Luisa Avanzo and MC) applicable throughout the Cinca valley, and (ii) to classify the existing stands in the region by productivity index.

Materials and Methods

Site description

The area of study is located in the Cinca valley (Huesca, Spain), with coordinates 0° 11' to 0° 08'E, 41° 55' to 41° 31'N (Fig. 16.1), with an average altitude of 188 m above sea level. Poplar plantations are located in alluvial soils formed principally by gravels. The groundwater table is about 2 m deep.

The forest management is characterized by the following steps: (i) 1-year-old poplars are established on flood irrigation lands following square planting (6 × 6 m, with a total of 278 trees/ha); (ii) tillage is performed yearly to eliminate competition by weeds; and (iii) the time period between irrigations is 15 days during summer at a rate of approximately 2000 m³/ha (approximately 15,000 m³/ha/year) (Padró, 1992).

The available data were collected throughout the inventoried areas of the Cinca valley to provide representative information for a variety of site conditions. For each felled tree, the following variables were measured: diameter at breast height (DBH), total height and stand age (all the inventoried areas have the same density; 278 trees/ha). Data originated from re-measured permanent plots (544 DBH × age combinations) and temporary plots (295 DBH × age combinations). Table 16.1 classifies the number of plots by type (permanent or temporary) for each clone, and Table 16.2 summarizes the data collected at the plot level. The number of measured trees by plot was four in the temporary plots and nine in the permanent plots. I-214, the least

Fig. 16.1. Area of the study.

Table 16.1. Summary of number of plots by type for each clone.

	I-214		MC		LA		Count	
Temporary	38	20.9%	141	40.5%	116	37.5%	295	35.16%
Permanent	144	79.1%	207	59.5%	193	62.5%	544	64.84%
Count	182		348		309		839	

Table 16.2. Summary of data collected at the plot level

	I-214		MC		LA	
	Age (years)	DBH (cm)	Age (years)	DBH (cm)	Age (years)	DBH (cm)
Count	182		347		309	
Average	5.357	14.748	6.541	18.043	5.543	15.914
SD	4.094	12.415	3.919	10.715	2.804	9.0735
Minimum	1.0	2.3	1.0	2.425	1.0	1.075
Maximum	16.0	47.5	16.0	42.35	12.0	39.15

abundant clone in the area, was present in only 182 plots (i.e. about half of the plots in which the other clones were represented).

Model fitting

The model proposed by Richards (1959) has been widely used in forestry. Its success can be explained by the inclusion, in its different forms, of the basic components of growth (Equation 1) (Bravo, 1988). Biologically based models, such as the one by Richards, have the advantage that the extrapolation of results out of the range of the original data is safer than that for empirical models (Vanclay, 1994). Different forms of Richards' model are derived for use in growth modelling; Equation 2 is an integral form, whereas Equations 2a, 2b and 2c are different forms of the model.

$$\frac{\delta DBH}{\delta t} = \alpha \cdot DBH^{n} - \beta \cdot DBH^{m} \qquad (1)$$

$$DBH = a \cdot \left(1 - \exp(-b \cdot t)\right)^c \tag{2}$$

$$DBH = K \cdot PI \cdot \left(1 - \exp(-b \cdot t)\right)^c \tag{2a}$$

$$DBH = \frac{PI}{\left(1 - \exp(-b \cdot t_r)\right)^c} \cdot \left(1 - \exp(-b \cdot t)\right)^c \tag{2b}$$

$$DBH = PI \left(\frac{1 - \exp(b \cdot t)}{1 - \exp(b \cdot t_r)}\right)^c \tag{2c}$$

where a is the asymptote of DBH (cm), b is a growth rate-related parameter, t is the age of the stand (years), t_r is the reference age (years), c is a shape parameter, K is a constant and PI is the productivity index (cm).

The procedure for fitting the productivity curve consists of the following steps: (i) determination of reference age; (ii) parameter estimation of the reference curve; and (iii) graphic representation of the different curves by PI.

Requisites for selecting an appropriate reference age are: (i) to be high enough to obtain a reliable expression of the quality, and (ii) to not be so high that a large number of plots cannot be found (Bengoa, 1999). The reference age is usually taken as the maximum of the average growth curve along the rotation period or, more straightforwardly, half of the rotation period (Pita, 1964). In poplar plantations for the region it fluctuates between 6 and 8 years and, for this work, we have chosen a reference age of 7 years.

In addition to fitting the productivity curves, we also fitted an empirical height–DBH relationship, based on a multiplicative model adapted by Cunia (1979). Its mathematical formulation is the following:

$$TH = a \cdot DBH^b \tag{3}$$

where a and b are parameters to estimate, DBH is the diameter at breast height (cm) and TH is the total height (m). A summary of the characteristics of the data set for fitting the height–DBH model is presented in Table 16.3.

Models 2 and 3 were estimated using the SAS software (1990). For model 2 (non-linear model), the non-linear regression (NLIN) procedure with the derivate-free iterative Marquardt method was applied, and multiple starting values for parameters were provided to ensure a least squares solution. Model 3 was fitted using a simple least square regression technique (the REG procedure).

The coefficient of determination (R^2) and the mean squared error (MSE) were calculated as:

$$R^2 = 1 - \frac{\sum_{1}^{n}\left(Y_{2i} - \hat{Y}_{2i}\right)^2}{\sum_{1}^{n}\left(Y_{2i} - \overline{Y}\right)^2} \tag{4}$$

$$\mathrm{MSE} = \frac{\sum_{1}^{n}\left(Y_{2i} - \hat{Y}_{2i}\right)^2}{n - m} \tag{5}$$

where Y_{2i}, $\hat{Y}_{2i}$, $\overline{Y}$, m and n are actual Y, predicted Y, observed average Y, number of parameters, and number of observations, respectively.

Table 16.3. Summary of the data set for fitting a height–DBH model (age in years, DBH and TH in cm)

	I-214			MC			LA		
	Age (years)	DBH (cm)	TH (m)	Age (years)	DBH (cm)	TH (m)	Age (years)	DBH (cm)	TH (m)
Count	809	809	809	1816	1816	1816	1415	1415	1415
Average	7.04	20.42	17.87	7.61	20.85	19.50	6.27	18.16	16.99
Variance	23.6	233.8	135.2	16.5	124.8	90.58	7.72	82.35	56.50
SD	4.86	15.29	11.63	4.06	11.17	9.517	2.77	9.075	7.516
Minimum	1.0	2.0	3.17	1.0	2.0	3.76	1.0	1.5	2.9
Maximum	16.0	66.84	44.6	16.0	48.8	40.2	12.0	42.3	36.8

Results

Model 2 was fitted separately for each clone. All parameters were statistically significant ($P < 0.05$). A family of anamorphic curves was constructed by changing PI values. A summary of the fit and the estimated parameters is given in Table 16.4.

A PI was calculated for each of the inventoried plots. We obtained a frequency distribution of the different productivities for each of the three clones studied. From these distributions it was observed that most of the plots fell within the class of average productivity, with few plots having very high or very low productivities.

Table 16.5 summarizes the number of plots by PI for each clone.

Model 3 was fitted separately for each clone and, again, all parameters were statistically significant. All three models exhibited a high R^2 and a very low MSE. Table 16.6 summarizes the estimated parameters and the statistics indicating the goodness of fit.

The graphical analyses of models 2 and 3 are shown in Fig. 16.2.

Table 16.4. Summary of the fit and the estimated parameters.

Clone	A	b	c	K	R^2	MSE
I-214	56.0046	0.111167	1.60662	2.6851	0.9242	11.67
Luisa Avanzo	51.7181	0.11277	1.49883	2.4777	0.8612	11.42
MC	46.0016	0.134302	1.60509	2.2143	0.9272	8.34

Table 16.5. Number of plots by productivity index (PI).

PI	I-214		MC		LA		Total
29	33	18.1%	38	10.9%	31	10.0%	102
25	25	13.7%	79	22.7%	58	18.8%	162
21	68	37.4%	148	42.5%	106	34.3%	322
17	47	25.8%	54	15.5%	52	16.8%	153
13	9	4.9%	29	8.3%	62	20.1%	100
Total	182		348		309		839

Table 16.6. Summary of the parameter estimates and fit statistics.

Clone	a	b	R^2	MSE
I-214	1.65552	0.801669	0.9784	0.013
Luisa Avanzo	1.57215	0.825299	0.9657	0.009
MC	1.71534	0.805359	0.9656	0.011

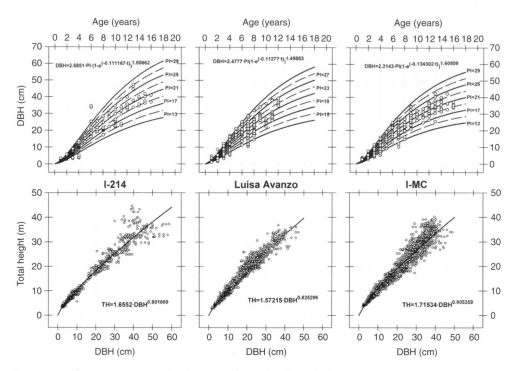

Fig. 16.2. Productivity curves and height–DBH relationship for each clone.

Discussion and Conclusions

In model 2, 'a' represents the asymptote of the curve, 'c' a shape parameter important for determining the lower inflection point of the curve, and 'b' a parameter reflecting the rate at which the asymptote is approached. Higher values of 'b' indicate greater intrinsic growth rates.

An advantage of the fitted model 2 is the ability to predict the DBH development without experimental determination of PI values. Nevertheless, prediction is advised only for short time periods.

The model has been constructed for a determinate plantation density (278 trees/ha), so that this model is valid for this range of density, an aspect that is common in all the poplar plantations analysed. When comparing the performance of the productivity curves for the three clones, it seems advisable to increase the available data for Luisa Avanzo at ages above 12 years, due to the scarcity of plots beyond this age for this clone. Likewise, it would also be necessary to validate all three models with independent data sets for each clone.

This study has demonstrated the existence of a great range of productivities for the region. Luisa Avanzo is a more demanding clone with regard to the site conditions (nearly 20% of the plots showed a PI of 13 or less).

We are now developing a stem taper function for each of the three clones. With the combination of productivity curves and DBH–height relationships with stem taper functions, we expect to be able to assess the yield of the different final products (peeler, sawnwood and pulpwood) of the tree.

Acknowledgements

The authors thank the Gobierno de Aragon for the data set of the permanent plots, and Jordi Voltas for helpful comments and suggestions on the manuscript. This research was partially supported by CICYT AGL2000-1255, Spain.

References

Bengoa, J.L. (1999) Curvas de calidad en altura para *Quercus pyrenaica* en La Rioja. *Congreso de Ordenación y Gestión Sostenible de Montes*. Santiago de Compostela, Spain.

Bravo, F. (1998) Modelos de producción para *Pinus sylvestris* L. en el Alto Valle del Ebro. Doctoral thesis, University of Valladolid, Spain.

Cunia, T. (1979) On sampling trees for biomass table construction: some statistical comments. In: *Proceedings of the Forestry Inventory Workshop*, SAF-IUFRO. Ft Collins, Colorado, pp. 643–664.

Monserud, R.A. (1984) Height growth and site index curves for inland Douglas fir based on stem analysis data and forest habitat type. *Forest Science* 30, 943–965.

Padró, A. (1982) Curvas de productividad del clon 'I-214' en regadío con planta R2T2 y marco 6x6 en el valle medio del Ebro. *Anales INIA Serie Forestal* 6, 63–73

Padró, A. (1992) *Clones de Chopo para el Valle Medio del Ebro*. Diputación General de Aragón, Zaragoza, 203 pp.

Pita, P.A. (1964) Clasificación provisional de las calidades de *Pinus sylvestris* L. *Anales Instituto Forestal de Investigación y Experimentación*, 7–25.

Richards, F.J. (1959) A flexible growth function for empirical use. *Journal of Experimental Botany* 10, 290–300.

Rodríguez, F., Serrano, L. and Aunós, A. (2001) Influencia del método de poda sobre el perfil del árbol en una chopera de Luisa Avanzo con 8 años de edad, en el valle medio del Cinca (Huesca). In: *I Simposio del Chopo*, Zamora, pp. 149–158.

Rueda, J., Cuevas, Y. and García-Jiménez, C. (1997) *Cultivo de Chopos en Castilla y León*. Junta de Castilla y León, Valladolid, 100 pp.

SAS (1990) SAS/STAT *User's Guide*, Version 6. SAS Institute, Cary, North Carolina.

Vanclay, J.K. (1994) *Modelling Forest Growth and Yield: Applications to Mixed Tropical Forest*. CAB International, Wallingford, UK, 312 pp.

Wang, G.G. and Huang, S. (2000) Height growth pattern of white spruce in natural subregions in Alberta, Canada. *Forest Ecology and Management* 134, 271–279.

17 Stand Growth and Productivity of Mountain Forests in Southern Siberia in a Changing Climate

N.M. Tchebakova[1] and E.I. Parfenova[1]

Abstract

Local-level, bioclimatic regression models that relate stand characteristics (forest composition, height, site quality class and wood stocking) to site climate (temperature sums, base 5°C, and dryness index) were developed to predict the stand structure of dark-needled forest (*Pinus sibirica* and *Abies sibirica*) climax successions and their transformations in a changing climate over the Sayan mountain range in southern Siberia.

The models explained up to 80% of the variation in forest growth and productivity characteristics. Productivity varied widely and depended on heat supply rather than moisture. Stand tree species composition depended on moisture: dark-needled species and light-needled tree species (*Pinus sylvestris*) were separated by a dryness index value of 1.0. Living phytomass was calculated from a wood stocking model. Tree heights and living phytomass were mapped over the mountain range under current climate conditions and a regional climate change scenario. The model predicts that total dark-needled forest phytomass will decrease by 17% in a warmed climate.

Introduction

The mountain forests of southern Siberia are of special interest due to their high diversity and productivity and especially due to the most valuable tree species of the Siberian taiga – Siberian cedar (*Pinus sibirica*). Two main tree species, cedar and fir (*Abies sibirica*), dominate the mountain taiga forests on the windward northern and north-western slopes (Nazimova, 1975; Smagin *et al.*, 1980). The climate of this region is moist, with annual precipitation varying from 500 mm at the lower elevation border of the cedar and fir forests to 1500 mm at their upper elevation border. The combination of sufficient heat supply and plentiful water favours the occurrence of forests composed of cedar and fir which are rich in biodiversity. These two species are called in Russian botanical and forestry literature 'dark-needled' tree species because of their high degree of shade tolerance. These forests were formed

[1] Institute of Forest, Siberian Branch, Russian Academy of Sciences, Russia
Correspondence to: ncheby@forest.akadem.ru

during the Holocene (Savina, 1986). Some relic species, such as the broadleaved tree species *Tilia cordata*, and groundcover species (e.g. *Asarum europaeum*, *Asperula odorata*, *Veronica officinalis*, *Dryopteris fillix-mas*) remain from the Tertiary in the lowland dark-needled ('*chern*' in Russian) forests which include various ferns and tall herbs. Chern forests covered vast areas of Siberia in the distant past (Shumilova, 1962). From the mid-1950s, cedar forests and especially productive chern forests underwent a significant decline due to intensive cutting. Since then, there have been many governmental decisions to organize sustainable forestry in these forests (Semechkin *et al.*, 1985). Pine (*Pinus sylvestris*) forests, which dominate the subtaiga and forest steppe, are found only in a narrow band of the foothills on a flat, climatically homogeneous area (Smagin *et al.*, 1980).

Evaluation of the forest stand transformations of a given site caused by both current land use and a changing climate can be performed by comparing current stand characteristics with those at climax stages. In many studies of mountain forests in southern Siberia, the climate was shown to be a principal environmental factor controlling forest composition and growth potential at the climax stage (Polikarpov, 1970; Polikarpov *et al.*, 1986; Parfenova and Tchebakova, 2000). Stand models based on climatic parameters are the tools to employ to evaluate transformations caused by the climate. To our knowledge, no such stand models have been developed for these valuable mountain forests. Our goals were to build stand regression models that predict forest composition and productivity based on site climates and to apply these models to climate change scenarios in order to evaluate possible changes in forest structure and productivity on a local scale.

Methods

Quadratic regression models were developed that related the site climate of a plot and stand productivity characteristics of the forests along a transect in the Kulumys Range of the West Sayan mountains (93° E and 53° N), an area 30 km long and 20 km wide (Fig. 17.1).

Stand data for uneven-aged, mature stands only (older than 160 years for *P. sibirica* and 120 years for *A. sibirica*) at quasi-climax stages were derived from 412 inventory plots. Each stand was characterized by tree species composition (percentage of wood volume), average tree height (m) and trunk wood stocking (m³/ha). Trunk wood stocking was analysed only for those stands which were 80% or more of one tree species. Additionally, stand living phytomass (t/ha) was calculated as a product of trunk wood stocking and the conversion coefficients representing ratios of the different stand fractions (bark, crown, roots and understorey) to the stem wood mass. The latter is calculated as a product of trunk wood volume and wood density. We derived appropriate conversion coefficients for our cedar and fir stands from Alexeyev and Birdsey (1998).

Two climatic indices – temperature sums, base 5°C (TS_5, heat supply), and a dryness index (DI, water supply) – were employed to characterize the site climate of a plot. In Russian climatology, temperature sums, base 5°C, are calculated as the sum of all positive temperatures (T) occurring for the period with daily temperature greater than 5°C:

$$TS_5 = \int T \, dt$$

integrated over the time period with $T > T_5$.

The dryness index is a ratio between available energy (radiation balance) and the energy required to evaporate annual precipitation. To calculate radiation balance,

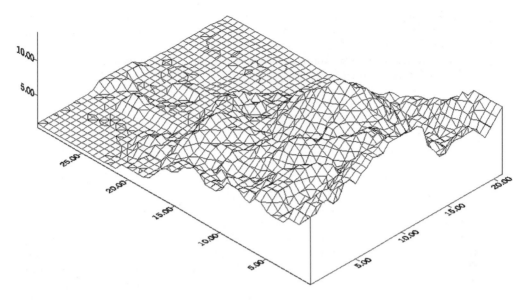

Fig. 17.1. A digital elevation model (m, z axis) of the test transect across the Kulumys Range, an area 20 km wide (x axis) and 30 km long (y axis).

data on temperature, humidity and cloudiness are required (Budyko, 1974; Tchebakova *et al.*, 1994). These climatic indices were calculated using the long-term climatic data of 10–15 weather stations adjacent to the test transect (Reference Books, 1967–1970). Climatic submodels relating TS_5 and DI to topography (elevation, slope and aspect when applicable) were developed to calculate climatic indices for each plot. Climatic submodels were coupled to a local digital elevation model (DEM) that we produced from a topographic map (1:100,000) to delineate current climatic layers of TS_5 and DI (Fig. 17.2a and b). With a regional climate change scenario of +1.5°C summer temperature and −4% summer precipitation (Hulme and Sheard, 1999), two new maps of climatic layers were recalculated by adding those departures to each pixel in the current climatic maps of TS_5 and DI. Stand models predicting tree heights and wood stocking were coupled with current and climate change TS_5 and DI layers to display heights and living phytomass over the test mountain range.

Results and Discussion

Stand models for predicting height, site quality class and wood stocking in cedar and fir forests along with some statistics are given in Table 17.1. Productivity characteristics were determined by a temperature factor, TS_5, with moisture, DI, being insignificant. The best fit of TS_5 to stand productivity characteristics was a quadratic regression with the argument expressed by the logarithm TS_5. The percentage of total variation explained for growth characteristics was as high as 80% for cedar stands and 40% for fir stands. It was less for wood stocking (Table 17.1).

In Fig. 17.3, the relationship between height and TS_5 is shown. On average, cedar trees are 8 m taller than fir trees. This difference is less in cold climates and greater in warm climates where cedar trees can grow to heights of 30 m or more. Although, on average, cedar trees are taller than fir trees, wood stocking of both fir

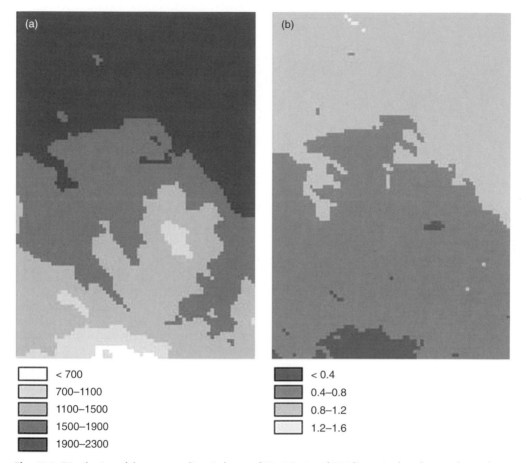

□	< 700	■	< 0.4
▨	700–1100	■	0.4–0.8
▨	1100–1500	▨	0.8–1.2
■	1500–1900	□	1.2–1.6
■	1900–2300		

Fig. 17.2. Distribution of the current climatic layers of TS_5 (°C, a) and DI (dimensionless, b) over the study area.

Table 17.1. Stand models for predicting growth and productivity characteristics of the Sayan mountain forests. All regression coefficients were statistically significant at $P < 0.05$.

Stand parameter	Cedar stand models	Fir stand models
Height (m)	$-2508.70+1571.70^*X-243.42^*X^2$ $R^2 = 0.79$; SD = 4.3; $n = 233$	$-1729.99+1086.43^*X-168.41^*X^2$ $R^2 = 0.40$; SD = 3.6; $n = 272$
Wood stocking (m³/ha)	$-48245.4+30504.5^*X-4797.0^*X^2$ $R^2 = 0.30$; SD = 72; $n = 184$	

X, log (temperature sums, base 5°C, TS_5); SD, standard deviation; n, sample size.

and cedar stands is about the same (Fig. 17.4) because the basal area of fir stands is usually greater than that of cedar stands (Nazimova, 1975). We developed a generalized wood stocking model for all stands regardless of forest composition. Wood stocking varied widely between 50 m³/ha in highlands and 400 m³/ha in lowlands.

Mountain forest composition of dark-needled stands along this transect was rather uniform. Because the ecology of cedar and fir is quite similar, these tree species compete for the same sites. Prevailing species occurrence depends on a forest succession stage rather than climate. As follows from the climatic ordination, DI

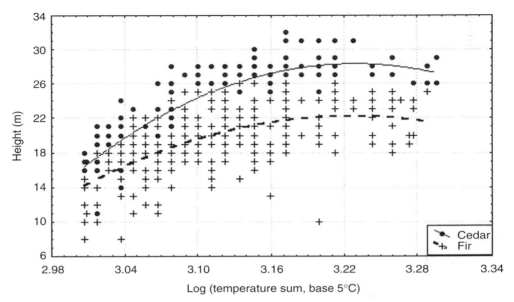

Fig. 17.3. Height (m) curves of cedar and fir depending on climate.

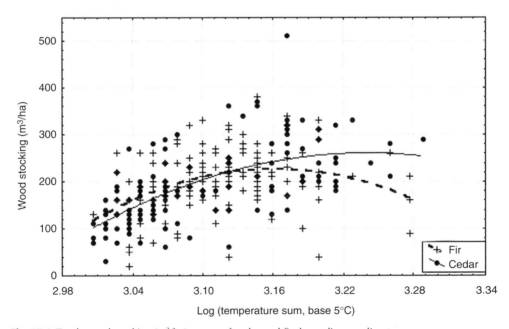

Fig. 17.4. Trunk wood stocking (m³/ha) curves of cedar and fir depending on climate.

separated light-needled pine from dark-needled cedar and fir: light-needled pine had DI values above 1.0 while the others were below 0.8 (Fig. 17.5). This border is quite abrupt and clearly seen in forest maps.

Cedar and fir tree height distributions across the study area are shown in Fig. 17.6. Currently, more than a half of the study area is occupied by cedar stands with trees higher than 25 m and by fir stands with trees higher than 20 m. In a warmed climate, the area will be almost fully covered by forests with such tall trees. From our estimates, the tree line will shift some 250 m up-slope. Current treeless areas in

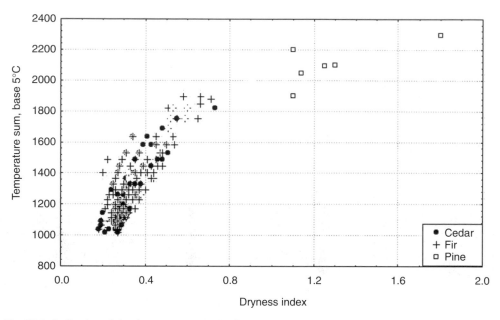

Fig. 17.5. Ordination of dominant tree species in climatic space.

highlands will be covered by forests with rather tall trees: 20–25 m cedar trees and 15–20 m fir trees. However, from our models we cannot say that trees could grow higher in a warmed climate than observed now. As seen from Fig. 17.3, in climates warmer than 1600°C of TS_5, heights of both tree species slowly start to decrease, because under these warmer conditions a greater than presently available water supply is required to favour tree growth. Tree growth in height may also be limited by species biology.

Living phytomass distribution calculated from the wood stocking model (Table 17.1) for the current and a warmed climate across a test transect is shown in Fig. 17.7. The data represented in the figures suggest that, under the current climate, less productive forests with a phytomass of <100 t/ha occupy only about 7% of the forest area. In general, the climate of this region favours productive cedar and fir forests with a phytomass >100 t/ha. In a warmed but drier climate, the area of these forests will decrease by about 8000 ha, which is about 20% of their current area. Dark-needled forests will shift up-slope by about 250–300 m. The pine sub-taiga and forest-steppe area will increase by the same area correspondingly. Although most productive forests will dominate across this area, under warming the total forest phytomass will decrease by 0.8 Pg from 4.6 Pg today to 3.8 Pg in the future because their total area will decrease in lowlands.

Conclusions

The productive dark-needled forests of the Sayan mountains are of special interest both from the theoretical aspect of preserving biodiversity and from the forestry practices aspect of sustainable forestry. The climax growth potential of these forests and their structural changes in a changing climate can be evaluated by using bioclimatic models where forest characteristics are defined as functions of climate.

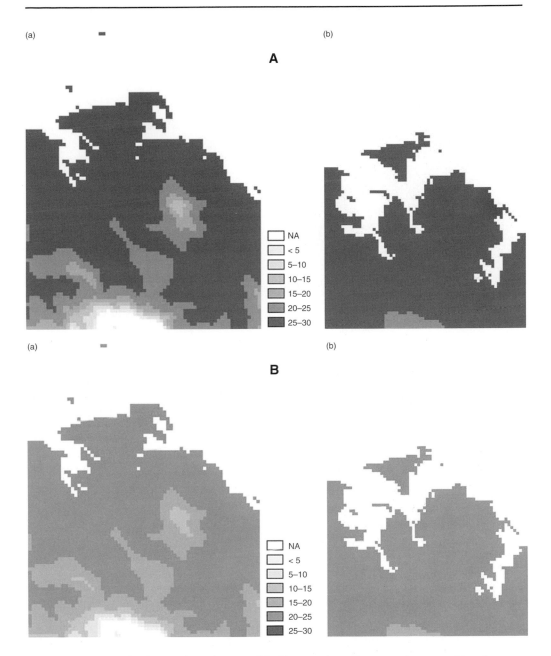

Fig. 17.6. Height (m) distributions for cedar (A) and fir (B) across the study area under current (a) and a warmed climate (b). NA means that a species is not available in a given pixel.

Bioclimatic stand models developed in this study allow us to evaluate growth and productivity characteristics from temperature parameters and tree species composition from moisture parameters. The dryness index separates the mountain, dark-needled (cedar and fir) forests and the light-needled, pine forests of the foothills by the dryness index value of 1.0. Because climates across the region are favourable for both cedar and fir, a dominant tree species in a stand is determined

(a)   (b)

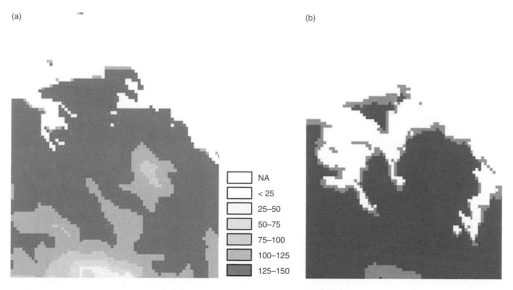

Legend:
- NA
- < 25
- 25–50
- 50–75
- 75–100
- 100–125
- 125–150

Fig. 17.7. Living phytomass (t/ha) of dark-needled forests across the study area under current (a) and a warmed climate (b). NA means that a species is not available in a given pixel.

by succession stage. The phases of cedar and fir forest successions caused by forest fires, droughts or insect pest outbreaks (Polikarpov, 1970) repeat in 200–250 years. Forest gap models (Shugart, 1984) may be used to specify the forest composition of these mixed cedar and fir stands at different succession stages.

Acknowledgement

This study was supported by the Russian Foundation for Basic Sciences No. 02-04-49888.

References

Alexeyev, V.A. and Birdsey, R.A. (eds) (1998) *Carbon Storage in Forests and Peatlands of Russia.* USDA Forest Service, Northeastern Research Station, General Technical Report NE-244, Radnor, Pennsylvania, 137 pp.

Budyko, M.I. (1974) *Climate and Life.* Academic Press, New York, 508 pp.

Hulme, M. and Sheard, N. (1999) *Climate Change Scenarios for the Russian Federation.* Climatic Research Unit, Norwich, 6 pp.

Nazimova, D.I. (1975) *Mountain Dark-needled Forests of West Sayan Mountains.* Nauka, Leningrad, 120 pp.

Parfenova, E.I. and Tchebakova, N.M. (2000) Possible vegetation change in the Altai mountains under climate warming. *Geobotanical Mapping* 1998–2000, 26–31.

Polikarpov, N.P. (1970) Complex investigations in mountain forests of West Sayan. In: Zhukov, A.B. (ed.) *Questions of Forestry,* Vol. 1. Institute of Forestry, Krasnoyarsk, pp. 26–79.

Polikarpov, N.P., Tchebakova, N.M. and Nazimova, D.I. (1986) *Climate and Mountain Forests of Southern Siberia.* Nauka, Novosibirsk, 225 pp.

Reference Books on Climate of the USSR (1964–1970) Issues: 17, 20–24. Gidrometeoizdat, Leningrad.

Savina, L.N. (1986) *Boreal Forests of Northern Asia during the Holocene*. Nauka, Novosibirsk, 192 pp.

Semechkin, I.V., Popova, Y.M. and Popov, V.E. (1985) Stand structure and productivity. In: Isaev, A.S. (ed.) Pinus sibirica *Forests of Siberia*. Nauka, Novosibirsk, pp. 117–132.

Shugart, H. (1984) *A Theory of Forest Dynamics: the Ecological Implications of Forest Succession Models*. Springer, New York, 278 pp.

Shumilova, L.V. (1962) *Botanical Geography of Siberia*. Tomsk University Press, Tomsk, 440 pp.

Smagin, V.N., Ilinskaya, S.A., Nazimova, D.I., Novoseltseva, I.F. and Cherednikova, J.S. (1980) In: Smagin, B.N. (ed.) *Forest Types of Mountains of Southern Siberia*. Nauka, Novosibirsk, 234 pp.

Tchebakova, N.M., Monserud, R.A. and Leemans, R. (1994) A Siberian vegetation model based on climatic parameters. *Canadian Journal of Forest Research* 24, 1597–1607.

Part 3

Estimation Processes

—————————

Parameter estimation is required prior to the building of a mathematical code that will be used by researchers, forest practitioners and managers, or decision makers and policy makers. Parameter estimates are necessary for process, empirical and hybrid models.

Data quality is a key issue affecting the prospective quality of the model. The sources of variability are related to the language communication process (the so-called linguistic variability), the intrinsic variability (intrinsic to the variable that is being measured) and the measurement variability (due to the measurement method).

Before estimating parameters, there are several decisions on data type and data structures (e.g. dependent and independent, re-measurements, independent, time and spatial series, class and continuous, and expert knowledge) and the estimation methods that may be used (e.g. qualitative (expert and literature), moments, maximum likelihood, marginal quasi-likelihood, penalized quasi-likelihood, parameter recovery, parameter prediction, percentile, Markov chains, OLS and NOLS, generalized OLS and NOLS, mixed linear and non-linear, SUR and NSUR, 2SLS and N2SLS, 3SLS and N3SLS, generalized methods, neural networks, geostatistical and Bayesian techniques).

After parameter estimation and model construction, the next step consists of the process of verifying the model. The use of simple or more complex tools in order to verify the statistical assumptions and the evaluation is essential to test the model's adherence to reality and the coherency of its results. A very important contribution to this process is the analysis of the biological interpretation of the parameters.

Parameter estimation is a dynamic task: reality is changing and the knowledge of the reality is also changing. All the priority research topics in this area have the same objective: reducing uncertainty in the ultimate model predictions.

The possibility of building web-based learning tools draws attention to the fact that many of the scientific papers submitted for publication are often not very clear, or even difficult to understand, especially with regard to parameter estimation. To understand the process it is often necessary to reproduce the methodology, and that can be difficult in many cases. The availability of a learning tool with some training examples and subsets of the original data, along with a more in-depth explanation of the methods used, could help in understanding the methods used and provide additional information that would certainly improve and validate the proposed procedures.

An existing model needs some updating – after some time it will need to be re-evaluated and modified: knowledge of the system evolves and the true situation may change. To accomplish this task, there are two main possibilities: (i) to start from the beginning (adding new data, re-evaluating the possibility of having new functional forms, new structures, new methods), or (ii) to develop and use tools to support model updating. Control engineers use variants of the Kalman filter to update the state of a system in real time, for example. As process measurements are recorded, they are optimally combined with model predictions in the filter update step. Such techniques were developed for real-time systems (i.e. short time steps) and may be very appropriate in process modelling. However, for traditional long-term growth modelling, the benefit of applying such techniques is questionable when there may be years between measurements. In the intervening time, it may be more efficient to re-fit the existing model, develop a completely new model, or use newly developed estimation techniques on existing model forms that were not known at the last calibration.

Tree mortality is one of the key processes in forest ecosystems. It has been modelled traditionally with logistic regression using the maximum likelihood estimation method. It is common for individual tree mortality data to include observations from several stands with more than one mortality tree per stand. As a consequence, the observations (trees in the same stand) are correlated. Ignoring data structure in mortality data leads to several consequences, such as underestimation of standard errors for higher-level fixed effects. The generalized linear mixed models are useful when data are not normally distributed or have been hierarchically collected. Methods of analysis and measures of model fit appropriate to use in these situations should be seriously considered.

The quasi-likelihood estimation method used in generalized linear mixed models is known for its ability to produce efficient estimates without exact information about the likelihood function. Both marginal quasi-likelihood (MQL) and penalized quasi-likelihood (PQL) estimation methods have been used. The quasi-likelihood methods are attractive because they are available in commonly used software. However, one should be aware of the possible limitations of these methods, e.g. in estimation, the often small number of observations at different levels may cause bias in the resulting estimates.

Multi-level mortality models can be improved by: (i) increasing data and variables in the model; (ii) estimating the mortality model simultaneously with the growth model; and (iii) developing goodness-of-fit measures in the case of generalized linear mixed models. Increasing information by obtaining more data can often be done, but it may be difficult to determine what level is needed. There are often unmeasured variables affecting tree mortality, and inclusion of this information can often improve the mortality model. However, if the number of variables increases, the model becomes more complex. The most recently developed mortality models often have the same independent variables used to describe tree vitality. To improve model estimates, the growth and mortality models may be estimated simultaneously so that the mortality curve asymptotically approaches the forest self-thinning limit. Developing goodness-of-fit measures in the case of correlated data is an issue that must be considered in the future.

Modelling methods are commonly assessed based on their properties of unbiasedness, asymptotic unbiasedness, consistency and efficiency (as related to a standard). They are also assessed as to their ability to hold properties under different types and distributions of data. Properties of fitting methods are affected by several variables, including sample size, the number of parameters to be estimated, the distribution of the model errors and the fitting method.

Some techniques only have desirable (usually asymptotic) properties for very large sample sizes, and may function poorly with small sample sizes. Often, techniques that are based on the normal distribution of model errors give good results for small samples when the model errors are indeed normal. For non-normal distributions when the distribution is known, maximum likelihood estimators give 'good' results. However, for very large numbers of parameters to be estimated, methods requiring search algorithms may become unstable, resulting in local minima, and may show large variations in parameter estimates for different sample data sets.

The following contributions examine these and other issues related to parameter estimation in forest models. The editors wish to acknowledge the efforts of Ana Amaro (Portugal) and Jeffrey Gove (USA) for coordinating this section, and thank them for their contributions to this volume.

18 Estimation and Applications of Size-biased Distributions in Forestry

Jeffrey H. Gove[1]

Abstract

Size-biased distributions arise naturally in several contexts in forestry and ecology. Simple power relationships (e.g. basal area and diameter at breast height) between variables are one such area of interest arising from a modelling perspective. Another, probability proportional to size (PPS) sampling, is found in the most widely used methods for sampling standing or dead and fallen material in the forest. Often it is desirable or necessary to estimate a parametric probability density model based on size-biased data. Traditional equal probability methods may not be appropriate, or may be less efficient in such circumstances, and estimation is better conducted utilizing size-biased theory. This chapter surveys some of the possible uses of size-biased distribution theory in forestry and related fields.

Introduction

Size-biased distributions are a special case of the more general form known as weighted distributions. First introduced by Fisher (1934) to model ascertainment bias, weighted distributions were later formalized in a unifying theory by Rao (1965). Such distributions arise naturally in practice when observations from a sample are recorded with unequal probability, such as from probability proportional to size (PPS) designs. Briefly, if the random variable X has distribution $f(x;\theta)$, with unknown parameters θ, then the corresponding weighted distribution is of the form

$$f^w(x;\theta) = \frac{w(x)f(x;\theta)}{\mathrm{E}\big[w(x)\big]}$$

where $w(x)$ is a non-negative weight function such that $\mathrm{E}[w(x)]$ exists.

A special case of interest arises when the weight function is of the form $w(x) = x^\alpha$. Such distributions are known as size-biased distributions of order α and are written as (Patil and Ord, 1976; Patil, 1981; Mahfoud and Patil, 1982):

---

[1] USDA Forest Service, Northeastern Research Station, USA
Correspondence to: jhgove@christa.unh.edu or jgove@fs.fed.us

$$f_\alpha^*(x;\theta) = \frac{x^\alpha f(x;\theta)}{\mu_\alpha'} \tag{1}$$

where $\mu_\alpha' = \int x^\alpha f(x;\theta)\,dx$ is the αth raw moment of $f(x;\theta)$. Denote X the original, or equal probability, random variable, and $X_\alpha^* \sim f_\alpha^*(x;\theta)$ the size-biased random variable. The most common cases of size-biased distributions occur when α=1 or 2; in the context of sampling, these special cases may be termed length- and area-biased, respectively.

Weighted distributions have numerous applications in forestry and ecology. Warren (1975) was the first to apply them in connection with sampling wood cells. Van Deusen (1986) arrived at size-biased distribution theory independently and applied it to fitting distributions of diameter at breast height (DBH) data arising from horizontal point sampling (HPS) (Grosenbaugh, 1958) inventories. Subsequently, Lappi and Bailey (1987) used weighted distributions to analyse HPS diameter increment data. More recently, weighted distributions were used by Magnussen *et al.* (1999) to recover the distribution of canopy heights from airborne laser scanner measurements. In ecology, Dennis and Patil (1984) use stochastic differential equations to arrive at a weighted gamma distribution as the stationary probability density function (PDF) for a stochastic population model with predation effects. In fisheries, Taillie *et al.* (1995) modelled populations of fish stocks using weighted distributions. In these last two examples, weighted distributions were not directly tied to the underlying sample selection method, but were simply convenient models for the observed data. Recognizing the fact that weighted distributions may be applied as convenient PDF models, Gove and Patil (1998) developed a compatible theory, unifying the DBH–frequency and basal area–DBH distributions based on the quadratic relationship between diameter and basal area. Lastly, Gove (2000) extended the work of Van Deusen (1986) by providing simulation experiments and guidelines for fitting size-biased distributions to data.

The purpose of this chapter is to review some of the more recent results on size-biased distributions pertaining to parameter estimation in forestry, with special emphasis on the Weibull family. In addition, some new results and avenues for possible future research will be presented. Finally, a new computer program with graphical user interface (GUI) developed by the author for fitting size-biased Weibull distributions will be briefly discussed.

Size-biased Weibull Distributions

Weibull distributions have found widespread use in forestry for modelling since they were first introduced by Bailey and Dell (1973). The two- and three-parameter Weibull PDFs are given as

$$f(x;\theta) = \left(\frac{\gamma}{\beta}\right)\left(\frac{x}{\beta}\right)^{\gamma-1} e^{-(x/\beta)^\gamma} \qquad x>0$$

$$f(x;\theta) = \left(\frac{\gamma}{\beta}\right)\left(\frac{x-\xi}{\beta}\right)^{\gamma-1} e^{-((x-\xi)/\beta)^\gamma} \qquad x>\xi$$

with $\theta = (\gamma,\beta)'$ and $\theta = (\gamma,\beta,\xi)'$, respectively. The unknown parameters $\gamma > 0$, $\beta > 0$ and $\xi > 0$ are the shape, scale and location parameters to be estimated for a given sample of data.

These PDFs can be easily converted to their size-biased counterparts using Equation 1, namely

$$f_\alpha^*(x;\theta) = \frac{x^\alpha}{\mu_\alpha'}\left(\frac{\gamma}{\beta}\right)\left(\frac{x}{\beta}\right)^{\gamma-1} e^{-(x/\beta)^\gamma} \qquad x > 0$$

$$f_\alpha^*(x;\theta) = \frac{x^\alpha}{\mu_\alpha'}\left(\frac{\gamma}{\beta}\right)\left(\frac{x-\xi}{\beta}\right)^{\gamma-1} e^{-((x-\xi)/\beta)^\gamma} \qquad x > \xi$$

for the two- and three-parameter versions, respectively, with the same restrictions on the parameters as for the equal probability PDFs. Gove and Patil (1998) have also shown that the size-biased two-parameter Weibull can be transformed, through change-of-variables techniques, to the standard gamma distribution. Such a transformation may be advantageous for simulation studies. For example, Gove (2000) used the standard gamma to draw probability-weighted samples to simulate the HPS tally distribution.

Because of their popularity in modelling the traditional DBH–frequency distribution, both the two- and three-parameter size-biased Weibull PDFs are appropriate as candidate probability models in all of the applications presented in this chapter.

Size-biased Weibulls: Moment Estimation

Size-biased two-parameter Weibull moment estimators

The development of moment estimators for the size-biased two-parameter Weibull distribution is given in Gove (2003a). There, a modified moment estimation scheme along the lines of Cohen (1965), using the coefficient of variation, is presented. Let $\tilde{\gamma}$ and $\tilde{\beta}$ represent the moment estimates for the shape and scale parameters, respectively; then the moment equations are

$$CV = \Gamma_\alpha \Gamma_{\alpha+1}^{-1} \sqrt{\frac{\Gamma_{\alpha+2}}{\Gamma_\alpha} - \frac{\Gamma_{\alpha+1}^2}{\Gamma_\alpha^2}} \qquad (2)$$

$$\tilde{\beta} = \frac{\bar{x}\tilde{\Gamma}_\alpha}{\tilde{\Gamma}_{\alpha+1}} \qquad (3)$$

where $\bar{x}$ and CV are the sample mean and coefficient of variation, respectively, with $\Gamma_\alpha = \Gamma(\alpha/\gamma + 1)$, $\tilde{\Gamma}_\alpha = \Gamma(\alpha/\tilde{\gamma} + 1)$ and $\Gamma(k) = \int_0^\infty x^{k-1} e^{-x} dx$, $k > 0$, the gamma function. Equation 2 is solved iteratively for the shape parameter, then the scale parameter can be found directly by substitution into Equation 3.

Size-biased three-parameter Weibull moment estimators

Unfortunately, the moment equations for the size-biased three-parameter Weibull are not easily couched in a modified scheme like that for the two-parameter where the coefficient of variation can be used. Thus, the moment equations for the first three raw moments are used; these moments can be built up from the moments of the equal probability three-parameter Weibull (Gove 2003a). Let $\mu_{\alpha,\zeta}^{*'} = \int d^\zeta f_\alpha^*(x,\theta)dx$ denote the ζth raw moment of the size-biased three-parameter Weibull distribution of order α. Then, it is straightforward to show that $\mu_{\alpha,\zeta}^{*'} = \frac{\mu_{\alpha+\zeta}'}{\mu_\alpha'}$. Now, since $\alpha = 1$ or

2 for the most common forestry applications, and $\zeta = 1,...,3$ for the first three raw moments, it is easy to see from the numerator of $\mu_{\alpha,\zeta}^{*'}$ that the first five raw moments of the three-parameter Weibull distribution are required for the estimating equations. The moments for the three-parameter Weibull are of the form

$$\mu_{\alpha}' = \beta^{\alpha}\Gamma_{\alpha} + \binom{\alpha}{1}\beta^{\alpha-1}\Gamma_{\alpha-1}\xi + \binom{\alpha}{2}\beta^{\alpha-2}\Gamma_{\alpha-2}\xi^2 + \cdots + \xi^{\alpha} \tag{4}$$

where the coefficients $\binom{\alpha}{i}$, $i = 1,...,\alpha$ follow Pascal's triangle. Thus, for example, the second raw moment from a length-biased three-parameter Weibull, $\mu_{1,2}^{*'}$ is

$$\frac{\mu_3'}{\mu_1'} = \frac{\beta^3\Gamma_3 + 3\beta^2\Gamma_2\xi + 3\beta\Gamma_1\xi^2 + \xi^3}{\beta\Gamma_1 + \xi}$$

It should be clear that the moment equations for the length- and area-biased versions differ. For comparison, the second raw moment from an area-biased three-parameter Weibull is given as $\mu_{2,2}^{*'}$, and is therefore more complicated:

$$\frac{\mu_4'}{\mu_2'} = \frac{\beta^4\Gamma_4 + 4\beta^3\Gamma_3\xi + 6\beta^2\Gamma_2\xi^2 + 4\beta\Gamma_1\xi^3 + \xi^4}{\beta^2\Gamma_2 + 2\beta\Gamma_1\xi + \xi^2}$$

The first three moment equations are set equal to the first three sample moments and solved simultaneously for the estimates $\tilde{\gamma}$, $\tilde{\beta}$, $\tilde{\xi}$. Further details are given in Gove (2003a,c).

Size-biased Weibulls: Maximum Likelihood Estimation

The maximum likelihood estimators (MLEs) for size-biased Weibulls can be found by building up from the equal probability likelihood, just as in the case of the three-parameter moment estimators in the previous section. The equal probability three-parameter Weibull log-likelihood is

$$\ln \mathcal{L} = n\ln\left(\frac{\gamma}{\beta^{\gamma}}\right) + (\gamma-1)\sum_{i=1}^{n}\ln(x_i - \xi) - \frac{1}{\beta^{\gamma}}\sum_{i=1}^{n}(x_i - \xi)^{\gamma}$$

and the two-parameter log-likelihood follows directly by setting $\xi = 0$.

The size-biased form was first given by Van Deusen (1986), where he noted that it was composed of the equal probability log-likelihood plus a constant and a correction term. He also noted that the purpose of the correction was to account for the fact that the observations are drawn with unequal probability. The general form of the size-biased log-likelihood is given as

$$\ln \mathcal{L}^* = \ln \mathcal{L} + \alpha\sum_{i=1}^{n}\ln x_i - n\ln\mu_{\alpha}'$$

where the second term is constant, depending only on the data, and thus may be dropped if desired.

In addition, the gradient vector and Hessian matrix of first- and second-order partial derivatives are also of the same form (Gove 2003a). For example, the gradient equations for the size-biased three-parameter Weibull follow the form

$$\frac{\partial\ln \mathcal{L}^*}{\partial\gamma} = \frac{\partial\ln \mathcal{L}}{\partial\gamma} - n\rho_{\gamma}(\alpha)$$

$$\frac{\partial\ln \mathcal{L}^*}{\partial\beta} = \frac{\partial\ln \mathcal{L}}{\partial\beta} - n\rho_{\beta}(\alpha)$$

$$\frac{\partial\ln \mathcal{L}^*}{\partial\xi} = \frac{\partial\ln \mathcal{L}}{\partial\xi} - n\rho_{\xi}(\alpha)$$

Notice that the correction term $(n\rho_{\theta_i}(\alpha))$ depends on the size-biased order α. Thus, there are unique corrections associated with length- and area-biased log-likelihoods. The Hessian matrix follows the same pattern, being composed of the equal probability and correction components. Detailed equations for the three-parameter gradient and Hessian are presented in Gove (2003a). In the two-parameter size-biased Weibull, the equations are much simpler, due to the simpler nature of the raw moment μ'_α in that distribution. The gradient equations for the two-parameter case are given in Gove (2000).

The Basal Area-size Distribution

As mentioned earlier, the basal area-size distribution (BASD) is the size-biased distribution of order $\alpha = 2$ of the traditional DBH–frequency distribution (Gove and Patil, 1998). The relationship can easily be shown algebraically and arises, not from sampling theory, but purely from the quadratic relationship between DBH and basal area. If the random variable X is tree diameter, then $X \sim f(x;\theta)$ is the DBH–frequency distribution. From it, we normally calculate the number of trees in the ith diameter class (N_i), once the parameters θ have been estimated from sample data

$$N_i = N \int_{x_{l_i}}^{x_{u_i}} f(x;\theta)dx \tag{5}$$

where x_{l_i} and x_{u_i} are the lower and upper diameter class limits, respectively, and N is the total number of trees per hectare.

The BASD comes about by redistributing the probability mass in terms of basal area, rather than tree frequency. The random variable in both cases is still DBH. The BASD can then be used to calculate the basal area (B_i) in the ith DBH class as

$$B_i = B \int_{x_{l_i}}^{x_{u_i}} f_2^*(x;\theta)dx$$

where B is the stand basal area per hectare. Thus, $X_2^* \sim f_2^*(x;\theta)$.

Gove and Patil (1998) presented several examples of stands fitted with a parameter recovery model, all with the same basal area and number of trees, but spanning a wide range of the two-parameter Weibull parameter space. As an example, the stand in their Figure 1d has been re-fitted with a three-parameter Weibull model and is presented in Fig. 18.1. This figure shows the empirical histogram for the DBH–frequency distribution along with the Weibull curve fitted by ML. Also shown is the corresponding BASD curve, which shares the same estimated parameter vector $\hat{\theta}$ from ML.

Estimation of Weibull Parameters under Size-biased Sampling

Arguably, the two most useful forms of size-biased distributions arising in forestry are the length- and area-biased models. Length-biased data arise from line intersect samples (LIS) (Kaiser, 1983), horizontal and vertical line samples (HLS, VLS) (Grosenbaugh, 1958) and transect relascope sampling (TRS) (Ståhl, 1998). Area-biased data arise naturally from HPS and vertical point sampling (VPS) (Grosenbaugh, 1958), and from point relascope sampling (PRS) (Gove et al., 1999) for coarse woody debris. In this section these links are explored in more detail, with special emphasis on the distribution of HPS tally tree diameters.

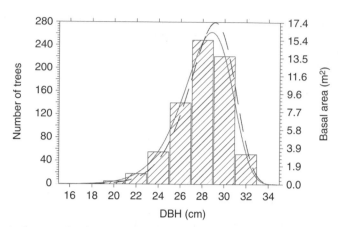

Fig. 18.1. Example diameter distribution (shaded) from Gove and Patil (1998) with $B = 45.9$ and $N = 741$ showing the estimated DBH–frequency distribution (solid) and associated BASD (dashed) for a three-parameter Weibull with MLEs: $\hat{\gamma} = 11.23$, $\hat{\beta} = 23.44$, $\hat{\xi} = 5.57$.

Because of the intrinsic link between basal area and HPS, it is not surprising that the distribution of tally diameters from a HPS turns out once again to be the size-biased distribution of order $\alpha = 2$ (Van Deusen, 1986; Gove, 2000). Thus, if the underlying population diameter distribution for a given stand is $f(x;\theta)$, then the corresponding HPS tally distribution is given by $f_2^*(x;\theta)$, where θ is a shared parameter set. Having sampled from $f_2^*(x;\theta)$ using a prism or suitable angle gauge with HPS, we next must estimate θ, usually by ML. In the following sections some strategies for estimation are discussed with regard to this problem.

Fitting single horizontal point samples

Van Deusen (1986) first discussed fitting Weibull distributions to diameter data arising from single horizontal point samples. The most common reason for doing this would be the subsequent fitting of parameter prediction models (Hyink and Moser, 1983). Later, Gove (2000) used simulations to address in more detail the possible problems with parameter estimation, using two-parameter Weibulls for illustration. The main results of the latter study are discussed in this section.

Briefly, it is possible to estimate θ either by fitting a Weibull to the estimated stand table (number of trees per hectare by DBH) diameters from a single HPS, or by fitting the area-biased Weibull directly to the tally diameters. However, in theory, θ is supposed to be a shared parameter set between $f(x;\theta)$ and $f_2^*(x;\theta)$. A problem arises because one can fit both distributions to their respective data for any given HPS and, in so doing, two different estimates of θ normally result in the process. Then the question becomes, which estimate is the best? This question does not arise when fitting distributions to diameters sampled on fixed area plots, because in either instance we are estimating $f(x;\theta)$ (Gove, 2000).

The simulations presented were extensive and will not be discussed in detail here. However, they were designed to assess the effects of both expected sample size (in terms of number of trees tallied) per point, and the shape of the population distribution $f(x;\theta)$ on estimation. The key findings were as follows. First, as the sample size per point increases, both parameter estimates tend to converge to the population values. However, the rate at which they do so depends in large part on the shape of the underlying population of diameters. In the case of fairly symmetrical population distributions, both parameter estimates converged at the same rate and had very similar root mean squared errors (RMSEs). However, as the population diameter distribution tended more towards a reverse J-shape, associated with typical uneven-aged stands, the parameter estimates from the size-biased distribution fit both converged more quickly and had lower RMSEs, often by more than half.

The reasons for the results are twofold. First, because the size-biased form is theoretically linked to the underlying sampling mechanism, its shape more nearly parallels that of the population distribution of HPS diameters and is therefore estimated more efficiently. This is particularly true, as illustrated in Figure 4a of Gove (2000), when the population diameter distribution is reverse J-shaped. As the population distribution of diameters becomes more symmetrical, the shapes of $f(x;\theta)$ and $f_2^*(x;\theta)$ tend to be more alike and estimation is therefore essentially equivalent for either density. Second, in the reverse J-shaped population, sampling with probability proportional to basal area is akin to sampling for rare events in terms of frequency. The vast majority of probability density for the associated tally distribution is confined to diameters of essentially merchantable size. Therefore, it is very difficult to realize a large enough sample of smaller diameter trees on any one point, to actually shift the estimated stand table from unimodel to reverse J-shaped. For example, the result of $m = 1000$ simulations from a reverse J-shaped distribution with population shape parameter $\gamma = 1.0$ was an estimated stand table shape parameter of $\hat{\gamma} = 1.54$ with $N^* = 40$ trees per point sampled. In contrast, the estimate for the size-biased shape parameter from the tally data for the same simulations was 1.07, with RMSE equal to one-quarter that of the stand table estimate for the shape parameter.

The most important conclusion that should be kept in mind from this study, is that concerning the overall purpose of the inventory. Horizontal point sample inventories are a rich reservoir of data for estimating forest characteristics. However, the normal recommendation of choosing an angle that selects 5–12 trees per point on average for estimating stand data (Avery and Burkhart, 1994: 218), generally will not suffice for parameter estimation of assumed diameter distributions. Therefore, the goals of parameter estimation and inventory may conflict and it is possible that, depending on the shape of the population diameter distribution, alternative inventory protocols may be required.

Fitting with multiple points

Fitting diameter distributions to a single HPS for use with parameter prediction model construction is undoubtedly a rather infrequent use of such data. It is probably more likely that one would be interested in fitting diameter distributions to sample data arising from more than one HPS point, say, for example, to a stand diameter distribution taken over n sample points. In this case, the questions posed in the previous study are still valid. However, the support for parameter estimation naturally increases with the increased sample size and one would expect that the ML estimates would continue to converge in both the stand and tally estimates to the respective population values. The problem can be viewed from two different perspectives based on the degree of homogeneity of the target population diameter distribution.

Homogeneous stands

In this case, one would envisage that the diameter distribution from one point to the next in an HPS inventory is relatively homogeneous within the population of interest. Thus, for parameter estimation purposes, the stand table can be computed directly from the sample of n points to estimate $f(x;\theta)$. Similarly, the tally from all n points can simply be pooled to estimate $f_2^*(x;\theta)$. Furthermore, let $\hat{\theta} = (\hat{\gamma},\hat{\beta})$ and $\hat{\theta}^* = (\hat{\gamma}^*,\hat{\beta}^*)$ be the MLEs for $f(x;\theta)$ and $f_2^*(x;\theta)$, respectively.

Two sets of simulations were conducted to extend the previous study to the multiple point case. The two populations chosen were those that showed the poorest convergence in the single HPS estimates: the reverse J-shaped and mild positively skewed populations. The expected number of trees sampled per point was fixed at $N^* = 10$, and the sample sizes ranged from $n = 5$ to 40 points for the simulations.

The results of the simulations are presented in Table 18.1. These results clearly show that in both cases, as the sample size increases, the parameter estimates converge to the population values more rapidly for the tally distribution. Not only is the overall bias less, but the RMSE is also significantly reduced. This is particularly true for the reverse J-shaped population, but also still holds rather convincingly for the mildly skewed population.

Because the reverse J-shaped case is closely linked to uneven-aged management, which seems to be gaining in popularity in the USA, it is of interest to look at this problematic case a little more closely. The results of the simulations are presented graphically in Fig. 18.2. In both cases, the population line is shown as solid, and the average density (dashed) lines generally approach it as the number of points increases. It is quite apparent from these graphs that the estimated densities for the stand table data are never quite able to estimate the true reverse J-shape. On the other hand, it takes in the neighbourhood of 25–30 HPS points to arrive at the correct estimates when using the tally data and $f_2^*(x;\theta)$ in such stands.

These simulations mirror the trends in the single HPS case exactly but, because of the increased sample size, show that convergence is better in the multiple-point scenario. The conclusions to be drawn, then, also parallel the single-point case: when sampling from stand conditions that approach that of the classic reverse J-shaped distribution, or show some degree of positive skewness, parameter estimation should be undertaken using the size-biased likelihood approach.

Table 18.1. Simulation results for 250 replications of $N^* = 10$ trees per point on n multiple HPS points drawn from a two-parameter Weibull population of tree diameters with $\theta = (\gamma,\beta)$.

θ	n	Average				% RMSE			
		$\hat{\gamma}^*$	$\hat{\beta}^*$	$\hat{\gamma}$	$\hat{\beta}$	$\hat{\gamma}^*$	$\hat{\beta}^*$	$\hat{\gamma}$	$\hat{\beta}$
(1,8)	5	1.05	8.50	1.48	11.16	19.1	34.2	75.0	71.7
	10	1.02	8.20	1.39	10.61	12.7	24.1	60.2	58.3
	20	1.02	8.23	1.32	10.20	8.9	16.6	51.1	49.1
	30	1.0	7.99	1.25	9.55	6.1	12.5	41.3	39.7
	40	1.0	7.95	1.22	9.21	5.6	10.9	38.7	36.9
(2,15)	5	2.08	15.18	2.32	15.57	16.4	11.4	32.6	16.0
	10	2.04	15.04	2.18	15.25	11.8	8.6	23.2	12.0
	20	2.02	15.07	2.15	15.32	7.3	5.7	19.4	10.5
	30	2.0	14.97	2.08	15.06	6.1	4.6	15.7	8.9
	40	2.01	15.04	2.13	15.24	5.2	3.9	16.8	10.4

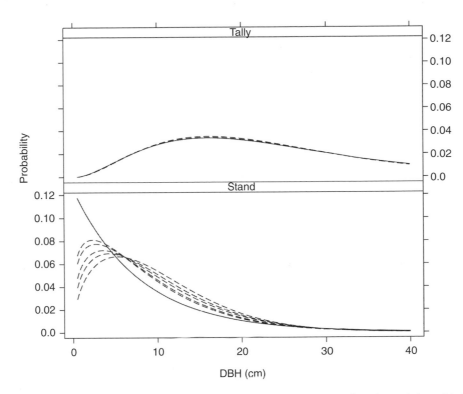

Fig. 18.2. Simulated average distribution results for the homogeneous reverse J-shaped population with the tally densities (top) and stand table densities (bottom); the mean densities (dashed) converge to the population curve (solid) with increasing sample size (see Table 18.1 for details).

Heterogeneous stands

Consider a stand (or larger area) where the diameter distribution varies, possibly considerably, throughout, but where it is still desired to estimate $f(x;\theta)$. In such cases it may or may not be feasible to stratify.

Stating that the diameter distribution varies is another way of saying that θ is not constant throughout. For example, consider an HPS with n points in which θ varies from point to point according to some stochastic process. Thus θ may be considered a random variable and may exhibit a spatial covariance structure between points. Such a scenario might possibly be modelled using continuous mixtures.

For illustration, assume that the conditional distribution of tallied diameters given θ is a two-parameter Weibull; $X_2^* | \theta \sim f_2^*(x | \theta)$. It would then make sense to use a bivariate distribution to model the variation in θ over the stand. One candidate probability model for the joint distribution of $\Theta \sim f(\theta;\mu,\Sigma)$ is bivariate normal with mean and covariance matrix μ and Σ, respectively. Other bivariate distributions could also be considered. With the bivariate normal, particular care must be taken to ensure that, for all practical purposes, $\theta > 0$. Thus, extreme variability between HPS points coupled with small-scale or shape parameters might argue against its use. However, for the sake of illustration it is a useful model.

With this modelling scheme, the bivariate normal is the mixing distribution, and the marginal stand tally distribution for X_2^* under HPS would be given by

$$f_2^*(x;\mu,\Sigma) = \iint f_2^*(x\mid\theta)f(\theta;\mu,\Sigma)d\gamma d\beta \tag{6}$$

Undoubtedly, the marginal distribution given in Equation 6 does not exist in closed form and the integration would require numerical methods. However, it does pose an interesting interpretation for the final density once γ and β have been integrated out. The only two remaining parameters are μ and Σ. Thus, the following method might be used to fit such a distribution based on the techniques discussed earlier in this chapter.

1. Fit a size-biased two-parameter Weibull PDF to each of n individual HPS points in the stand using the methods in previous sections.
2. Calculate the sample mean vector $\hat{\mu}$ and respective sample covariance matrix S from the parameter estimates on the n individual HPS sample points, as estimates of and μ and Σ, respectively.

The mixture density $f_2^*(x;\mu,\Sigma)$ may now be estimated by $f_2^*(x;\hat{\mu},S)$. However, it must be kept in mind that the above has in no way proved that $\hat{\mu}$ and S have any of the desirable properties of say, MLEs, for μ and Σ. It is simply a possible model for a heterogeneous stand parameter estimation scenario.

Discussion

The discussion on estimation and applications of size-biased distributions to this point demonstrates that they both have a solid theoretical underpinning and practical use in forestry. Well-known relationships between basal area and horizontal point sampling, for example, are preserved under this theory. It should not be surprising then that other results will also hold for size-biased distributions. For example, Gove (2003b) has shown that the relationship between the quadratic mean stand diameter and the harmonic mean basal area from an HPS holds for area-biased distributions; the result is shown to apply also to the BASD.

In fact, size-biased distributions can also bring new insight to previously unknown relationships. For example, Gove and Patil (1998) showed that the third raw moment of the DBH–frequency distribution has an intuitive and consistent interpretation through the BASD – a result that had been missed prior to the application of this theory. Similarly, it can be shown analytically (Gove, 2003b) that $f(x;\theta)$ and $f_2^*(x;\theta)$ will *always* cross at the quadratic mean stand diameter ($\bar{D}_q$). To illustrate, refer back to Fig. 18.1, for this stand $\bar{D}_q = 28.08$ cm, and this is exactly where the two PDFs cross.

A new computer program (BALANCE) (Gove, 2003c) has been developed to facilitate the use of size-biased distributions in forestry. BALANCE was written in FOR-TRAN-90, and is fully integrated with a graphical user interface and runs under Microsoft Windows© operating systems. Currently, BALANCE allows the user to fit two- and three-parameter equal probability Weibull distributions. In addition, both length- and area-biased versions of these PDFs can also be fitted. BALANCE computes the moment estimates and then uses these as starting values for ML. Results are presented in three windows; a listing of the input data in a grid window, a summary report window with fit statistics, and a graphics window with various graphical displays. The latter may be exported in encapsulated PostScript format, an example of which is shown in Fig. 18.1. Notice in this figure that, even though the equal probability density was estimated for the DBH-frequency distribution, BALANCE also shows the related BASD.

Clearly, size-biased distributions provide a useful paradigm for sampling and modelling in forestry research. The availability of computer programs such as BALANCE to make fitting such distributions easier should serve to increase their application.

References

Avery, T.E. and Burkhart, H.E. (1994) *Forest Measurements*, 4th edn. McGraw-Hill, New York.

Bailey, R.L. and Dell, T.R. (1973) Quantifying diameter distributions with the Weibull function. *Forest Science* 19, 97–104.

Cohen, A.C. (1965) Maximum likelihood estimation in the Weibull distribution based on complete and on censored samples. *Technometrics* 7, 579–588.

Dennis, B. and Patil, G. (1984) The gamma distribution and weighted multimodal gamma distributions as models of population abundance. *Mathematical Biosciences* 68, 187–212.

Fisher, R.A. (1934) The effects of methods of ascertainment upon the estimation of frequencies. *Annals of Eugenics* 6, 13–25.

Gove, J.H. (2000) Some observations on fitting assumed diameter distributions to horizontal point sampling data. *Canadian Journal of Forest Research* 30, 521–533.

Gove, J.H. (2003a) Moment and maximum likelihood estimators for Weibull distributions under length- and area-biased sampling. *Ecological and Environmental Statistics* (in press).

Gove, J.H. (2003b) A note on the relationship between the quadratic mean stand diameter and harmonic mean basal area under size-biased distribution theory. *Canadian Journal of Forest Research* (in press).

Gove, J.H. (2003c) *Balance: a System for Fitting Diameter Distribution Models*. General Technical Report. NE-xx, USDA Forest Service (in press).

Gove, J.H. and Patil, G.P. (1998) Modeling the basal area-size distribution of forest stands: a compatible approach. *Forest Science* 44(2), 285–297.

Gove, J.H., Ringvall, A., Ståhl, G. and Ducey, M.J. (1999) Point relascope sampling of downed coarse woody debris. *Canadian Journal of Forest Research* 29, 1718–1726.

Grosenbaugh, L.R. (1958) *Point Sampling and Line Sampling: Probability Theory, Geometric Implications, Synthesis*. Occasional Paper 160, USDA Forest Service, Southern Forest Experiment Station.

Hyink, D.M. and Moser, J.W. Jr (1983) A generalized framework for projecting forest yield and stand structure using diameter distributions. *Forest Science* 29, 85–95.

Kaiser, L. (1983) Unbiased estimation in line-intercept sampling. *Biometrics* 39, 965–976.

Lappi, J. and Bailey, R.L. (1987) Estimation of diameter increment function or other tree relations using angle-count samples. *Forest Science* 33, 725–739.

Magnussen, S., Eggermont, P. and LaRiccia, V.N. (1999) Recovering tree heights from airborne laser scanner data. *Forest Science* 45(3), 407–422.

Mahfoud, M. and Patil, G.P. (1982) On weighted distributions. In: Kallianpur, G., Krishnaiah, P. and Ghosh, J. (eds) *Statistics and Probability: Essays in Honor of C.R. Rao*. North-Holland, New York, pp. 479–492.

Patil, G.P. (1981) Studies in statistical ecology involving weighted distributions. In: Ghosh, J.K. and Roy, J. (eds) *Applications and New Directions*, Proceedings of the Indian Statistical Institute Golden Jubilee. Statistical Publishing Society, Calcutta, pp. 478–503.

Patil, G.P. and Ord, J.K. (1976) On size-biased sampling and related form-invariant weighted distributions. *Sankhyā, Series B* 38(1), 48–61.

Rao, C.R. (1965) On discrete distributions arising out of methods of ascertainment. In: Patil, G.P. (ed.) *Classical and Contagious Discrete Distributions*. Statistical Publishing Society, Calcutta, pp. 320–332.

Ståhl, G. (1998) Transect relascope sampling: a method for the quantification of coarse woody debris. *Forest Science* 44(1), 58–63.

Taillie, C., Patil, G.P. and Hennemuth, R. (1995) Modeling and analysis of recruitment distributions. *Ecological and Environmental Statistics* 2(4), 315–329.

Van Deusen, P.C. (1986) Fitting assumed distributions to horizontal point sample diameters. *Forest Science* 32, 146–148.

Warren, W. (1975) Statistical distributions in forestry and forest products research. In: Patil, G.P., Kotz, S. and Ord, J.K. (eds) *Statistical Distributions in Scientific Work*, Vol. 2. D. Reidel, Dordrecht, The Netherlands, pp. 369–384.

19 The SOP Model: the Parameter Estimation Alternatives

Ana Amaro[1]

Abstract

The SOP model is a growth and yield model for eucalyptus stands planted in different regions in Portugal. The model's functional form is based on a non-linear difference system of two equations (dominant height and basal area).

In 1997, the first time that SOP was parameterized, the ordinary least squares (OLS), two- and three-stage least squares criteria were the methods applied, using more than 3000 re-measurements for the endogenous and exogenous variables corresponding to 3 years of stand measurement observations. Due to the system characteristics, three-stage least squares was considered the best procedure for the estimation process.

Using more than 8000 re-measurements, gathered over 5 years, and a more intense regional reality observation, the SOP model was re-parameterized using the same set of estimation criteria but using a sequential steps methodology.

The OLS criteria application assumes, for the validity of the parameter hypothesis testing, a number of characteristics for the stochastic part of the model, namely zero mean, homoscedasticity, non-autocorrelation and desirable normality. Beyond these, of course, the functional form should reflect the reality, and the equation's right-hand-side (RHS) variables must be deterministic.

The SOP model parameterization process violates some assumptions. In this chapter a description of the methodology and an indirect analysis of the true meaning of the assumptions are presented.

Introduction

The growth and yield of forest stands are often modelled using empirical formulations based on the logistic growth behaviour of all living organisms and data that allow the validation of this biological assumption. In most cases, these data are re-measurements of the state of the stand, namely dominant height and basal area. Both of the measures will allow the estimation of the stand volume, with the latter being used as one of the relevant tools during the decision-support process.

[1] Department of Mathematics, Instituto Superior de Gestão, Portugal
Correspondence to: aamaro@isg.pt

The functional model used to simulate the growth of eucalyptus stands belonging to Soporcel, a pulp and paper company in Portugal, is a system of two related equations based on the Richards' and Lundqvist–Korf single-equation models (Amaro, 1997):

$$\begin{cases} dh_2 = (A_0 + A_1 I)^{1 - \frac{\ln(1 - e^{-(k_0 + k_1)t_2})}{\ln(1 - e^{-(k_0 + k_1)t_1})}} dh_1^{\frac{\ln(1 - e^{-(k_0 + k_1)t_2})}{\ln(1 - e^{-(k_0 + k_1)t_1})}} \\ ba_2 = (B_0 + B_1 I)(\frac{ba_1}{B_0 + B_1 I})^{(\frac{t_1}{t_2})^{n_{00} + n_{01} npl_1 + (n_{10} + n_{11} npl_1)I}} \end{cases} \quad (1)$$

with $B_0 = B_{00} + B_{01} dh_{t2=10}$ and $B_1 = B_{10} + B_{11} dh_{t2=10}$, and where dh_2 and dh_1 represent dominant height (m), ba_2 and ba_1 represent basal area (m^2/ha) at times t_2 and t_1 (years) and npl_1 represents the number of live trees at t_1.

The model was parameterized for two different regions of productivity (inner and coastal) using data from the forestry inventory from 1992 to 1994: I is a binary spatial function that assumes a value of 0 if the region in which the stand is located is in the inner part of Portugal and 1 if it is in the coastal part of Portugal, and npl_1 is the number of live trees per hectare when $t = t_1$. The coefficients A_0, A_1, B_{00}, B_{01} B_{10}, B_{11}, k_0, k_1, n_{00}, n_{01}, n_{10} and n_{11} were estimated (Amaro, 1997), an allometric relationship to generate volume (Amaro, 1997) was added (Tomé et al., 1995) and the model – SOP$_1$ – was validated and implemented in the company's decision-support system.

Although the mean residuals of the SOP$_1$ model are not significantly different from zero, the high residual variance observed, especially in the basal area submodel and in the inner region of Portugal, contributed to the need for re-parameterizing SOP for different subregions in order to better control the variance, possibly due to variations in environmental factors (Amaro, 1997).

To implement this project and increase SOP model accuracy, additional data were used and a different stand classification was considered. The applied methodology for SOP re-parameterization is described.

Data

The data refer to eucalyptus stands belonging to Soporcel and were gathered within the forest inventory from 1992 to 1996. Validation procedures, based on coherency criteria, were applied to the rough database, initially with around 17,000 observations. These correspond to all pairs of possible re-measurements for the same stand, as retrieved from the forestry inventory database.

Although in previous analysis, for the same modelling objective, some conclusions were drawn in order to select a structure based on the longest length interval (Amaro and Reed, 2001), this was done for a special difference equation (Lundqvist–Korf with the k parameter free) and no significantly different results were obtained for all possible data structure intervals. Due to the fact that another equation is present (based on Richards[2]) and to the fact that this last structure contains the full information on it and has proved to produce consistent estimates, this will be considered.

The validation criteria process released 8577 re-measurements of all possible intervals to be considered in the parameter-estimation process. Due to climatic characteristics (especially water availability and frost risk) the stands were classified into seven different regions with different productivity potentials (Amaro et al., 1994) and seven productivity regions considered to be relevant by Soporcel (Table 19.1).

[2] Information Management Unit (IMU) classification was on the basis of the modelling process for SOP's first version.

Methodology

To describe stand growth, the same two equations used previously were established. In order to estimate the equations' parameters, several steps were considered, with the objectives of identifying the initial estimates for the parameters, testing the need for regional reclassification and finally finding the best estimates. The number of regional submodels was defined during this process.

The non-linear model that describes the stand growth through dominant height (ALDO) is a difference equation, based on a Richards' function with the following formulation:

$$dh_2 = A_{dh}^{1-\frac{\ln(1-e^{-kt_2})}{\ln(1-e^{-kt_1})}} dh_1^{\frac{\ln(1-e^{-kt_2})}{\ln(1-e^{-kt_1})}} + \varepsilon_{dh} \tag{2}$$

where (t, year) refers to stand age and (dh, m) to dominant height. Indices 1 and 2 refer to two different time moments, and A_{dh} (the asymptote) and k (related to growth rate) are the parameters to be estimated.

To simulate basal area growth (ALBA model) the difference equation, derived from a Lundqvist–Korf modified function, is considered:

$$ba_2 = \left(A_{bal} + A_{ball}\, dh_{10}\right)\left(\frac{ba_1}{A_{bal} + A_{ball}\, dh_{10}}\right)^{\left(\frac{t_1}{t_2}\right)^{n_0+n_1 npl}} + \varepsilon_{ba} \tag{3}$$

where (ba, m^2/ha) is basal area, (dh_{10}, m) dominant height at 10 years and (npl, per ha) the number of live trees. A_{bal}, A_{ball} (combined represent the asymptote), n_0 and n_1 (combined represent a growth rate measure) are the parameters to be estimated. The error terms in each equation are represented by ε_k.

To estimate volume (V, m^3 ha^{-1}) the following allometric equation is used:

$$V = 0.3636\, ba^{0.9171}\, dh^{1.1025} \tag{4}$$

The system of equations is non-linear and, apparently, recursive as the term dh_{10} in the right-hand side of Equation 3 will be estimated with Equation 2. In this special case of systems, if the correlation between the endogenous variables and the errors of the other equation equals zero, the ordinary least squares (OLS) method could apply (Gujarati, 1995). However the difference equation nature of the system is responsible for the existence of the same variables in the left-hand side (LHS) and right-hand side (RHS) of every equation (although at different times).

Table 19.1. Number of re-measurements within each information management unit (IMU) and productivity region (PR), related to the region qualitative classification of water availability (water) and frost risk (frost); very high (+++), average (0), very low (---).

IMU	Water	Frost	Poor	Marginal	Low	Medium	Good	Very good	Excellent	Total
1	+++	0	0	20	0	19	74	164	4	281
2	++	-	0	0	0	0	28	219	31	278
3	+	++	52	1	12	115	105	1	0	286
4	0	---	4	0	0	2	55	1592	21	1674
5	-	+++	849	2045	125	146	5	0	0	3170
6	---	---	0	15	383	1824	0	4	0	2226
7	---	+	538	58	32	24	2	8	0	662
Total			1443	2139	552	2130	269	1988	56	8577

Not only may the cross-equation correlations differ from zero, as dominant height and basal area growth are correlated, but the endogenous variables are also exogenous variables (DH_2 is DH_1 in another moment and BA_2 is BA_1). Thus the system is not recursive and some additional problems arise: the RHS variables are stochastic and no preliminary information is available about the homoscedasticity of the error components and expected zero values.

Several authors have discussed similar problems (e.g. Murphy, 1983; Borders, 1989; Huang and Titus, 1999). The non-linearity will be responsible for the asymptotic behaviour of the estimators that may be consistent (but implying large data samples). To be consistent, several restrictions must be observed, namely the errors must be identically and independently distributed with zero mean and constant variance and the model specification must be correct.

SOP-specific non-linearity will not allow an analytical solution to the estimators that will be computed through a numerical method using, as convergence criteria, the least squares loss function minimization. The Levenberg–Marquardt (Bates and Watts, 1988) convergence algorithm was selected for its recognized efficiency: based on the Gauss–Newton algorithm, it does overcome the estimation impossibility in cases where the Hessian matrix is singular. On the other hand, this algorithm is very sensitive to local minima, explaining why the initial estimates were derived with one of these alternative algorithms: Quasi-Newton, Simplex or Rosenbrock (StatSoft, 2001). The convergence was determined by the difference between sequential parameter estimates to be less than 10^{-6}.

Due to the complexity of the econometric model (especially in respect of the regionalization definition objective) and the need to 'build knowledge' about the system's behaviour, several sequential methods were used with the support of two commercial applications: STATISTICA (StatSoft, 2001) and SPSS (SPSS, 1996):

- Non-linear ordinary least squares (NOLS) for ALDO initial parameterization considering the following structure for parameters A_{dh} and k:

$$A_{dh} = A_1 + \sum_{i=2}^{n} A_i RA_i \tag{5}$$

  where A_1 refers to A_{dh} for region 'Poor' (Table 19.1) and A_i the difference from the corresponding parameter for ALDO region RA_i (binary variable); n represents the number of regions and

$$k = k_1 + \sum_{i=2}^{n} k_i RA_i \tag{6}$$

  with the same correspondences to parameter k.
- Non-linear two-stage least squares (N2SLS) for submodel ALBA in which DH_{10} is estimated through ALDO submodel (Equation 2), and A_{baI}, A_{baII}, n_0 and n_1 with the same structure considered in Equation 6 for parameter k:

$$A_{baX} = A_{baX1} + \sum_{i=2}^{m} A_{baXi} RB_i \tag{7}$$

$$n_X = A_{X1} + \sum_{i=2}^{m} n_{Xi} RB_i \tag{8}$$

  X refers to the A_{ba} parameter (I or II) and the n parameter (0 or 1), RB_i is a binary variable and m is the number of regions.
- Seven initial parameterization regions will be considered, based on productivity regions classification (Table 19.1) establishing 7-term Equations (Equations 2–4). Whenever inconsistencies were detected, through parameter estimate

values or non-convergent solutions, a re-classification based on splits through IMUs or merges of different initial regions was performed. The final classification was based on a step-by-step analysis using the statistical and biological meanings of the parameter estimates criteria.

- After the number of regions was considered stable, and due to the non-recursive nature of the system and the correlation between errors of the different equations involved, to eliminate the simultaneous equation bias, non-linear three-stage least squares (N3SLS) was applied.
- To evaluate the $\psi + 1$ parameter model significance compared with a ψ parameter model, the likelihood ratio test was performed.
- Some usual evaluation indicators are used to assess the model estimation quality, comparing the two N2SLS and N3SLS methods (namely mean residuals and residual mean sum of squares). *F*-Snedecor homogeneity variance tests were also performed in order to assess the validity of the error variance homogeneity but in particular graphical analyses were done in order to detect serious violations of the assumptions.

Results and Discussion

The preliminary NOLS estimates for the ALDO and ALBA submodels are considered in Table 19.2 (computed individually for each region). For the ALBA submodel, the endogenous variable DH_{10} was substituted by the value suggested by the inventory data. Due to the non-significance of some estimates (possibly due to small samples or high intrinsic variation) and the lack of meaning of some significant estimates for the asymptote parameters, a step-by-step reorganization was done (using every piece of information and especially considering the IMU climatic characteristics), crossing over IMU by PR units, generating the SOP regions as considered in Table 19.3.

This was done with a simultaneous estimation performed using Equations 2–8 and the NOLS method. The dominant height growth pattern is simultaneously correlated with both parameters on the equation, which generates an objective classification of the new regions using a productivity criterion. The same is true for basal

Table 19.2. Initial estimates for the ALDO submodel within the seven IMUs and seven PRs (italic figures are non-significant at the 0.05 level).

IMU	1	2	3	4	5	6	7
A_{dh}	73.8629	*42.5668*	*59.9684*	*36.4943*	*36.5916*	35.6854	27.9614
k	0.025023	0.072635	*0.010764*	0.060505	0.055140	0.080311	0.091759
A_{bal}	56.40669	16.83116	214.5511	14.74440	24.54613	19.16068	23.59806
A_{ball}	−0.46084	1.00848	−6.5269	0.83015	−0.06501	0.65952	−0.19106
n_0	0.51684	0.30733	0.2758	0.65820	1.00851	0.76304	0.56774
n_1	0.00032	0.00075	0.0003	0.00028	0.00018	0.00020	0.00072

PR	Poor	Marginal	Low	Medium	Good	Very good	Excellent
A_{dh}	27.0519	35.0363	40.9820	42.4488	38.4873	32.8762	*60.2579*
k	0.075046	*0.062980*	*0.061284*	0.049435	*0.082352*	*0.086783*	*0.037605*
A_{bal}	61.66494	19.60075	20.47559	42.26702	23.06307	5.571351	*637.1737*
A_{ball}	−1.88888	0.18185	0.51062	−0.14242	0.34265	1.074174	*−17.1955*
n_0	0.64334	1.04146	0.62242	0.68841	0.37064	0.532283	0.5212
n_1	0.00038	0.00014	0.00035	0.00015	0.00081	0.000577	*−0.0000*

Table 19.3. NOLS estimates for ALDO and ALBA (PR, productivity region and IMU, information management unit).

Regional characterizations (PR and IMU based)	SOP region	A_{dh}	k	A_{bal}	A_{ball}	n_0	n_1
PR = Poor OR PR = Medium AND IMU ≠ 6	Poor	29.51789	0.06241	72.7638	−2.2746	0.6330	0.0003
Marginal PR = Medium AND IMU = 6	A	35.18502	0.06241	19.9833	0.1512	0.9910	0.0002
	B	35.18502	0.08157	19.9833	0.5743	0.8072	0.0002
PR = Good AND IMU = 3	C	29.51789	0.12376	19.9833	0.1512	0.6330	0.0007
Low	D	40.62639	0.06241	19.9833	0.5743	0.6330	0.0003
(PR = Very good OR PR = Excellent) AND (IMU ≠ 1 AND IMU ≠ 2)	E	29.51789	0.10485	19.9833	0.5743	0.6330	0.0003
(PR = Good AND IMU ≠ 3) OR ((PR = Very good OR PR = Excellent) AND (IMU = 1 OR IMU = 2))	Good	45.08593	0.06241	41.3486	0.1512	0.6330	0.0003

area. This is why a nominal classification was adopted. Exceptions were made for two regions that were, due to the base characteristics of the regions (supported by the IMU climatic classification) and to the asymptote estimates, readily classified as being Poor and Good productive quality (Table 19.3).

Using the NOLS estimates as initial values on the iterative estimation process, the N2SLS values were estimated (Table 19.4). The sample correlation between the residuals of the two equations, 0.41, is significant. In order to overcome the simultaneous bias that may occur due to the correlation, N3SLS was run (Table 19.4).

The asymptotic standard error estimates of both N2SLS and N3SLS parameter estimates are similar. The error variance estimate for the N2SLS model is 1.34977, being 1.34964 for the N3SLS model (not significantly different at $P < 0.05$). All the estimates were considered significant at the 0.05 level and the mean residual was not significantly different from zero in both cases. A graphical analysis was run is order to detect patterns on the mean residual sum of squares, crossing this information with the endogenous variable range, the dummy variables and age of stands. Although not statistically tested, no relevant patterns were observed.

In order to better interpret the parameter estimates, meanings and relationships between the stand classifications in the respective SOP regions, the average values for age stands, estimate of dominant height at 10 years of age and number of plants per hectare were computed (Table 19.5) and a spatial distribution of SOP regions for the bi-dimensional IMU classification criteria is considered (Fig. 19.1).

Dominant height analysis

The Poor region stands are located in reduced water availability and medium to very high frost-risk regions. The dominant height asymptote (29.73424 m) is the

Table 19.4. N2SLS and N3SLS estimates for ALDO and ALBA (PR, productivity region and IMU, information management unit).

Regional characterizations (PR and IMU based)	SOP region	A_{dh}	k	A_{bal}	A_{ball}	n_0	n_1
PR = Poor OR	Poor	29.51796	0.06241	79.7839	-3.2124	0.6395	0.0003
PR = Medium AND		(29.73424)	(0.06325)	(80.3384)	(-3.2262)	(0.6392)	(0.0003)
IMU ≠ 6							
Marginal	A	35.18511	0.06241	15.8431	0.3476	1.0256	0.0002
		(34.94904)	(0.06325)	(18.8100)	(0.3490)	(1.0257)	(0.0002)
PR = Medium AND	B	35.18511	0.08157	15.8431	0.7921	0.8053	0.0002
IMU ≠ 6		(34.94904)	(0.08316)	(18.8100)	(0.7932)	(0.8049)	(0.0002)
PR = Good AND	C	29.51796	0.12376	15.8431	0.3476	0.6395	0.0007
IMU ≠ 3		(29.73424)	(0.12178)	(18.8100)	(0.3490)	(0.6392)	(0.0007)
Low	D	40.62649	0.06241	15.8431	0.7921	0.6395	0.0003
		(40.40774)	(0.06325)	(18.8100)	(0.7932)	(0.6392)	(0.0003)
(PR = Very good OR	E	29.51796	0.10485	15.8431	0.7921	0.6395	0.0003
PR = Excellent) AND		(29.73424)	(0.10319)	(18.8100)	(0.7932)	(0.6392)	(0.0003)
(IMU ≠ 1 AND IMU ≠ 2)							
(PR = Good AND IMU ≠ 3)	Good	45.08606	0.06241	36.1187	0.3476	0.6395	0.0003
OR ((PR = Very good OR		(44.77771)	(0.06325)	(36.0894)	(0.3490)	(0.6392)	(0.0003)
PR = Excellent) AND							
(IMU = 1 OR IMU = 2))							

N2SLS estimates (N3SLS estimates).

lowest observed (Table 19.4), also assigned to regions C (regions with medium water availability but high frost risk) and E (very low frost risk and medium–high water availability). The differentiation of the three regions is achieved by the k growth rate parameter that assumes a higher value for regions C and E. Within the dominant height scenario, and due to the climatic characteristics of both regions, the k parameter is expected to be generally higher for region E: the main difference between C and E is the frost risk, which is almost non-existent for region E. It is also true that the stands in region C are generally younger than those in region E, and this may bias the k estimate.

Due to the observed estimates for regions C and E, it seems that frost risk does not restrict growth at the dominant height level. The highest k estimates are assigned to these two regions, showing that, although with low final potential (probably due to water availability level), the rate at which this potential is achieved is high.

When comparing regions D and Good, both with high dominant height asymptotes (although D is lower) and the same k values, the most relevant climatic difference between the two regions is water availability and not frost risk.

Although region D has higher potential ($A = 40.40774$ m) than regions C and E (29.73424 m), the growth rate is very low (0.06241) when compared with both growth rates for the other regions ($k_C = 0.12178$ and $k_E = 0.10319$). This may even mean that the variation assigned to growth in region D is so high that mathematically 'the best way to solve this problem' was to push the asymptote up and allow a very soft growth rate (in realistic time periods, e.g. 15 years, the potential is far from being attained!) (see Fig. 19.2).

The productivity classification is, then, dependent on both values of the asymptote and growth rate, although the dominant height asymptote seems to be correlated with water availability.

Table 19.5. Estimates for mean stand age (years), mean dominant height at 10 years (SOP estimate m) and mean number of trees (per ha).

Regional characterizations (PR and IMU based)	SOP region	(Water availability; frost risk)	t_1 (year)	dh_{10} (m)	Npl (per ha)
PR = Poor OR PR = Medium AND IMU ≠ 6	Poor	(- ; +++) (--- ; +)	5.3	29	950
Marginal	A	(- ; +++)	6.8	25	1050
PR = Medium AND IMU = 6	B	(--- ; ---)	7.6	35	1200
PR = Good AND IMU = 3	C	(+ ; ++)	4.4	26	950
Low	D	(- ; +++)	7.1	34	1100
(PR = Very good OR PR = Excellent) AND (IMU ≠ 1 AND IMU ≠ 2)	E	(0 ; ---)	7.1	36	1200
(PR = Good AND IMU ≠ 3) OR ((PR = Very good OR PR = Excellent) AND (IMU = 1 OR IMU = 2))	Good	(++ ; ---) (+++; 0)	5.3	44	1200

Note: qualitative classification of water availability (water) and frost risk (frost): very high (+++), average (0), to very low (---).

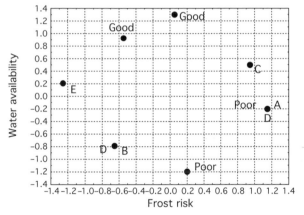

Fig. 19.1. SOP$_2$ spatial distribution using the bi-dimensional IMU classification criteria (water availability and frost risk).

Basal area analysis

Based on the comments on both regions C and E, looking at the asymptote for basal area it is clear that this parameter (18.8 + 0.79 DH$_{10}$ average = 47.2 m^2/ha) is higher for region E: the average estimate for region C is given by 18.8 + 0.35 DH$_{10}$ average = 27.9 m^2/ha.

Although with some variation, it may be observed that as frost risk becomes higher, the value for the potential basal area (asymptote) becomes lower. This variation is partly due to water availability, which, for example, may explain the difference between the basal area asymptotes in regions Good and B that are not very exposed to frost risk but have low water availability.

The stands located in region Good are clearly the most productive stands: both asymptotes, for dominant height and basal area, have the maximum values observed. The growth rate parameter for dominant height is the lowest, whereas for basal area, based on the average number of trees per hectare, the value is around average.

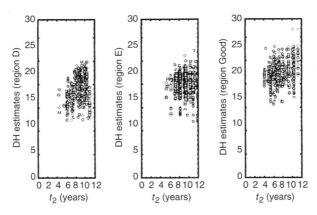

Fig. 19.2. Dominant height estimates (m) for regions D, E and Good.

Statistical analysis

All the more usual quality evaluation estimation indicators (e.g. mean residual and root mean residual sum of squares) were computed during all stages of the process, showing that no strong violations of the zero mean error and constant error variance assumptions were detected: the functional forms (tested previously against a relevant number of different functional forms (Amaro, 1997)) are appropriate to express the growth pattern of eucalyptus stands in Portuguese conditions. The large data set has helped to generate estimates that are asymptotically unbiased.

However, the estimation process was performed with potential statistical violations: the RHS component of both equations is not deterministic and is equal to the endogenous variable at different times. The past estimation procedures (Amaro, 1997), similar projects and their positive outcomes (e.g. Amaro and Reed, 2001), along with the effective parameter interpretability, help to validate the procedure and to attempt the hypothesis that the statistical violations are not at all relevant.

SOP$_1$ was re-estimated considering the new data set with N2SLS and N3SLS. The same irrelevance on the quality indicators was observed. The residual error variance estimate for SOP$_1$ compared with SOP$_2$ shows a clear reduction in uncertainty (Table 19.6).

The observed reduction is relevant although non-significant (likelihood test). Although the base functional form is exactly the same, the new structure of the dummy variables, especially the regions that are assigned to each of these categories, has a different spatial distribution. The aggregation of regions (from SOP$_2$ to SOP$_1$) is not direct (e.g. region 0 in SOP$_1$ corresponds to IMU 3, 5 and 7 that are spread out and mixed in the seven proposed new regions).

Conclusions

This work started from a eucalyptus stand growth model, SOP$_1$, that considers two spatial submodels. SOP$_1$ is being used by a pulp and paper company, Soporcel, within their decision-support process. Uncertainty has been detected within some projections and SOP$_2$ was designed in order to reduce this uncertainty.

Table 19.6. Residual mean sum of squares for both SOP$_1$ and SOP$_2$ ($n = 8577$), considering both N2SLS and N3SLS methods.

SOP model	Number of parameters	N2SLS	N3SLS
SOP$_1$ (Amaro, 1997)	12	1.42223	1.45898
SOP$_2$	20	1.34977	1.34964

A new data set was obtained, filtered and validated and an estimation procedure has been defined.

Due to the empirical nature of the SOP system of equations, the step-by-step approach, using different methods in order to estimate parameters, is probably more appropriate. Although it is much more time-consuming, it allows, as the analyses evolve, the addition of more knowledge to the system's behaviour and understanding.

The NOLS individual estimation allowed the initial estimate assessment and a rough idea of its magnitude. The use of less heavy algorithms (Quasi-Newton and Rosenbrock's), although possibly not as efficient as Levenberg–Marquardt's, speeded up the individual estimation process.

The several possibilities for stand aggregation were used as an argument for the simultaneous estimation with NOLS, using the regional dummy variables, built on the basis of the less productive region (Poor). This procedure allowed the aggregation of regions without any concern for the verification of statistical assumptions: seven different regions were selected as modelling units.

Due to the simultaneous nature of the equations, N2SLS was run and, further, due to the correlation between errors on both equations and in order to prevent bias and subsequently run significance tests, N3SLS was also run.

The estimates obtained during the three main steps on the seven final regions do not differ in a relevant way, although the final estimates obtained through N3SLS are most probably the best estimates, and in particular produce the most efficient standard deviations parameter estimates that permit the significance tests.

The residual error variance estimate for SOP$_1$, actually being used by Soporcel, was reduced when compared with SOP$_2$.

The statistical violations, namely the non-deterministic characteristic of both the dominant height and basal area variables on the RHS of the correspondent equations, were considered non-relevant based on past analysis, parameter meaningfulness and impact of violation.

The non-significance of the parameter tests would lead to a region merge or separation task that by no means invalidates the actual values: the process is unfinished and more work needs to be done in order to not only validate the SOP$_2$ model (which Soporcel will eventually do and will identify the real negative points of the model) but also to continuously reduce the observed uncertainty (e.g. by modifying the system's functional forms – the most accurate way of targeting the statistical assumptions – or by creating additional submodels). All of this must be assisted by an accurate inventory database.

Finally and re-stating what has already been said (Amaro, 1997), water availability and frost risk are both limiting factors to eucalyptus growth patterns! Water seems to help to establish the potential and frost risk more as a rate-to-potential regulator. In basal areas, however, the potential seems to be more influenced by frost risk.

Acknowledgements

This study was supported by project Sapiens PCTI/1999/MGS/36578 and data provided by Soporcel.

References

Amaro, A. (1997) Modelação do crescimento de povoamentos de *Eucalyptus globulus* Labill de 1ª rotação em Portugal. PhD dissertation, IST, UTL, 241 pp.

Amaro, A. and Reed, D. (2001) Forest re-measurements and modelling strategies. In: *Proceedings of the Conference on Forest Biometry, Modelling and Information Science*. University of Greenwich, 25–29 June. Available at: cms1.gre.ac.uk/conferences/iufro/proceedings/AmaroReed1.pdf

Amaro, A., Themido, I. and Tomé, M. (1994) A definição de unidades de gestão de informação num sistema de apoio à decisão para a gestão florestal. In: *III Congresso Florestal Nacional*. Actas 1, Figueira da Foz, pp. 74–82.

Bates, D.M. and Watts, D.G. (1988) *Nonlinear Regression Analysis and its Applications*. John Wiley, New York, 365 pp.

Borders, B.E. (1989) System of equations in forest stand modelling. *Forest Science* 35, 548–556.

Gujarati, D.N. (1995) *Basic Econometrics*. McGraw-Hill International, Economic Series, 838 pp.

Huang, S. and Titus, S. (1999) Estimating a system of nonlinear simultaneous individual tree models for white spruce in boreal mixed-species stands. *Canadian Journal of Forestry Research* 29, 1805–1811.

Murphy, P.A. (1983) A nonlinear timber yield equation system for loblolly pine. *Forest Science* 29, 582–591.

SPSS (1996) *SPSS Base System Syntax Reference Guide*. Release 7.5.1.

StatSoft (2001) *STATISTICA (Data Analysis Software System)*, version 6. Available at: www.statsoft.com

Tomé, M., Falcão, A., Carvalho, A. and Amaro, A. (1995) A global growth model for eucalypt plantations in Portugal. *Lesnictví-Forestry* 41(4), 197–205.

20 Evaluating Estimation Methods for Logistic Regression in Modelling Individual-tree Mortality

Virpi Alenius,[1] Hannu Hökkä,[1] Hannu Salminen[1] and Sylvain Jutras[2]

Abstract

In this study we compared individual-tree mortality models for peatland Scots pine (*Pinus sylvestris* L.) in Finland constructed using different estimation methods. We applied standard logistic regression with the maximum likelihood (ML) method by ignoring the data structure, and alternatively accounted for the data hierarchy using generalized linear mixed models with either the marginal quasi-likelihood (MQL) or penalized quasi-likelihood (PQL) estimation method. We evaluated the models on the basis of traditional logistic regression goodness-of-fit criteria including the χ^2 test, sensitivity, specificity, rate of correct classification, bias and the receiver operating characteristic (ROC) curves with subsequent R^2. The interpretation of the fit measures appeared to be complicated. The ML and MQL methods resulted in models with high sensitivity, a high rate of correct classification and low bias. Despite the good fit measures, the Hosmer–Lemeshow test suggested rejection of the models. The graphical expression of the models' ROC curves did not give additional information to make a selection between any of the models, but the R^2 showed that the models obtained with the ML and MQL methods were slightly better than that obtained with the PQL method.

Introduction

When modelling the development and survival of forest stands, the status of an individual tree at a certain time is described by two possible values; the tree is either alive or dead, i.e. the response is typically binary. Logistic regression has generally been used as the method to describe the relationship between a binary outcome and continuous explanatory variables. Consequently, logistic regression has been applied in modelling the probability of individual trees dying along the forest succession. During recent years, other methods such as binary classification tree (CART) (Dobbertin and Biging, 1998) and neural computing (Guan and Gertner, 1991; Hasenauer *et al.*, 2001) have also been successfully introduced into forestry modelling, but are not considered in this chapter.

[1] Finnish Forest Research Institute, Rovaniemi Research Station, Finland
Correspondence to: Virpi.Alenius@metla.fi
[2] Département des Sciences du Bois et de la Forêt, Université Laval, Canada

When modelling and classifying mortality it is necessary to determine to what extent causality and regularities are taken into account and what phenomena are regarded as stochastic. Subsequently, mortality can be classified as regular (non-catastrophic) or irregular (catastrophic) (Vanclay, 1995). Regular mortality reflects lowering vigour and decreasing growth rate, which may originate from ageing or lack of resources and increasing competition and, as such, is at least partially predictable (Murty and McMurtrie, 2000). Irregular mortality is typically caused by storms, forest fires, epidemics of pathogenes or insect population explosions, where the survival probability is unpredictable at tree level but not necessarily at landscape level. Several logistic mortality models accounting for regular tree-level mortality mainly due to competition have been developed (e.g. Monserud, 1976; Hamilton, 1986; Avila and Burkhart, 1992; Monserud and Sterba, 1999; Shen *et al.*, 2000; Eid and Tuhus, 2001; Yao *et al.*, 2001). Such models are mainly applied to control the predicted stocking level in stand simulators (e.g. Crookston, 1990; Botkin, 1993; Van Dyck, 2001; Hynynen *et al.*, 2002).

It is common for individual tree mortality data to include several stands with more than one tree in each stand. As a consequence, the observations (trees in the same stand) are always correlated. This fact has generally not been accounted for in modelling individual tree mortality, although it calls for different methods of model construction, and consequently leads to different procedures in parameter estimation. Ignoring data structure in two- or more level mortality data leads to, for example, too small standard errors for higher-level fixed effects which are measured at plot or stand levels, a problem similar to any multivariate normal regression of multilevel data. Therefore, when logistic regression is applied to multilevel data, methods of analysis and measures of model fit which are appropriate especially for these situations should be seriously considered. In this study, we compared logistic mortality models for peatland Scots pine (*Pinus sylvestris* L.), which were developed both with and without random stand effect and by alternative estimation methods using two-level data. We discussed the methods of parameter estimation and the use of different goodness-of-fit measures in the case of correlated data.

State of the Art

Traditional logistic regression

In logistic regression (Hosmer and Lemeshow, 2001), the dependent variable may be binary or expressed in proportions. With the logit transformation, the linear model is as follows

$$g(x) = \ln\left[\frac{\pi(x)}{1-\pi(x)}\right] = \alpha + \beta x \tag{1}$$

where $\pi(x) = \dfrac{e^{\alpha+\beta x}}{1+e^{\alpha+\beta x}}$, α is an intercept and β is a vector of unknown parameters. The distribution of the errors is supposed to be binomial with zero mean and variance $\pi(x)[1-\pi(x)]$. The error term has two possible values: if $E(y) = 1$ then $\varepsilon = 1-\pi(x)$ with probability $\pi(x)$, and if $E(y) = 0$ then $\varepsilon = -\pi(x)$ with probability $1-\pi(x)$. The conditional distribution of y follows the binomial distribution with probability given by the conditional mean, $\pi(x)$.

The maximum likelihood (ML) method is used to estimate the parameters of the logistic regression model, because the unknown parameters are non-linearly related to $\pi(x)$. The model's statistical evaluation is done on the basis of several criteria developed to guide in choosing the best model from different candidates. Assessing the fit is usually done with the Pearson χ^2 statistic and the deviance. The Hosmer–Lemeshow test looks for the correctness of the model in grouping probabilities in risk deciles. The classification table describes the model's goodness of fit with the correct classification rate. The receiver operating characteristic (ROC) curve (Hosmer and Lemeshow, 2001) is based on sensitivity and specificity measures. A high ROC curve describes the good prediction ability of the model. The area under the curve is interpreted as the model's R^2.

Multilevel logistic regression with random effect

Multilevel logistic regression models for hierarchically structured data have become more common in recent studies of social sciences and medicine (e.g. Hox, 1995; Goldstein and Rasbash, 1996; Carlin *et al.*, 1999; Fielding, 2000). Forestry applications are few (Jalkanen, 2001), and published tree mortality models, for example, have been constructed from multilevel data, but in parameter estimation the data structure has been ignored.

Multilevel binary data can be analysed using a logistic regression model which is composed of fixed covariates from different hierarchy levels and random effects (Hosmer and Lemeshow, 2001; McCulloch and Searle, 2001). The random effects not related to the lowest level of hierarchy are generally assumed to follow normal distribution with mean zero and constant variance. Here the random stand effect describes the unmeasured variance between the stands in case of mortality. The error variance (tree-level variance) is assumed to be:

$$\phi^2 \pi(x)(1 - \pi(x)) \tag{2}$$

where ϕ is the dispersion scale (or extra-binomial variance). If $\phi = 1$, then the error variance is purely binomial. If $\phi > 1$, the situation is called over-dispersion. Unlike under-dispersion ($\phi < 1$), over-dispersion is fairly common, and may occur when data are collected from experiments with a hierarchical design (McCullagh and Nelder, 1989). If the dispersion scale is forced to be equal to 1, the potential over-dispersed variance is added to the random effect variance.

The ML and restricted maximum likelihood (REML) methods are relevant when analysing linear multilevel models with normal distribution, but logistic regression models with random effects need special quasi-likelihood estimation methods. The key point is to estimate the effect of the covariates on the binary outcome with awareness of the random effect(s). The idea of quasi-likelihood methods is that only the model for the mean (response and covariates) and the relationship between the mean and the variance is necessary for the estimation. Two quasi-likelihood methods have been developed for estimation: the marginal quasi-likelihood (MQL) and the penalized quasi-likelihood (PQL) methods. Today these methods are available in most statistical software packages. However, Rodriguez and Goldman (1995) have pointed out that in certain circumstances the first-order MQL estimates are badly biased, sometimes more biased than those obtained by ML without random effects. The PQL method has also been noticed to have a major problem in practice if the cluster size is small (McCulloch and Searle, 2001). Alternatives to the ML approach are, for instance, GEE, generalized estimating equations (Zeger *et al.*, 1988) or Bayesian methods (MCMC, bootstrapping) (e.g. Goldstein, 1995; Browne and Draper, 2002), which will not be addressed in this chapter.

There is a basic problem with the goodness-of-fit measures in the case of multi-level logistic regression; all tests based on likelihood are approximations, because the likelihood is also an approximation. According to Hosmer and Lemeshow (2001), many of the goodness-of-fit statistics have not been evaluated for hierarchical data, but they do recommend some of these tests instead of dismissing any model checking.

Methodology

Multilevel model

In this study, the response variable is binary, i.e. the observed responses y_{ij} of the tree status may have two values: 0 = live or 1 = dead. With the standard assumptions of the dependent variable, the distribution for the response is $y_{ij} \sim Bin(1, \pi_{ij})$ with binomial variance $\pi_{ij}(1 - \pi_{ij})$, where π_{ij} is the probability of the jth tree to die in stand i.

A two-level model with a random effect can be written in the general form:

$$g(X_{ij}, \beta_{ij}, \mu_i) = logit(\pi_{ij}) = \beta_0 + X_{ij}\beta_{ij} + u_i \tag{3}$$

where π_{ij} is as above, term $X_{ij}\beta_{ij}$ is the component of the linear predictor, which has fixed coefficients (i.e. tree diameter at breast height, basal area of larger trees divided by stand basal area, stand basal area, proportion of birch of total basal area, quadratic mean diameter), and u_i represents random departure from the average probability for a tree to die in the ith stand at logit scale. As stated earlier, the random variable u is normally distributed with zero mean and variance σ_u^2. The residual error variance is a function of π_{ij} as in Equation 2.

Estimation

The log-likelihood function for the logistic regression with random effect is defined as follows:

$$L(\beta) = \sum_{i=1}^{m} \int_{-\infty}^{\infty} \left[\prod_{j}^{n} \frac{e^{y_{ij} \times (\beta_0 + x_y\beta + u_i)}}{1 + e^{\beta_0 + x_y\beta + u_i}} \right] \frac{1}{\sqrt{2\pi}} \frac{1}{\sigma_u} \exp\left(-\frac{u_i^2}{2\sigma_u^2} \right) du_i \tag{4}$$

For multilevel binary models, the quasi-likelihood method is attractive in parameter estimation due to its ability to generate highly efficient estimators (McCulloch and Searle, 2001). The MQL method utilizes the *score functions* of log-likelihood as the model for the mean and the model for the relationship between mean and variance. This is not a very reasonable way to estimate the variance–covariance matrix, because in the hierarchical collected data there is a correlation between observations. In MQL this relationship between mean and variance is based on an assumption of equal variance (simple variance–covariance structure), which is not always reasonable. Goldstein and Rasbash (1996) showed that improved estimation is obtained by applying the second-order PQL method, in which the second-order Taylor expansion for the random part is calculated based on current residuals u_i with variance–covariance matrix D. The corresponding equations for β are:

$$MQL = \frac{\partial}{\partial \beta} \sum Q_{ij} = 0 \tag{5}$$

and

$$PQL = \frac{\partial}{\partial \beta} \sum Q_{ij} - \frac{1}{2}u'D^{-1}u = 0 \qquad (6)$$

where Q_{ij} is constructed using only information about how the variance changes as a function of the mean. $Q_{ij} = \int_{y_{ij}}^{u_{ij}} \frac{y_{ij} - t}{\tau^2 \vartheta(t)} dt$, where τ is a constant of proportionality relating var(y_i) to the linear predictor variance $v(\mu_i)$.

To demonstrate the differences that result from ignoring the data structure in the analysis, we first estimated the standard logistic regression model with maximum likelihood for 5-year mortality. In the multilevel analysis, first-order MQL and second-order PQL methods (Goldstein and Rasbash, 1996) were used. The tree-level error variance was also tested for over- and under-dispersion (Goldstein, 1995). All parameters of the multilevel logistic models were estimated simultaneously. We used SAS (SAS Institute, 1994) and MLwiN (Goldstein *et al.*, 1998) software to estimate the models.

Goodness of fit

In traditional logistic regression, the difference between observed and fitted values is measured with Pearson residuals:

$$r\left(y_j, \hat{\pi}_j\right) = \frac{\left(y_j - m_j \hat{\pi}_j\right)}{\sqrt{m_j \hat{\pi}_j \left(1 - \hat{\pi}_j\right)}} \qquad (7)$$

where y_j denotes the number of dead trees among m_j subjects with x_j covariates. The fitted values $\hat{y}_j = m_j \hat{\pi}_j$ in logistic regression are calculated for every independent variable and are called the covariate patterns m_j $j = 1, 2, ..., J$. The Pearson χ^2 test is based on adjusted residuals.

$$X^2 = \sum_{j=1}^{J} r\left(y_j, \hat{\pi}_j\right)^2 \qquad (8)$$

χ^2 – under the assumption that the fitted model is correct – follows χ^2 distribution with degrees of freedom equal to $J-(p+1)$, where p is the number of parameters. It should be noticed that in χ^2, y_j is considered as a relative frequency and m_j is assumed to be large. One of the goodness-of-fit measures is deviance, which is based on so-called adjusted deviance residuals. The difference between the likelihood of saturated and estimated models, the likelihood ratio test, is equal to deviance and also follows χ^2 distribution. However, it is only an approximation.

The Hosmer–Lemeshow test groups the estimated probabilities from smallest to largest into groups (g), which are called the 'deciles of risk'. The Hosmer–Lemeshow goodness-of-fit $\hat{C}$ is based on the Pearson χ^2 test from the $g \times 2$ table of observed and estimated expected frequencies:

$$\hat{C} = \sum_{k=1}^{g} \frac{\left(O_k - n'_k \bar{\pi}_k\right)^2}{n'_k \bar{\pi}_k \left(1 - \bar{\pi}_k\right)} \qquad (9)$$

where n'_k is the total number of subjects in the kth group, c_k denotes the number of covariate patterns in the kth group, $o_k = \sum_{j=1}^{c_k} y_j$ is the number of responses among c_k

covariate patterns, and $\bar{\pi}_k = \sum_{j=1}^{C_k} \frac{m_j \hat{\pi}_j}{n'_k}$ is the average estimated probability. $\hat{C}$ is approximated by the χ^2 distribution with $g - 2$ degrees of freedom. Evans (1998, cited in Hosmer and Lemeshow, 2001) has shown that with some restrictions the test can be applied to correlated data.

The *classification table* is an informative way to describe the results. The model predicts the continuous probabilities; to convert these back to dichotomous results a cutpoint (c) must be defined. If the estimated probability is less than *c*, the outcome value is 0; otherwise it is 1. One way to determine the cutpoint is to choose it in such a way that the *rate of correct classification* is maximized. The rate of correct classification can be determined as the sum of the diagonal elements of the classification table divided by the total number of observations.

Sensitivity and *specificity* can be obtained from the classification table. These measures describe how correct the model is in classifying dead and living trees. Sensitivity is the frequency of observed '1' which the model predicted as '1' divided by all observed '1'. Specificity is the frequency of observed '0', which the model has predicted as '0', divided by all observed '0'. Evaluating the cutpoint in such a way that sensitivity and specificity are as high as possible, the estimated probabilities in risk deciles frequency should have large frequencies close to 0 and 1. Sensitivity and specificity form the basis of describing the model's *ROC curve*, which is drawn as sensitivity against (1−specificity) with all possible cutpoints. The height of the curve describes the model's ability to predict the phenomenon. The area under the curve represents the model's R^2.

We calculated χ^2, the classification table, specificity, sensitivity, bias, the Hosmer–Lemeshow test, ROC curves and R^2 as the basis of evaluation and discussed their use for this purpose in multilevel data. The bias was calculated in two ways: (i) tree-level bias as the average difference of tree-level mortality observations and predictions, and (ii) stand-level bias as an average difference of stand-level observed and predicted mortalities.

Data

We used the permanent peatland growth plots (SINKA) as the modelling data. The SINKA stand is composed of a cluster of three circular plots located 40 m apart. In order to avoid too-laborious measurements in dense sapling stands, the radii of the circular sample plots were adjusted according to the stand density in such a way that the whole SINKA cluster contained approximately 100 tally trees. For those, the minimum diameter at breast height (DBH) was 4.5 cm if the stand was past pole age, and 2.5 cm otherwise. The first re-measurement of the SINKA data was done in 1988–1994 following a period of five growing seasons on each plot. For a more detailed description of the data, see, for example, Hökkä *et al.* (1997). From the structure of the data, it is clear that trees within stands are correlated.

For the purposes of this study, both pure Scots pine and mixed pine–birch stands (an admixture of *Betula pubescens* Ehrh.) were included in the data if at least one pine was present in the cluster. The data were restricted to stands where no management operations (thinning, ditching) had been carried out during the last 5-year period preceding the first measurement or between the measurement occasions. Altogether, the numbers of stands and trees were 295 and 17,293, respectively. The mortality proportion for pine was 2.73%. The average tree DBH for pine was 9.1 cm (range 2.5–44.5 cm). The proportion of large trees in the data was low; there were only 27 pines with DBH larger than 30 cm.

Results and Discussion

Model assumptions and estimation methods

In the basic logistic regression approach without random variable (ML), the assumption of independent residuals was violated. This can be seen in the lower values of the standard errors of stand-level variables compared with those of the multilevel models (Table 20.1). The parameter estimates were not far from those of the multilevel models, the greatest difference being in the coefficients of DBH^{-1} and DW (18% and 16% underestimates compared with the MQL estimates, but the difference was non-significant when the standard errors were considered). In the MQL estimation, the tree-level variable DBH^{-1} influenced mortality more when compared with the basic logistic regression, while the opposite was true with the stand-level variables. The variance component related to the between-stand variation in mortality was clearly significant in all multilevel models. The advantage of the model with random effect (MQL and PQL) is that the model can be calibrated to a specific stand, if mortality from the preceding 5-year period has been observed (see McCulloch and Searle, 2001). For a more detailed description of the construction and interpretation of the models, see Jutras *et al.* (2003).

For the multilevel models, the MQL and PQL estimation methods also showed somewhat different results (Table 20.1). There was no problem of convergence with PQL, which has been the case in many other studies. In our data, PQL estimates in the fixed part were always larger than MQL estimates, but the difference was statistically significant only for *G*, stand basal area (m²/ha). The opposite was true for the random stand effect. This suggested that with MQL, less variation was accounted for by the fixed part. With MQL, minor overdispersed variation was detected. When the scale effect was fixed at 1 in MQL, the overdispersed variance was transformed to the between-stand variance. With PQL, the assumption of binomial variance appeared to be valid in our data.

Table 20.1. Models for peatland Scots pine mortality estimated by different methods. Standard errors are given in parentheses.

Variables[a]	ML, without random effect	1st order MQL	1st order MQL, with extra dispersion	2nd order PQL	2nd order PQL, with extra dispersion
Constant	−5.719 (0.307)	−5.855 (0.465)	−4.515 (0.227)	−6.808 (0.515)	−6.808 (0.515)
DBH^{-1}	30.884 (10.553)	37.64 (12.78)	37.155 (13.356)	42.1 (14.731)	42.098 (14.736)
BALRAT	2.091 (0.319)	2.020 (0.355)	2.027 (0.373)	2.137 (0.403)	2.137 (0.403)
G	0.111 (0.010)	0.107 (0.022)	0.108 (0.021)	0.132 (0.023)	0.132 (0.023)
G_B	2.133 (0.231)	2.007 (0.435)	2.011 (0.434)	2.170 (0.474)	2.170 (0.474)
DW	−0.128 (0.024)	−0.110 (0.043)	−0.112 (0.043)	−0.125 (0.046)	−0.125 (0.046)
σ_u^2	–	1.985 (0.231)	1.842 (0.197)	1.579 (0.251)	1.579 (0.251)
Dispersion scale ϕ	1.000 (0.000)	1.000 (0.000)	1.112 (0.012)	1.000 (0.000)	1.001 (0.011)

[a]DBH^{-1}, tree diameter at breast height (mm); BALRAT, basal area of larger trees/stand basal area; *G*, stand basal area, (m²/ha); G_B, proportion of birch of total basal area (%); and DW, quadratic mean diameter (cm).

Browne and Draper (2002) pointed out that in many cases both MQL and PQL methods produce clearly biased parameter estimates, PQL being generally more recommendable than MQL. Differences in estimates became evident in situations where the number of observations within clusters was small (see McCulloch and Searle, 2001). According to Goldstein and Rasbash (1996), the low number of observations at the lowest level and especially the ratio of the level-1 units to level-2 units may be the main source of bias in estimates. It is clear that with few observations the estimation of covariance parameters is unstable. This is also the case with PQL, as shown by Browne and Draper (2002). In our data, the ratio (average number of trees in stand : number of stands) was 59 : 295, and was probably sufficient, since all the methods appeared to give parameter estimates that were close to each other. Figure 20.1 illustrates the effect of cluster size on estimates with the help of the difference of stand-wise-calculated average mortality and corresponding predicted probability. Figure 20.1 shows that when the number of trees per stand is less than 30, there is large variability in residuals. When there are fewer than 10 trees per cluster, the residuals may be anything between 0.0 and ±0.8. In a way this is trivial in data where the average probability is low, because for clusters with a low number of observations, one tree is representing high relative proportion in such clusters. The model, however, predicts low probabilities in these stands, too.

Another factor that has been assumed to influence the first-order MQL estimates is the magnitude of the higher-level variance. This was not a problem with our data either, despite the fact that the two-level variance was relatively high (1.6–2.0) compared with the value of 0.5 which Rodriguez and Goldman (1995) and Goldstein and Rasbash (1996) gave as the limit for large variance.

In cases where PQL and MQL appear to give biased results, both Goldstein (1999) and Browne and Draper (2002) suggest the use of Bayesian methods. However, these are computationally much more demanding (Browne and Draper, 2002).

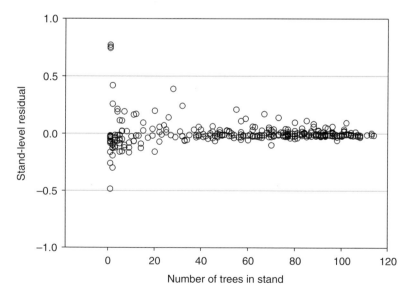

Fig. 20.1. Stand-level mortality residuals (average observed mortality in the stand – average predicted (MQL method) probability) as a function of the number of trees in the stand.

Fitting measures

The logistic regression goodness-of-fit measures are generally based on Pearson residuals and deviance. They can be used only with the assumption of the independence of observations. In hierarchical data, the residuals are correlated in clusters, i.e. stands, so these fitting measures are biased and may be used only for speculation. Table 20.2 shows a lot of instability in the Pearson χ^2 test among different models. As discussed by Hosmer and Lemeshow (2001), Evans (1998) has improved the Pearson χ^2 test to make it applicable to correlated data, but we are not able to consider it in more detail here.

The classification table can be used as an overall measure of goodness-of-fit also in cases of correlated data, since it is not affected by the data structure. The result of the ordinary Hosmer–Lemeshow test is not reliable because it is assuming independence among the observations. According to the results, these models did not fit into the data.

Our multilevel mortality models appeared to be more specific than sensitive. Sensitivity was highest with ML, followed by MQL, and lowest with the PQL method. Mortality models need to be sensitive, i.e. more accurate, to find dead trees than living ones. With all models there were always some living trees predicted as dead. With PQL, the rate of correct classification was highest (73%), but it predicted

Table 20.2. Goodness-of-fit measures calculated on the basis of the fixed part of the models.

Measure of fit	Models		
	ML	1st order MQL	2nd order PQL
Pearson X^2	15149.87	19079.26	17306.53
Hosmer–Lemeshow	44.7	45.1	233.9
P value χ^2_8	(<0.0001)	(<0.0001)	(<0.0001)
Sensitivity (%)	73.7	64.2	60.2
Specificity (%)	56.2	69.1	73.5
Rate of correct classification	56.7%	68.5%	73.1%
Area under the ROC curve (R^2)	0.7434	0.7462	0.7291
Tree-level bias	+0.00031	−0.00104	+0.01105
Relative (%)	+1.1	-3.8	+40.5
Stand-level bias	+0.001572	+0.001392	+0.01901
Relative (%)	+3.0	+2.8	+36.5

Tree-level bias :

$$bias_j = \frac{\sum \left(y_{ij} - \hat{\pi}_{ij} \right)}{N}$$

where y_{ij} = tree j status, 0 or 1 in stand i; π_{ij} = predicted probability that tree j in stand i will die; and N = total number of tree observations.
Stand-level bias :

$$bias_j = \frac{\sum \left(Y_i - \hat{Y}_i \right)}{S}$$

where Y_i = observed mortality in stand i, $\dfrac{\sum y_{ij}}{n_i}$; $\hat{Y}_i$ = predicted mortality in stand i, $\dfrac{\sum \hat{\pi}_{ij}}{n_i}$; S = total number of stands in data; and n_i = total number of trees in stand.

living trees more accurately than dead ones, resulting in clearly lower sensitivity than ML and MQL. With MQL, the rate of correct classification was 68%, but it predicted dead trees more accurately than the PQL method.

The ROC curve can also be used to select the best model in correlated data, because it is based on sensitivity and specificity. Visually, the curves differed upwards from the 45° straight line, indicating that the models had rather good ability to predict tree mortality, but there was virtually no difference among curves (Fig. 20.2). However, the area under the curve (R^2) was highest for the MQL method and lowest for the PQL method (Table 20.2). This suggested that the model obtained by the MQL method had the highest discrimination ability. Stand-wise-calculated ROC curves illustrated the model's (MQL) ability to predict dead trees in specific stands in the data (Fig. 20.2).

The logistic model (ML) without random effect gave the lowest tree-level bias, which is to be expected because it is a population average model and should work well marginally. The stand-level bias was lowest for the MQL method, which is a cluster-specific model accounting for the hierarchical data structure and should work well at cluster (stand) level.

In general, to be chosen as the best model, all measures need to be good. The MQL and ML methods appeared to result in models with good fit measures, acceptable bias and high R^2. Because the MQL method is theoretically correct, the corresponding model should be chosen. The reasons for the poor performance of the PQL model remain unclear, but the result is not necessarily unusual (see discussion in McCulloch and Searle, 2001). Here it may be partly due to the strongly skewed data, with low overall mortality.

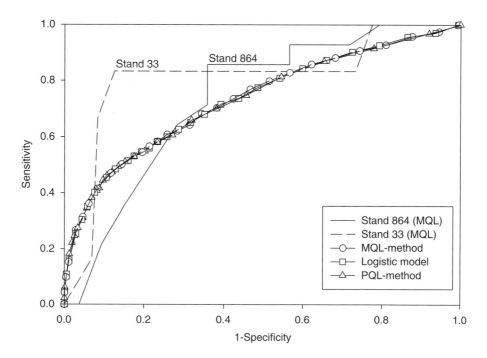

Fig. 20.2. The average ROC curves of different models. Additional curves are drawn for two example stands, 33 and 864 (on the basis of the MQL method).

Conclusions

Our results suggest that with relatively balanced data the marginal quasi-likelihood (MQL) method produced consistent model estimates for multilevel binary mortality models. However, the model's goodness-of-fit is problematic to measure in correlated binary data. The classification tables, specificity, sensitivity, correct classification rate, bias and R^2 may be used to evaluate the model's performance. Applying the traditional Hosmer–Lemeshow statistics cannot be recommended. Here the test suggested rejection of all models, although the models produced by the ML and MQL methods behaved well in terms of sensitivity, the correct classification rate, and bias. Compared with the other measures, which actually measure the marginal performance of the models, the ROC curve analyses the predicted probability distribution, and thus is better at describing the variability in the predictions. The graphical expression did not give additional information to make a selection among the models, but the R^2 showed that the models obtained by the ML and MQL methods were slightly better than that obtained by the PQL method.

When the binary multilevel models are applied, the continuous outcome predicted by the fixed part is back-transformed into binary values. If the overall cutpoint is applied, the data structure will not be utilized in selecting '1's and '0's. However, in stand simulators the model can be used in a deterministic way (Hynynen *et al.*, 2002). The stem frequencies of each DBH class will be decreased by the amount of the predicted continuous mortality probability, i.e. it is not necessary to specify a cutpoint and determine which individual will die.

Acknowledgement

We wish to thank Dr Juha Lappi for his valuable comments.

References

Avila, O.B. and Burkhart, H.E. (1992) Modeling survival of loblolly pine trees in thinned and unthinned plantations. *Canadian Journal of Forest Research* 22, 1878–1882.

Botkin, D.B. (1993) *Forest Dynamics: an Ecological Model*. Oxford University Press, Oxford, 309 pp.

Browne, W.J. and Draper, D. (2002) A comparison of Bayesian and likelihood methods for fitting multilevel models. *Journal of the Royal Statistical Society*, submitted.

Carlin, J.B., Wolfe, R., Coffey, C. and Patton, G.C. (1999) Analysis of binary outcomes in longitudinal studies using weighted estimating equations and discrete time survival methods: prevalence and incidence of smoking in an adolescent cohort. *Statistics in Medicine* 18, 2655–2679.

Crookston, N.L. (1990) *User's Guide to the Event Monitor: Part of Prognosis Model Version 6*. USDA Forest Service, Intermountain Research Station, Odgen, Utah, General Technical Report INT-275, 21 pp.

Dobbertin, M. and Biging, G.S. (1998) Using the non-parametric classifier CART to model forest tree mortality. *Forest Science* 44, 507–516.

Eid, T. and Tuhus, E. (2001) Models for individual tree mortality in Norway. *Forest Ecology and Management* 154, 69–84.

Evans, S.R. (1998) Goodness-of-fit in two models for clustered binary data. PhD thesis, University of Massachusetts at Amherst, Amherst, Massachusetts, 145 pp.

Fielding, A. (ed.) (2000) *Generalized Linear Mixed Models in Multilevel and Other Complex Data Structures in the Social Sciences*. Social Science Methodology in the New Millennium.

Proceedings of the Fifth International Conference on Logic and Methodology, Cologne, Germany, 3–6 October. TT-Publikaties (on CD-ROM).

Goldstein, H. (1995) *Multilevel Statistical Models*. Kendall's Library of Statistics, Arnold Statistics, London, 178 pp.

Goldstein, H. (1999) *Multilevel Statistical Models*. Kendall's Library of Statistics 3, Arnold, London, 140 pp.

Goldstein, H. and Rasbash, J. (1996) Improved approximation for multilevel models with binary responses. *Journal of the Royal Statistical Society* A 159, 505–514.

Goldstein, H., Rasbash, J., Plewis, I., Draper, D., Browne, W., Yang, M., Woodhouse, G. and Healy, M. (1998) *A User's Guide to MLwiN*. Institute of Education, University of London, 140 pp.

Guan, B.T. and Gertner, G. (1991) Modeling red pine tree survival with an artificial neural network. *Forest Science* 37, 1429–1440.

Hamilton, D.A. (1986) A logistic regression model of mortality in thinned and unthinned mixed conifer stands of northern Idaho. *Forest Science* 32, 989–1000.

Hasenauer, H., Merkl, D. and Weingartner, M. (2001) Estimating tree mortality of Norway spruce stands with neural networks. *Advances in Environmental Research* 5, 405–414.

Hökkä, H., Alenius, V. and Penttilä, T. (1997) Individual-tree basal area growth models for Scots pine, pubescent birch and Norway spruce on drained peatlands in Finland. *Silva Fennica* 31, 161–178.

Hosmer, D.W. and Lemeshow, S. (2001) *Applied Logistic Regression*. Wiley Series in Probability and Statistics, John Wiley & Sons, New York, 375 pp.

Hox, J.J. (1995) *Applied Multilevel Analysis*. TT-Publikaties, Amsterdam, 119 pp.

Hynynen, J., Ojansuu, R., Hökkä, H., Siipilehto, J., Salminen, H. and Haapala, P. (2002) Models for predicting stand development in MELA System. *Metsäntutkimuslaitoksen Tiedonantoja. Finnish Forest Research Institute, Research Paper* 835, 116 pp.

Jalkanen, A. (2001) The probability of moose damage at the stand level in southern Finland. *Silva Fennica* 35, 159–168.

Jutras, S., Hökkä, H., Alenius, V. and Salminen, H. (2003) Modeling mortality of individual trees in drained peatland sites in Finland. *Silva Fennica* 37, 235–251.

McCullagh, P. and Nelder, J. (1989) *Generalized Linear Models*. Chapman and Hall, London, 511 pp.

McCulloch, C.E. and Searle, S.R. (2001) *Generalized, Linear, and Mixed Models*. Wiley Series in Probability and Statistics, Wiley, New York, 358 pp.

Monserud, R.A. (1976) Simulation of forest tree mortality. *Forest Science* 22, 438–444.

Monserud, R.A. and Sterba, H. (1999) Modeling individual tree mortality for Austrian forest species. *Forest Ecology and Management* 113, 109–123.

Murty, D. and McMurtrie, R.E. (2000) The decline of forest productivity as stands age: a model-based method for analysing causes for the decline. *Ecological Modelling* 134, 185–205.

Rodriguez, G. and Goldman, N. (1995) An assessment of estimation procedures for multilevel models with binary responses. *Journal of the Royal Statistical Society* A 158, 73–89.

SAS Institute (1994) *SAS/STAT User's Guide*, Vol. 2, 4th edn. SAS Institute, Cary, North Carolina.

Shen, G., Moore, J.A. and Hatch, C.R. (2000) The effect of nitrogen fertilization, rock type, and habitat type on individual tree mortality. *Forest Science* 47, 203–213.

Van Dyck, M.G. (2001) *Keyword Reference Guide for the Forest Vegetation Simulator*. USDA Forest Service. Forest Management Service Center, 93 pp.

Vanclay, J.K. (1995) Growth models for tropical forests: a synthesis of models and methods. *Forest Science* 41, 7–42.

Yao, X., Titus, S.J. and MacDonald, S.E. (2001) A generalized logistic model of individual tree mortality for aspen, white spruce, and lodgepole pine in Alberta mixedwood forests. *Canadian Journal of Forest Research* 31, 283–291.

Zeger, L.S., Liang, K.-Y. and Albert, P.S. (1988) Models for longitudinal data: a generalized estimating equation approach. *Biometrics* 44, 1049–1060.

21
Using Process-dependent Groups of Species to Model the Dynamics of a Tropical Rainforest

Nicolas Picard,[1] Sylvie Gourlet-Fleury[1] and Plinio Sist[1]

Abstract

The high tree species diversity in tropical forests is difficult to take into account in models. The usual solution consists of defining groups of species and then adjusting a set of parameters for each group. In this study, we address this issue by allowing a species to move from one species group to another, depending on the biological process that is concerned. We developed this approach with a matrix model of forest dynamics, for a tropical rainforest in French Guiana, at Paracou, focusing on the methodological aspects. The forest dynamics is split into three components: recruitment, growth and mortality. We then built five recruitment groups, five growth groups and five mortality groups. One species is characterized by a combination of the three groups, thus yielding in total $5 \times 5 \times 5 = 125$ possibilities, out of which 43 are actually observed. The resulting matrix model provides a better view of the floristic composition of the forest, and does not have more parameters than it would have with five global species groups. However, its predictions are no more precise than those of the matrix model based on five global groups.

Introduction

When trying to model the dynamics of a tropical rainforest, one is confronted with the huge diversity of tree species. Even though some authors treat every single species separately (e.g. Shugart *et al.*, 1980), the usual solution consists of building *ad hoc* groups (generally by cluster analysis) from species characteristics that are in use in the model, and then adjusting a set of parameters for each group (Favrichon, 1998; Köhler and Huth, 1998; Finegan *et al.*, 1999; Köhler *et al.*, 2000, 2001; Huth and Ditzer, 2001). Groups may be built from ecological characteristics of the species, thus providing so-called 'functional groups', but they may also take into consideration extraneous information such as commercial categories (Wan Razali, 1986; Vanclay, 1989; Boscolo and Vincent, 1998). Still, defining groups in relation to a model of forest dynamics remains difficult, as two species may appear similar with respect to a biological function, and at the same time different with respect to another function.

[1] Cirad-forêt, Montpellier, France
Correspondence to: nicolas.picard@cirad.fr

For instance, species groups that are homogeneous with respect to growth may be built, but it is likely that these groups will be heterogeneous with respect to recruitment or mortality.

In this chapter, we address this issue by defining three distinct species groupings: one is growth-specific, that splits the species along their growth characteristics; the second is recruitment-specific, that gathers species with similar recruitment characteristics; and the third is mortality-specific. The key point is that the model of forest dynamics allows a species to shift from one species group to another depending on the biological process (growth, recruitment or mortality) that is concerned. A species may thus take its recruitment parameters from one species group, its growth parameters from another and its mortality parameters from a third. The resulting model thus has as many parameters as in the classical approach, but their combinations give a much greater richness of modelled species.

We developed this approach with a matrix model of forest dynamics, for a tropical rainforest in French Guiana, at Paracou. We have deliberately focused on the methodological aspects in this chapter, leaving the study of the ecological relevance for future work. We analysed the Paracou data to build five growth-specific groups, five recruitment-specific groups, five mortality-specific groups and five comprehensive species groups that simultaneously rely on growth, recruitment and mortality. Two matrix models were then developed: one is classical and is based on the five comprehensive species groups; the other one illustrates the proposed new approach and makes use of the three process-specific groupings into five groups. The predictions of the two matrix models are then compared.

Materials and Methods

The Paracou plots

The matrix models were built from the data of the Paracou forest, in French Guiana, 40 km west of Kourou (5° 15′ N, 52° 55′ W). The climate is equatorial with an average annual rainfall of 3160 mm and an average temperature of 26°C. On this site twelve 6.25-ha plots of natural rainforests were settled in 1984 by the Cirad-Forêt. Nine plots underwent silvicultural treatments in 1987/88, and three plots were left as controls. Each plot is divided into four subplots, which yields 48 subplots in total. The girth of every tree greater than 10 cm diameter at breast height (DBH) was measured, and taxonomic information was also noted. As the botanical inventory did not allow every tree to be identified at the species level, a tree was characterized either by its species or by a group of species, which we shall refer as a pseudo-species. Measurements have been made annually from 1984 to 1995, and every 2 years since. In total, more than 46,000 trees have been inventoried and 202 pseudo-species have been identified. More details about the Paracou experimental station can be found in Schmitt and Bariteau (1990).

Characterizing pseudo-species

Each pseudo-species was first characterized by a set of parameters that describe its growth, its recruitment and its mortality. On each of the 48 subplots and at years $t =$ 1984, 1988, 1990 and 1992, we computed the average diameter increment, the mortality rate and the recruitment flux between year t and $t + 2$. Let us denote a_{nts} the average diameter increment of the pseudo-species s (= 1 . . . 202) between year t and

$t + 2$ in subplot n, y_{nts} the number of trees of the pseudo-species s that are alive in subplot n at year t, z_{nts} the number of trees of the pseudo-species s that die in subplot n between year t and $t + 2$, and r_{nts} the number of trees of the pseudo-species s that are recruited (i.e. whose DBH surpasses 10 cm) in subplot n between year t and $t + 2$.

Following Favrichon (1998)'s work, growth and recruitment were related to the stocking of the plot, whereas mortality was not. The stocking of a subplot at year t was quantified by the ratio b of its basal area at year t over its basal area in 1984 (assuming that 1984 characterizes the steady state of the subplot):

$$b_{nt} = \frac{B_{nt}}{B_{n,1984}} \tag{1}$$

where B_{nt} is the basal area of subplot n at year t. The average diameter growth of the pseudo-species s was then characterized by the parameters α_{0s} and α_{1s} of the regression: $a_{nts} = \alpha_{0s} - \alpha_{1s}b_{nt} + \varepsilon_{nts}$, where ε_{nts} are the residuals. Similarly the recruitment of the pseudo-species s was characterized by the parameters β_{0s} and β_{1s} of the regression: $r_{nts} = \beta_{0s} - \beta_{1s}b_{nt} + \varepsilon_{nts}$. As the diameter increments or the recruitment fluxes on the same subplot at 2 consecutive years are not independent, a standard linear regression (which would assume that the ε_{nts} are independent) is not appropriate. A regression for repeated measurements (Diggle et al., 1996) is required. A preliminary analysis showed that an exponential model (Diggle et al., 1996: 57) could adequately fit the variance–covariance structure of the residuals. The parameters α_0, α_1, β_0, β_1 were then estimated by maximizing the log-likelihood (Diggle et al., 1996: 63).

As mortality was assumed to be density-independent, the mortality rate m_s of the pseudo-species s was simply estimated as: $m_s - (\Sigma_{n,t}\, z_{nts})/(\Sigma_{n,t}\, y_{nts})$. The computations were actually achieved for all pseudo-species with at least 15 individuals (on average between 1984 and 1992) in at least three subplots. At this step, we have thus got the set of dynamics parameters $(\alpha_0, \alpha_1, \beta_0, \beta_1, m)$ for a subset of pseudo-species, which will be used to build species groups by cluster analysis.

Building species groups

In a comprehensive approach, the dissimilarity $d_{ss'}$ between any two pseudo-species s and s' was defined as the Euclidian distance between the standardized vector of their dynamics parameters: $d_{ss'} = [(\alpha_{0s}^* - \alpha_{0s'}^*)^2 + (\alpha_{1s}^* - \alpha_{1s'}^*)^2 + (\beta_{0s}^* - \beta_{0s'}^*)^2 + (\beta_{1s}^* - \beta_{1s'}^*)^2 + (m_s^* - m_{s'}^*)^2]^{0.5}$, where for any parameter x, $x_s^* = (x_s - \bar{x})/\sqrt{s_x}$, $\bar{x}$ is the empirical mean of x_s over all pseudo-species and s_x is the empirical variance of x_s. A hierarchical cluster analysis using Ward's minimum variance method was then used to define five comprehensive species groups. Species groups that are specific to a process of the forest dynamics were also defined by restricting the parameters used to compute the dissimilarity between pseudo-species. Five growth-specific groups were thus obtained in the same way from the following dissimilarity: $d_{ss'} = [(\alpha_{0s}^* - \alpha_{0s'}^*)^2 + (\alpha_{1s}^* - \alpha_{1s'}^*)^2]^{0.5}$. Five recruitment-specific groups were also obtained from the dissimilarity: $d_{ss'} = [(\beta_{0s}^* - \beta_{0s'}^*)^2 + (\beta_{1s}^* - \beta_{1s'}^*)^2]^{0.5}$, and five mortality-specific groups were obtained from the dissimilarity: $d_{ss'} = |m_s^* - m_{s'}^*|$.

Let g_s $(s = 1 \ldots 5)$, g_s^{gr}, g_s^{rc} and g_s^{mt} be respectively the comprehensive species group to which the pseudo-species s belongs, the growth-specific group to which it belongs, the recruitment-specific group to which it belongs, and the mortality-specific group to which it belongs. The dynamic behaviour of a species can be characterized either by its comprehensive species group g_s, or by the combination $(g_s^{gr},$

g_s^{rc}, g_s^{mt}) of its process-specific groups. The former characterization encompasses five categories, whereas the latter potentially encompasses $5 \times 5 \times 5 = 125$ categories.

Matrix model

The model is an Usher model with species groups and density-dependent coefficients. Its principles come from Buongiorno and Michie (1980), Favrichon (1998) and Usher (1969). The trees of the stand are broken down by diameter class and species group. Time is discrete with a time step Δt. Between time t and $t + \Delta t$, a tree of species group s and diameter class i has three possibilities: (i) it dies, with probability $q_{si}(t)$; (ii) it stays alive and moves up to the next diameter class, with probability $u_{si}(t)$; or (iii) it stays alive in the same diameter class, with probability $p_{si}(t) = 1 - q_{si}(t) - u_{si}(t)$.

Let $N_{si}(t)$ be the number of trees of species group s in diameter class i at time t. Its equation of evolution is:

$$N_{si}(t+\Delta t) = p_{si}(t)N_{si}(t) + u_{si-1}(t)N_{si-1}(t) \quad (i>1) \tag{2}$$

$$N_{s1}(t+\Delta t) = p_{s1}(t)N_{s1}(t) + R_s(t) \tag{3}$$

where $R_s(t)$ is the recruitment flux. A linear relationship between the u_{si} values and the ratio $b(t)$ of the subplot basal area at year t over its basal area in 1984 (see Equation 1) was selected: $u_{si}(t) = \delta_{0si} - \delta_{1si}b(t)$. For recruitment, a linear relationship between $b(t)$ and either the recruitment flux $R_s(t)$ or its log-transform was selected: $R_s(t) = \gamma_{0s} - \gamma_{1s}b(t)$ or $\ln R_s(t) = \gamma_{0s} - \gamma_{1s}b(t)$.

Parameter estimation

Given two inventories at year t and $t + \Delta t$, the upgrowth transition probability u_{si} can readily be estimated as the proportion of trees of species group s and diameter class i that move up to diameter class $i + 1$. Similarly, the mortality rate q_{si} can readily be estimated as the proportion of trees of species group s and diameter class i that die. Let u_{sint} be the estimate of u_{si} and q_{sint} be the estimate of q_{si} obtained from the subplot n (= 1 . . . 48) and from the inventories t and $t + \Delta t$.

To reduce the number of parameters of the model and to ensure the smoothness of the transition probabilities, the following regressions were actually performed:

$$u_{sint} = \delta_{0s} + \delta_{1s}D_i + \delta_{2s}D_i^2 + \delta_{3s}D_i^3 - \delta_{4s}b_{nt} + \varepsilon_{sint} \tag{4}$$

$$q_{sint} = \mu_{0s} + \mu_{1s}D_i + \mu_{2s}D_i^2 + \varepsilon_{sint} \tag{5}$$

where D_i is the average diameter of diameter class i, and b_{nt} is given by Equation 1.

As in the work of Favrichon (1998), we selected a time step $\Delta t = 2$ years. We then used the data from the 48 subplots at $t =$ 1984, 1988, 1990, 1992. Each subplot thus appears four times in Equations 4 or 5, so that the residuals ε_{sint} cannot be considered as independent, and a longitudinal data analysis is again required. Again an exponential model was selected for the variance–covariance structure:

$$\mathrm{Cov}(\varepsilon_{sint}, \varepsilon_{si'n't}) = 0 \quad \text{if } n \neq n' \text{ or } i \neq i', \forall(t,t') \tag{6}$$

$$\mathrm{Cov}(\varepsilon_{sint}, \varepsilon_{sint'}) = \sigma_s^2 \rho_s^{|t-t'|} \tag{7}$$

and the parameters were estimated by maximizing the log-likelihood (Diggle *et al.*, 1996).

Similarly for recruitment: let R_{snt} be the number of trees of species group s that are recruited between years t and $t + \Delta t$ on subplot n (= 1 . . . 48). The following regressions were performed: $R_{snt} = \gamma_{0s} - \gamma_{1s}b_{nt} + \varepsilon_{snt}$ or $\ln R_{snt} = \gamma_{0s} - \gamma_{1s}b_{nt} + \varepsilon_{snt}$. As each subplot appears four times, a regression for repeated measurements using the same variance–covariance structure as before was achieved.

In regressions 4 and 5, all the subplots have the same weight. This can lead to some distortions, as a subplot with a low number of trees (and thus imprecise estimates of u_{sint} and q_{sint}) will have the same weight as a subplot with a high number of trees (and thus accurate estimates of u_{sint} and q_{sint}). For a few species groups containing few species and few individuals, we indeed obtained very unrealistic mortality rates. In those cases we used in place of Equation 5 an a posteriori estimate of the mortality rates that is derived from Equation 2 (Houde and Ledoux, 1995).

A set of parameters $(\delta_{0s}, \delta_{1s}, \delta_{2s}, \delta_{3s}, \delta_{4s}, \mu_{0s}, \mu_{1s}, \mu_{2s}, \gamma_{0s}, \gamma_{1s})_{s=1\,...\,5}$ is associated with each species grouping. The complete parameter set was estimated for the comprehensive species grouping. The parameters $(\delta_{0s}, \delta_{1s}, \delta_{2s}, \delta_{3s}, \delta_{4s})_{s=1\,...\,5}$ were then estimated for the growth-specific grouping, $(\mu_{0s}, \mu_{1s}, \mu_{2s})_{s=1\,...\,5}$ were estimated for the mortality-specific grouping, and $(\gamma_{0s}, \gamma_{1s})_{s=1\,...\,5}$ were estimated for the recruitment-specific grouping.

Results

Species groups

Of the 202 pseudo-species present in the database, 152 (75%) had enough data for their parameters to be estimated. The five comprehensive groups that are derived from all parameters may be described as follows (Table 21.1): Group 1: low αs, medium βs; this group thus includes slow-growing species that are rather insensitive to a change in the stand stocking; Group 2: medium αs and medium βs; this group thus includes species with an intermediate behaviour in every respect; Group 3: high αs, medium βs; this group thus includes fast-growing trees that are sensitive to a change in the stand stocking; Group 4: low βs, medium αs and high mortality rate; this group thus includes the species that require a stand stocking large enough to recruit young trees; Group 5: very high αs and very high βs; this group thus includes the species that grow very fast and have a high recruitment when the stand is open (pioneer species). Groups 4 and 5 include species with outlying characteristics and thus encompass few species.

The growth-specific groups discriminate the species along a gradient from slow-growing, density-independent species (Group 1 with low αs) to fast-growing density-dependent species (Group 5 with high αs), see Table 21.2. Similarly the

Table 21.1. Mean values of the growth parameters α_0, α_1, the recruitment parameters β_0, β_1, and the mortality rates m in the five comprehensive species groups. S is the number of species in the groups.

g	α_0	α_1	β_0	β_1	m	S
1	0.64	0.40	0.70	0.56	0.024	46
2	1.33	1.04	0.28	0.20	0.010	75
3	2.76	2.24	0.90	0.84	0.016	25
4	0.95	0.76	−0.54	−1.05	0.184	4
5	3.15	1.57	24.29	25.09	0.027	2

Table 21.2. Mean values of the growth parameters α_0, α_1 in the five growth-specific groups, the recruitment parameters β_0, β_1 in the five recruitment-specific groups, and the mortality rates m in the five mortality-specific groups. S is the number of species.

g^{gr}	α_0	α_1	S	g^{rc}	β_0	β_1	S	g^{mt}	m	S
1	0.42	0.17	35	1	0.15	0.08	117	1	0.014	34
2	1.67	1.31	46	2	1.01	0.81	23	2	0.005	71
3	1.04	0.80	51	3	2.83	2.43	9	3	0.025	32
4	2.73	2.19	17	4	7.80	8.03	1	4	0.060	11
5	5.67	4.90	3	5	24.29	25.09	2	5	0.184	4

recruitment-specific groups discriminate the species along a gradient from low to high recruitment (Table 21.2), and the mortality-specific groups discriminate the species along a gradient from low to high mortality. Each pseudo-species is then characterized by a triplet (g^{gr}, g^{rc}, g^{mt}).

Forty-three distinct triplets are observed; that is, one-third of the theoretical maximum (125). This simply follows from biological trade-offs between the dynamics traits. For instance, no tree species can be at the same time fast-growing with a high recruitment and low mortality. Indeed, the triplet (5, 5, 1) does not correspond to any pseudo-species. In other words, the process-specific groups are not independent. For instance, a χ^2 test reveals a relationship between the growth-specific and the recruitment-specific groups ($P = 0.04$).

Model predictions

As in the work of Favrichon (1998), 11 diameter classes were defined, ranging from 10 to 60 cm with a constant width of 5 cm, the last diameter class grouping all the trees greater than 60 cm DBH. The parameters for the classical matrix model based on comprehensive species groups are given in Table 21.3. A posteriori estimates of the mortality rates were used for species groups 4 and 5: for species group 4, $q_i = 0.0224$, $\forall i$; for species group 5, $q_1 = 0.2381$, $q_2 = 0.0505$, $q_3 = 0.1258$ and $q_i = 1$, $\forall i \geq 4$.

The parameters for the matrix model based on growth-, recruitment- or mortality-specific groups are given in Table 21.4. Each of the 43 species groups that result from the crossing of these three groupings is specified by a triplet (g^{gr}, g^{rc}, g^{mt}), so that the species group (g^{gr}, g^{rc}, g^{mt}) takes its upgrowth transition parameters from group g^{gr} for growth (Table 21.4), its recruitment parameters from group g^{rc} for recruitment, and its mortality parameters from group g^{mt} for mortality. The resulting matrix model thus behaves like a matrix model with 43 species groups.

Both matrix models were used to project each of the 48 subplots from 1988 (at the end of the silvicultural treatments) to 1994. Evaluation is made at the end of the growing period by comparing the observed and predicted values of number of trees. This procedure does not validate the models as it does not rely on independent data. Let N_{sin} be the observed and $\hat{N}_{sin}$ the predicted number of trees in the ith diameter class of species group s in subplot n in 1994. The quality of the prediction for the ith diameter class of species group s is quantified by the percentage of explained variance: $\text{PEV}_{si} = 1 - \text{MSS}_{si}/S(N_{si})$, where $S(N_{si})$ is the empirical variance of N_{si} over the 48 subplots and $\text{MSS}_{si} = \Sigma_{n=1}^{48}(N_{sin} - \hat{N}_{sin})^2 / 48$ is the mean sum of squares. The higher the PEV is, the better the predictions are.

Table 21.5 shows the PEV values for each diameter class and each of the five comprehensive groups, according to the classical matrix model. As $\text{MSS}_{si} = S(N_{si} - \hat{N}_{si}) + (\bar{N} - \hat{\bar{N}})^2$, where $\bar{N}$ is the empirical mean of the N_{si} and $\hat{\bar{N}}$ is the empirical

Table 21.3. Selected regression equations and parameter values of the matrix model for the comprehensive species grouping.

g	Growth	ρ	R² (%)	Recruitment	ρ	R² (%)	Mortality	ρ	R² (%)
1	$u_i = \delta_1 D_i + \delta_2 D_i^2$	0.17	3.5	$R = \gamma_0 - \gamma_1 b$	−0.03	41.5	$q_i = \mu_0 + \mu_1 D_i$	0.13	7.1
2	$u_i = \delta_0 + \delta_1 D_i + \delta_2 D_i^2 + \delta_3 D_i^3 - \delta_4 b$	0.02	12.4	$R = \gamma_0 - \gamma_1 b$	−0.01	30.6	$q_i = \mu_0 + \mu_1 D_i$	−0.11	0.5
3	$u_i = \delta_0 + \delta_1 D_i + \delta_2 D_i^2 + \delta_3 D_i^3 - \delta_4 b$	−0.09	7.9	$\ln R = \gamma_0 - \gamma_1 b$	0.24	40.2	$q_i = \mu_0 + \mu_1 D_i$	−0.05	0.2
4	$u_i = \delta_0 + \delta_1 D_i + \delta_2 D_i^2 + \delta_3 D_i^3$	−0.04	2.1	$\ln R = -\gamma_1 b$	−0.44	8.3	$q_i = \mu_0 + \mu_1 D_i$	0.00	1.3
5	$u_i = \delta_0 + \delta_1 D_i + \delta_2 D_i^2 - \delta_4 b$	0.13	14.6	$\ln R = \gamma_0 - \gamma_1 b$	0.12	43.7	$q_i = \mu_0 + \mu_1 D_i + \mu_2 D_i^2$	−0.03	3.7

Group 1

	Estimate	SE	P
δ_0	–	–	–
δ_1	$3.347\ 10^{-3}$	$1.908\ 10^{-4}$	<0.001***
δ_2	$-5.358\ 10^{-5}$	$3.955\ 10^{-6}$	<0.001***
δ_3	–	–	–
δ_4	–	–	–
γ_0	32.169	1.795	<0.001***
γ_1	25.680	2.161	<0.001***
μ_0	$-4.840\ 10^{-2}$	$1.125\ 10^{-2}$	<0.001***
μ_1	$3.361\ 10^{-3}$	$2.968\ 10^{-4}$	<0.001***
μ_2	–	–	–

Group 2

	Estimate	SE	P
δ_0	$2.702\ 10^{-1}$	$2.843\ 10^{-2}$	<0.001***
δ_1	$-9.369\ 10^{-3}$	$2.667\ 10^{-3}$	<0.001***
δ_2	$3.458\ 10^{-4}$	$7.754\ 10^{-5}$	<0.001***
δ_3	$-3.671\ 10^{-6}$	$6.870\ 10^{-7}$	<0.001***
δ_4	$1.415\ 10^{-1}$	$1.054\ 10^{-2}$	<0.001***
γ_0	21.265	1.364	<0.001***
γ_1	15.207	1.640	<0.001***
μ_0	$5.783\ 10^{-3}$	$3.223\ 10^{-3}$	0.036*
μ_1	$2.789\ 10^{-4}$	$8.038\ 10^{-5}$	<0.001***
μ_2	–	–	–

Group 3

	Estimate	SE	P
δ_0	$5.292\ 10^{-1}$	$6.922\ 10^{-2}$	<0.001***
δ_1	$-1.976\ 10^{-2}$	$6.573\ 10^{-3}$	0.001**
δ_2	$7.265\ 10^{-4}$	$1.957\ 10^{-4}$	<0.001***
δ_3	$-7.470\ 10^{-6}$	$1.767\ 10^{-6}$	<0.001***
δ_4	$2.900\ 10^{-1}$	$2.782\ 10^{-2}$	<0.001***
γ_0	3.844	0.271	<0.001***
γ_1	3.140	0.330	<0.001***
μ_0	$1.295\ 10^{-2}$	$5.782\ 10^{-1}$	0.013*
μ_1	$2.854\ 10^{-4}$	$1.588\ 10^{-4}$	0.036*
μ_2	–	–	–

Group 4

	Estimate	SE	P
δ_0	$1.288\ 10^{-1}$	$3.846\ 10^{-2}$	<0.001***
δ_1	$-1.286\ 10^{-2}$	$4.018\ 10^{-3}$	<0.001***
δ_2	$4.514\ 10^{-4}$	$1.230\ 10^{-4}$	<0.001***
δ_3	$-4.475\ 10^{-6}$	$1.129\ 10^{-6}$	<0.001***
δ_4	–	–	–
γ_0	–	–	–
γ_1	-0.822	$7.717\ 10^{-2}$	<0.001***
μ_0	$4.907\ 10^{-1}$	$2.805\ 10^{-2}$	<0.001***
μ_1	$-3.112\ 10^{-3}$	$8.453\ 10^{-4}$	<0.001***
μ_2	–	–	–

Group 5

	Estimate	SE	P
δ_0	1.350	$1.796\ 10^{-1}$	<0.001***
δ_1	$-4.595\ 10^{-2}$	$1.103\ 10^{-2}$	<0.001***
δ_2	$5.197\ 10^{-4}$	$1.681\ 10^{-4}$	<0.001***
δ_3	–	–	–
δ_4	$3.509\ 10^{-1}$	$1.605\ 10^{-1}$	0.014*
γ_0	5.549	0.422	<0.001***
γ_1	5.276	0.610	<0.001****
μ_0	$1.479\ 10^{-1}$	$5.214\ 10^{-2}$	0.002**
μ_1	$-1.410\ 10^{-2}$	$5.847\ 10^{-3}$	0.008**
μ_2	$3.441\ 10^{-4}$	$1.555\ 10^{-4}$	0.013*

Levels of significance: *** 0.1%, ** 1%, * 5%. ρ is the correlation coefficient between any two successive estimates on the same subplot, as defined in Equation 7.

Table 21.4. Selected regression equations and parameter values of the matrix model for the growth-, recruitment- and mortality-specific groupings.

g	Growth	ρ	R^2 (%)	Recruitment	ρ	R^2 (%)	Mortality	ρ	R^2 (%)
1	$u_i = \delta_0 - \delta_4 b$	0.14	1.5	$\ln R = \gamma_0 - \gamma_1 b$	0.18	16.5	$q_i = \mu_0$	—	—
2	$u_i = \delta_0 + \delta_1 D_i + \delta_2 D_i^2 + \delta_3 D_i^3 - \delta_4 b$	0.02	10.4	$\ln R = \gamma_0 - \gamma_1 b$	0.18	41.2	$q_i = \mu_0$	—	—
3	$u_i = \delta_0 + \delta_1 D_i + \delta_2 D_i^2 - \delta_4 b$	0.01	9.0	$\ln R = \gamma_0 - \gamma_1 b$	0.18	43.2	$q_i = \mu_1 D_i$	−0.04	1.3
4	$u_i = \delta_0 + \delta_1 D_i + \delta_2 D_i^2 - \delta_4 b$	−0.02	7.1	$\ln R = \gamma_0 - \gamma_1 b$	0.01	41.8	$q_i = \mu_1 D_i$	0.11	3.3
5	$u_i = \delta_0 + \delta_1 D_i + \delta_2 D_i^2 + \delta_3 D_i^3 - \delta_4 b$	0.01	14.8	$\ln R = \gamma_0 - \gamma_1 b$	0.12	43.7	$q_i = \mu_0 + \mu_1 D_i$	0.00	1.3

	Group 1			Group 2			Group 3		
	Estimate	SE	P	Estimate	SE	P	Estimate	SE	P
δ_0	$9.589\ 10^{-2}$	$1.260\ 10^{-2}$	$<0.001^{***}$	$3.419\ 10^{-1}$	$4.076\ 10^{-2}$	$<0.001^{***}$	$8.228\ 10^{-2}$	$1.094\ 10^{-2}$	$<0.001^{***}$
δ_1	—	—	—	$-1.059\ 10^{-2}$	$3.831\ 10^{-3}$	0.003^{**}	$2.354\ 10^{-3}$	$5.210\ 10^{-4}$	$<0.001^{***}$
δ_2	—	—	—	$3.969\ 10^{-4}$	$1.116\ 10^{-4}$	$<0.001^{***}$	$-4.374\ 10^{-5}$	$6.885\ 10^{-6}$	$<0.001^{***}$
δ_3	—	—	—	$-4.309\ 10^{-6}$	$9.913\ 10^{-7}$	$<0.001^{***}$	—	—	—
δ_4	$6.555\ 10^{-2}$	$1.480\ 10^{-2}$	$<0.001^{***}$	$1.803\ 10^{-1}$	$1.521\ 10^{-2}$	$<0.001^{***}$	$5.945\ 10^{-2}$	$8.111\ 10^{-3}$	$<0.001^{***}$
γ_0	3.186	$2.042\ 10^{-1}$	$<0.001^{***}$	3.965	$2.076\ 10^{-1}$	$<0.001^{***}$	4.148	$2.271\ 10^{-1}$	$<0.001^{***}$
γ_1	1.343	$2.420\ 10^{-1}$	$<0.001^{***}$	2.644	$2.463\ 10^{-1}$	$<0.001^{***}$	3.030	$2.706\ 10^{-1}$	$<0.001^{***}$
μ_0	$1.522\ 10^{-2}$	$5.428\ 10^{-2}$	—	$9.293\ 10^{-3}$	$5.617\ 10^{-2}$	—	—	—	—
μ_1	—	—	—	—	—	—	$8.867\ 10^{-4}$	$6.810\ 10^{-5}$	$<0.001^{***}$

	Group 4			Group 5		
	Estimate	SE	P	Estimate	SE	P
δ_0	$3.120\ 10^{-1}$	$4.360\ 10^{-2}$	$<0.001^{***}$	2.219	$2.744\ 10^{-1}$	$<0.001^{***}$
δ_1	$8.030\ 10^{-3}$	$2.050\ 10^{-3}$	$<0.001^{***}$	$-1.643\ 10^{-1}$	$2.905\ 10^{-2}$	$<0.001^{***}$
δ_2	$-1.169\ 10^{-4}$	$2.815\ 10^{-5}$	$<0.001^{***}$	$5.335\ 10^{-3}$	$9.716\ 10^{-4}$	$<0.001^{***}$
δ_3	—	—	—	$-5.044\ 10^{-5}$	$9.547\ 10^{-6}$	—
δ_4	$3.213\ 10^{-1}$	$3.337\ 10^{-2}$	$<0.001^{***}$	$5.043\ 10^{-1}$	$1.206\ 10^{-1}$	$<0.001^{***}$
γ_0	3.785	$4.612\ 10^{-1}$	$<0.001^{***}$	5.549	$4.223\ 10^{-1}$	$<0.001^{***}$
γ_1	4.138	$7.215\ 10^{-1}$	$<0.001^{***}$	5.276	$6.105\ 10^{-1}$	$<0.001^{****}$
μ_0	—	—	—	$4.904\ 10^{-1}$	$2.805\ 10^{-2}$	$<0.001^{***}$
μ_1	$2.932\ 10^{-3}$	$1.438\ 10^{-4}$	$<0.001^{***}$	$-3.113\ 10^{-3}$	$8.453\ 10^{-4}$	$<0.001^{***}$

Levels of significance: *** 0.1%, ** 1%, * 5%. ρ is the correlation coefficient between any two successive estimates on the same subplot, as defined in Equation 7.

mean of the $\hat{N}_{si}$, a systematic bias of the model predictions may lead to high values of MSS, and thus negative values of PEV. We indeed got negative values of PEV, mainly for species groups 4 and 5. The number of trees in these groups was thus badly predicted by the classical matrix model on a short run after disturbance. On the other hand, the number of trees in the other species groups was quite well predicted, with PEV values ranging from 28 to 92% (excluding negative values).

Table 21.5 also shows the PEV values according to the matrix model based on process-specific groups, after aggregating the 43 groups into the five growth-specific groups. Fewer negative values were obtained than with the classical matrix model, but the positive values were on average smaller than with the former model. Similar results were obtained if the 43 groups were aggregated into the recruitment-specific or the mortality-specific groups.

Note that a comparison of the PEV values according to the two matrix models in Table 21.5 is not relevant, as they use different definitions of groups. As the 43 process-specific groups *cannot* be aggregated into the five comprehensive species groups (the former are not a partition of the latter), a direct comparison of the two matrix models is only possible at a higher level of aggregation, namely the stand level. The PEV of the total number of trees per hectare is 77% for the classical matrix model and 68% for the matrix model based on process-specific groups. The PEV of the total basal area per hectare is 62% for the classical matrix model and 66% for the matrix model based on process-specific groups.

Discussion and Conclusion

Allowing the species to shift from one species grouping to another, depending on the biological process of interest, presents a better description of the species richness. With each species grouping is a set of parameters, and this approach may also be seen as a crossing over of parameter sets. The resulting model has as many parameters as in the classical approach where species belong to a unique species

Table 21.5. Percentage of explained variance[a] (PEV) of the number of trees in each diameter class and each species group in 1994. The projections are made using either the classical matrix model (clas.) or the matrix model based on process-specific groups (proc.); for the former, the five groups (g) are the comprehensive groups whereas for the latter, the five groups (g^{gr}) are the growth-specific groups. PEV[a] cannot be estimated when $N_{sin} = 0$ for all n (dashed cells).

Model	g or g^{gr}	Diameter class										
		1	2	3	4	5	6	7	8	9	10	11
Clas.	1	83	75	74	70	62	56	60	66	51	57	< 0
Clas.	2	92	75	82	80	61	70	78	85	70	73	< 0
Clas.	3	< 0	34	59	59	36	40	28	46	49	36	37
Clas.	4	< 0	< 0	< 0	< 0	–	< 0	< 0	–	–	–	< 0
Clas.	5	< 0	< 0	< 0	< 0	< 0	–	–	–	–	–	< 0
Proc.	1	34	17	61	72	83	81	24	77	35	67	< 0
Proc.	2	73	62	69	55	41	68	49	70	60	67	40
Proc.	3	70	26	43	63	25	62	48	68	50	40	19
Proc.	4	< 0	19	46	40	32	36	< 0	38	33	< 0	61
Proc.	5	37	9	< 0	< 0	< 0	–	< 0	< 0	6	< 0	< 0

[a] For species group s and diameter class i, $\text{PEV}_{si} = 1 - \text{MSS}_{si}/S(N_{si})$, where $S(N_{si})$ is the empirical variance of N_{si} over the 48 subplots, $\text{MSS}_{si} = \sum_{n=1}^{48}(N_{sin} - \hat{N}_{sin})^2/48$ is its mean sum of squares, and N_{sin} is the number of trees in species group s, diameter class i and subplot n.

group. This approach is quite general and could be applied to several types of models.

In the case of the Paracou forest, the matrix model based on process-specific groups does not yield better short-run (6 years) predictions after disturbance than the classical matrix model based on comprehensive species groups. This raises questions about: (i) the construction of the species groups, and (ii) the estimation of the model parameters.

First, species groups could be built from parameters other than the ones that we used. The parameter α_0 can be interpreted as the diameter growth rate in an empty plot (when $b = 0$), whereas α_1 represents the sensitivity of the diameter growth rate to the plot stocking. Similarly, β_0 represents the recruitment rate for an empty plot and β_1 represents the sensitivity of the recruitment to the plot stocking. A cluster analysis based on (α_0, α_1) (or on (β_0, β_1)) thus favours species groups that behave similarly in large openings and have a similar sensitivity to the stocking. However no empty plot ($b = 0$) is observed at Paracou, and α_0 and β_0 result from an extrapolation. It may be safer to use parameters that correspond to situations that are observed at Paracou. In particular the cluster analysis could be based on $(\alpha_0 - \alpha_1, \alpha_1)$ or $(\beta_0 - \beta_1, \beta_1)$, where $\alpha_0 - \alpha_1$ represents the diameter growth rate in the stationary state ($b = 1$) and $\beta_0 - \beta_1$ represents the recruitment rate in the stationary state. Species groups that are homogeneous with respect to $\alpha_0 - \alpha_1$ or $\beta_0 - \beta_1$ are then more likely to make accurate predictions of the stationary state (or long-run predictions) than species groups based on α_0 or β_0. Moreover, expert knowledge of the autecological traits of the Guyanese species should be consulted to validate the species groups that are obtained with the cluster analysis.

Secondly, the parameter estimation is a more delicate step for the model with shifting species groups than for the classical models based on fixed species groups. In the classical approach, a change in a parameter of a given species group only affects this group. The model predictions for a group can then be used to detect anomalous parameter values and correct them. With shifting species groups, the change in a parameter will affect several groups. All the 43 species groups that we obtained at Paracou are interconnected through their parameters: a species group with its parameters cannot be isolated and treated separately from the other groups. As a consequence, a diagnosis of anomalous parameter values from the model predictions is much more difficult. The interconnection between parameters could, however, be investigated with an elasticity analysis (de Kroon *et al.*, 1986) by computing quantities such as $\partial \ln P_s / \partial \ln p_{s'}$ where P_s is the predicted number of trees or basal area of species group s, and $p_{s'}$ is one of the model parameters relative to group s'.

Eventually, to deal with the different weights of the subplots in regressions 4 and 5, one could replace regressions 4 and 5 by a two-stage weighted least-square regression (Anderson *et al.*, 1985). The first stage would be identical to the longitudinal data regression that is described in this chapter and would yield initial estimated $\hat{u}_{sint}$ and $\hat{q}_{sint}$. The second stage would be a longitudinal data regression with the following variance–covariance structure:

$$\text{Cov}(\varepsilon_{sint}, \varepsilon_{si'n't}) = 0 \quad \text{if } n \neq n' \text{ or } i \neq i', \forall(t, t') \tag{8}$$

$$\text{Cov}(\varepsilon_{sint}, \varepsilon_{sint'}) = \hat{\sigma}_{sint} \hat{\sigma}_{sint'} \rho_s^{|t-t'|} \tag{9}$$

where $\hat{\sigma}_{sint} = \hat{u}_{sint}(1 - \hat{u}_{sint})/N_{sint}$ for regression 4, $\hat{\sigma}_{sint} = \hat{q}_{sint}(1 - \hat{q}_{sint})/N_{sint}$ for regression 5, and N_{sint} is the observed number of trees of species group s in diameter class i and subplot n at year t. The latter relationships follow from the fact that the numbers of trees of species group s in diameter class i and subplot n that remain in the same class, grow up or die follow a multinomial law with parameters (N_{sint}, p_{sint}, u_{sint}, q_{sint}).

In conclusion, the use of shifting species groups in models of forest dynamics offers a greater richness of modelled species behaviours, without an increase of the number of parameters of the model. However it raises specific problems for model correction that would warrant further investigation.

Acknowledgements

We thank two reviewers for their helpful comments on an earlier version of the manuscript.

References

Anderson, D.R., Burnham, K.P. and Crain, B.R. (1985) Estimating population size and density using line transect sampling. *Biometrical Journal* 27, 723–731.

Boscolo, M. and Vincent, J.R. (1998) *Promoting Better Logging Practices in Tropical Forests: a Simulation Analysis of Alternative Regulations*. Development Discussion Paper 652, The Harvard Institute for International Development, Harvard University, Cambridge, Massachusetts.

Buongiorno, J. and Michie, B.R. (1980) A matrix model of uneven-aged forest management. *Forest Science* 26, 609–625.

de Kroon, H., Plaisier, A., van Groenendael, J. and Caswell, H. (1986) Elasticity: the relative contribution of demographic parameters to population growth rate. *Ecology* 67, 1427–1431.

Diggle, P.J., Liang, K.Y. and Zeger, S.L. (1996) *Analysis of Longitudinal Data*. Oxford Statistical Science Series No. 13, Clarendon Press, Oxford, 253 pp.

Favrichon, V. (1998) Modeling the dynamics and species composition of tropical mixed-species uneven-aged natural forest: effects of alternative cutting regimes. *Forest Science* 44, 113–124.

Finegan, B., Camacho, M. and Zamora, N. (1999) Diameter increment patterns among 106 tree species in a logged and silviculturally treated Costa Rican rain forest. *Forest Ecology and Management* 121(3), 159–176.

Houde, L. and Ledoux, H. (1995) Modélisation en forêt naturelle: stabilité du peuplement. *Bois et Forêts des Tropiques* 245(3), 21–26.

Huth, A. and Ditzer, T. (2001) Long-term impacts of logging in a tropical rain forest: a simulation study. *Forest Ecology and Management* 142(1–3), 33–51.

Köhler, P. and Huth, A. (1998) The effects of tree species grouping in tropical rainforest modelling: simulations with the individual-based model Formind. *Ecological Modelling* 109, 301–321.

Köhler, P., Ditzer, T. and Huth, A. (2000) Concepts for the aggregation of tropical tree species into functional types and the application to Sabah's lowland rain forests. *Journal of Tropical Ecology* 16, 591–602.

Köhler, P., Ditzer, T., Ong, R.C. and Huth, A. (2001) Comparison of measured and modelled growth on permanent plots in Sabahs rain forests. *Forest Ecology and Management* 144(1–3), 101–111.

Schmitt, L. and Bariteau, M. (1990) Gestion de l'écosystème forestier guyanais: etude de la croissance et de la régénération naturelle – Dispositif de Paracou. *Bois et Forêts des Tropiques* 220, 3–23.

Shugart, H.H., Hopkins, M.S., Burgess, I.P. and Mortlock, A.T. (1980) The development of a succession model for subtropical rain forest and its application to assess the effects of timber harvest at Wiangaree State Forest, New South Wales. *Journal of Environmental Management* 11, 243–265.

Usher, M.B. (1969) A matrix model for forest management. *Biometrics* 25, 309–315.

Vanclay, J.K. (1989) A growth model for north Queensland rainforests. *Forest Ecology and Management* 27(3–4), 245–271.

Wan Razali, B.W.M. (1986) Development of a generalized forest growth and yield modelling system for mixed tropical forests of peninsular Malaysia. PhD thesis, University of Washington.

22 Modelling Current Annual Height Increment of Young Douglas-fir Stands at Different Sites

Pero J. Radonja,[1] Milos J. Koprivica[1] and
Vera S. Lavadinovic[1]

Abstract

Generally, modelling the non-linear and complex process of current annual height increment of any timber species is significant both in dendrometric studies and in practical forest exploitation. Using the methods of artificial intelligence based on neural networks, we attempted to extract the non-linear process of height increment from the observed data sets and to generate a prediction as accurately as possible. The first part of the chapter analyses height increment of different provenances of young Douglas-fir (*Pseudotsuga menziesii* (Mirb.) Franco) stands at different sites. After that, the corresponding data-based models of height increment are evaluated. The models of suitable sites, standard sites and unsuitable sites for Douglas-fir fast growth and development, as well as the models for superior provenances and inferior provenances, are proposed.

Introduction

Forest models can be used as very successful research and management tools. The models designed for research require many complicated and not readily available data, whereas the models designed for management use simpler and more readily accessible data (Johnsen *et al.*, 2001). On the other hand, forest models are commonly divided into process-based and empirical.

Process-based models have intellectual and scientific advantages compared with empirical models. Process models deal with deep scientific understanding of the considered processes and are associated with a large number of analysed processes, in the case of statistical models. However, in spite of that, some researchers believe that process models will never be of practical use (Zeide, 1997).

On the other hand, all models used in forest management are data-based or empirical. In our opinion, the process-based models are a very good basis for understanding the acquired data, i.e. for understanding the physiological and ecological sides of a number of processes in forestry. Note that measurement data, in fact, integrate all relevant physiological processes, including those still unknown to us.

[1] SE Serbiaforest, Institute of Forestry, Yugoslavia
Correspondence to: inszasum@EUnet.yu

It is known that increased complexity of a model reduces the generality of the considered model. Because of that, we shall omit the analyses of climate factors (Zhang et al., 2000) The reduced model is based on the widely used assumption that the relationship between height increment and age of dominant trees depends solely on global site properties (Baldwin *et al.*, 2001).

Based on multiannual measurements, this chapter presents the results of research on young Douglas-fir stand development at different sites. More precisely, we present the analyses and modelling of current annual height increment of young Douglas-fir stands of different provenances at different sites. Ten different sites were analysed in Serbia: Jelova Gora, Goc, Kosmaj, Jastrebac, Bogovadja, Majdanpek, Zlatibor, Crnoljeva Mt. (Vrcelj-Kitic, 1982), Tanda (in eastern Serbia, near the town Bor) and Juhor (in central Serbia) (Lavadinovic, 1995), and four provenances, Nos 03, 09, 17 and 30 (Lavadinovic and Koprivica, 1996, 1997). The models were generated by averaging individual models obtained for different locations (sites).

Analysis of Height Increment of Young Douglas-fir Stands at Different Sites

One of the best indicators of site properties and provenance success is the value of current annual height increment. However, unfortunately, the ages of the study stands are insufficient to determine the years of culmination and the maximum values of height increment.

Figure 22.1 illustrates the measured data and curves of height increment in the sample plots on Goc, data denoted by 'x', and Jelova Gora, data denoted by '+'. The sample plot on Goc is at an altitude of 400–500 m. The altitude of stands on Jelova Gora is about 950 m. It can be seen that the development of Douglas-fir stands on Jelova Gora is slower, which is correct. The slower development results from the higher altitude of stands on Jelova Gora (Stamenkovic and Vuckovic, 1988). The sites on Mt Goc and Jelova Gora can be considered to be very favourable for Douglas-fir development and introduction.

Figure 22.2 presents the curves of height increment and measured data of sample plots on Kosmaj (dashed line and '*'), and sample plots on Jastrebac (solid line and 'o'). The altitudes of sample plots of Kosmaj and Jastrebac are 300 and 700 m, respectively. In spite of that, the increment at the age of 12 years on Jatrebac is some-

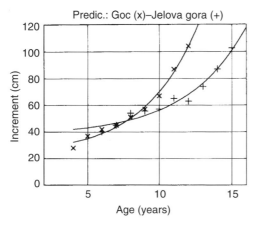

Fig. 22.1. Curves of height increment and the corresponding data.

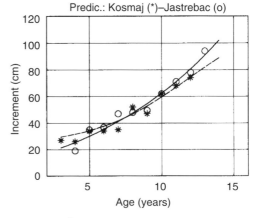

Fig. 22.2. Curves of height increment and the corresponding data.

what greater than on Kosmaj. Unfortunately we do not have the measurement at age 13 years for Kosmaj. In the considered age period particularly from 5 to 12 years, the curves of height increment for the stands on Kosmaj and Jastrebac are very similar. The sites on Kosmaj and Jastrebac can be considered as standard sites for Douglas-fir cultivation and development.

Note that fitting and prediction of all measured data in this chapter are performed using a method of artificial intelligence. The fitting and prediction are also performed by two-layer neural networks. The application of neural networks will be further explained in a later section of this chapter.

Figure 22.3 shows the measured data and the height increment curves for the sites at Bogovadja ('*' and dashed line) and Majdanpek ('o' and solid line). The variations in the height increment due to climate factor variation are very great. It is known that the effects of the climate factors are greater on the more unsuitable sites. Based on Fig. 22.3 we can say that the culmination of height increments for Bogovadja and Majdanpek will probably be after the age of 15 years and will amount to about 80 cm. For example: the culmination of height increment on location Goc (Fig. 22.1) will be probably at about the age of 15 years and about 130 cm. Because of the unfavourable conditions for Douglas-fir cultivation at Bogovadja and Majdanpek, it is very difficult to define the model for these sites.

Analysis of Height Increment of Young Douglas-fir Stands at Tanda and Juhor

This section presents the results from the sample plot Tanda in eastern Serbia and the sample plot Juhor in central Serbia. As has been stated, a good indicator of site properties and provenance success is the value of height increment. Because of that, we also deal with height increment in this section. Note that all the measurements were taken at the end of the vegetation period each year.

The geographical position of the sample plot Tanda is 44° 14' N, 22° 09' E. Its altitude is 370 m, exposure south-east, on a site of Hungarian oak and Turkey oak (*Quercetum farnetto-cerris* Rud.). Parent rock is granite, and the soil is acid brown, shallow, sandy and moderately dry. Douglas-fir seed was obtained via the FAO, and collected by the Center of Forest Seed in Mackon, USA (Lavadinovic and Koprivica, 1997).

The measured data and the corresponding height increment curves for the provenances with the greatest height increment (Nos 03 and 30) are presented in Fig. 22.4. It can be seen that the increment value is about 90 cm at the age of 16 years. The measurement data and the curves of height increment of the least favourable provenances (Nos 09 and 17) are shown in Fig. 22.5.

The sample plot Juhor is situated between 43° 47' and 43° 55' N and between 18° 52' and 18° 58' E, at an altitude of 660–700 m, on a beech site (*Acetum submontanum* Jov.) (Lavadinovic and Koprivica, 1996).

The more successful provenances (03 and 30, see Fig. 22.6), have a greater height increment than the provenances 09 and 17 (Fig. 22.7), at the location Juhor. In general, there are difficulties in comparing the Tanda and Juhor sites. To illustrate this, the height increments for the best and worst provenances at the age of 10 years on Juhor have values of 60 and 30 cm, respectively. The corresponding values at Tanda are only 30 and 20 cm, respectively. However, at age 15 years, the height increments for the worst provenances on Tanda are greater (55–65 cm) than on Juhor (45–60 cm).

As we have seen, the best results for height increment on a beech site (*A. submontanum* Jov.) in Serbia were obtained in the provenances Nos 03 and 30, and the worst were Nos 09 and 17. The common characteristic of the best provenances is

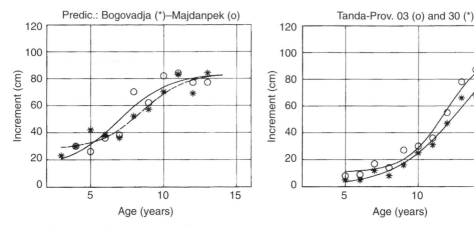

Fig. 22.3. Curves of height increment and the corresponding data.

Fig. 22.4. Current annual height increment.

that they originate from an altitude between 300 and 450 m, latitude between 45° and 47° 7′ and longitude between 122° 4′ and 123° 8′. As the study stands grow at an altitude of about 700 m, in the case of stands on Juhor, it seems that the altitude was the decisive factor in that case.

It is known that inferior provenances are more sensitive to altitude than superior provenances. Because of that, height increment at the age of 15 years, for the provenances 09 and 17 (Fig. 22.5) at the site Tanda, which was about 60 cm, differs from the increment at the site Juhor, which was about 50 cm (Fig. 22.7). For the good provenances (Nos 03 and 30), height increment was practically the same, about 90 cm, both at Tanda and at Juhor (Figs 22.4 and 22.6, respectively).

Modeling Based on Neural Networks

The modelling of current annual height increment, in the conditions of different sites, is difficult to implement using traditional linear or non-linear regression approaches. With the recent advances in the technology of artificial neural networks

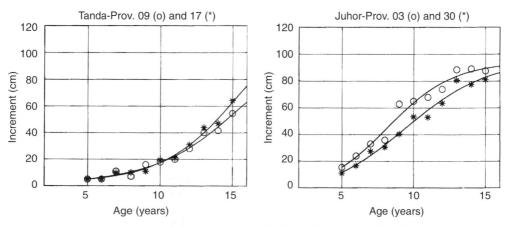

Fig. 22.5. Current annual height increment.

Fig. 22.6. Current annual height increment provenances Nos 03 and 30.

(ANN; Haykin, 1994), it is now possible to develop very successful non-linear ANN models to examine complex increment–age relationships. Note that the problem of generating the model which represents the increment–age relationship for a study site (location) is in fact, in the simplest case, a problem of curve fitting. However, the application of neural networks (NN) provides important advantages of experiential knowledge acquisition. Because of that, NN are able to extract a non-linear model from the observed data sets and to generate predictions more easily than non-linear regression approaches.

Generalization is the major attractive property of NN. The generalization is to use information that NN have collected during the training period in order to synthesize input–output mapping with novel data. The small mean squared error of the obtained model with test data means that the good generalization capability is achieved. Note that the very accurate approximation of a function class can lead to poor generalization capability (Haykin, 1994).

In our case, we can make predictions using the trained neural network. The training or learning of NN is performed by measured data.

The main difficulty of applying NN techniques in many scientific areas is the problem of over-learning and losing the ability to generalize. The common procedures followed to avoid over-learning mainly include adding a priori knowledge into the model through, for example, reducing the size of the neural network by decreasing the number of neurons in the hidden layer or stopping the training process early (Haykin, 1994; Zhang *et al.*, 2000). The early stopping of the learning or training process can be realized by increasing the sum-squared goal error or simply by decreasing the maximum number of epochs to train.

The best model of the considered non-linear biological process is obtained when a two-layer NN, based on the Levenberg–Marquardt algorithm, is used (NN Toolbox, 2000). The commonly used activation (or transfer) functions are linear functions for output neurons, or for neurons in the output layer, and logistic sigmoid functions for hidden neurons or for neurons in the hidden layer (Zhang *et al.*, 2000). In the cases studied, one neuron in the output layer and 1–2 neurons in the hidden layer are used. The process of learning and the obtained error of modelling in the case of application of one neuron in the output layer and two neurons in the hidden layer are presented in Figs 22.8 and 22.9, respectively.

Note that, in the case when we use only one neuron in the hidden layer, the risk of over-fitting does not exist. Although this software (NN Toolbox, 2000) was origi-

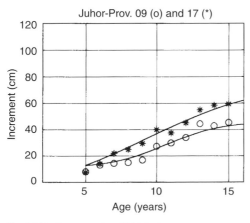

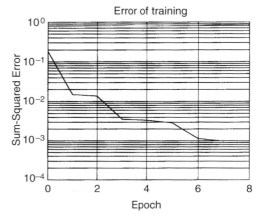

Fig. 22.7. Current annual height increment provenances Nos 09 and 17.

Fig. 22.8. Training the network.

nally developed for electrical engineering purposes, it is user-friendly and can readily be applied in dendrometric studies and analyses (Radonja, 2000, 2001; Radonja *et al.*, 2000).

Generating Data-based Models of Height Increment

Potential uses of the study model depend on the purpose for which it was developed and the assumptions made during its development (Reed, 1997). The most common uses of forest-growth models are to predict the growth at a particular site to enable the manager to make plans for harvesting, i.e. to determine the level of wood exploitation.

Also, the growth model is usually used to produce economic information to facilitate comparison of a number of investment options. For example, which species of conifer or which provenance should we introduce to a particular site?

It is known that forest stand dynamics involve numerous physical, chemical, biochemical, physiological and ecological processes. Besides that, there is sometimes a conceptual problem: it is doubtful that the processes of a higher level (global site properties) can be explained completely in terms of processes at lower levels, for example only using the curve of height increment (Zeide, 1997). Because of that, all the existing process models of forest dynamics are obviously limited to one class of processes. In our case, based on information on the suitability of study sites, we decide which model of height increment will be used. The process of modelling in our case is performed in two steps. In the first step, the measured data are fitted and predicted by neural networks and, in the second step, the model is generated by averaging the obtained curves of height increment. The model shown in Fig. 22.10 is based on data presented in Fig. 22.1.

This model (Model A) is evidently for a very good site, i.e. for a site favourable for Douglas-fir cultivation and introduction. The second model, Model B (Fig. 22.11), is based on data presented in Fig. 22.2 and this model is for standard sites. For unsuitable sites, for which the curves of height increment are shown in Fig. 22.3, we can use the third model, Model C, presented in Fig. 22.12.

Model D1 (data denoted by 'x', Fig. 22.13) is based on data presented in Fig. 22.4, and model D2 (data denoted by '+') is based on data presented in Fig. 22.5. Model D2 is in fact model D1 translated for 3 years. It can be seen that the better

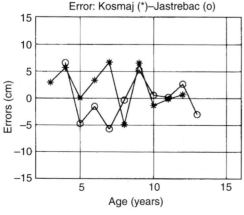

Fig. 22.9. Errors of modelling.

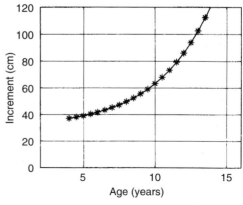

Fig. 22.10. Model A.

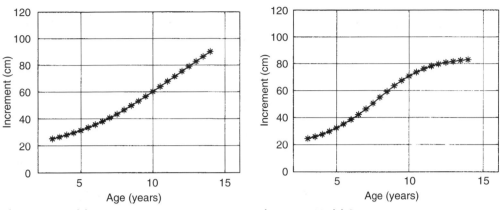

Fig. 22.11. Model B.   **Fig. 22.12.** Model C.

(more successful) provenances (Model D1) reach the culmination of height increment earlier than the less favourable provenances (Model D2).

Model E1, data denoted by 'x', Fig. 22.14, is based on data presented in Fig. 22.6, and model E2, data denoted by '+', is based on data presented in Fig. 22.7. Models E1 and E2 are, in fact, the same model, but with different coefficients which determine height increment at the age of 15 years. In the first case, its value is 90 cm and in the second case, its value is 50 cm. Furthermore, at the age of 10 years, the best provenances have twice as large height increments (Fig. 22.14, 60 cm, data denoted by 'x', model E1) as the unfavourable provenances (30 cm, data denoted by '+', Model E2).

At Tanda, Models D1 and D2, and Juhor, Models E1 and E2, we studied the same provenances. At the very beginning, until the age of 15 years, height increment at Juhor was greater. After that, the height increment at Tanda was greater. There are some difficulties in comparing the sites that correspond to Models A, B and C with the sites Tanda and Juhor, because different provenances were used in these different sample plots.

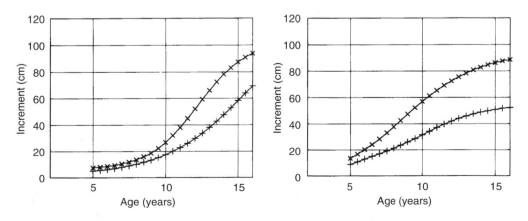

Fig. 22.13. Models D1 (×) and D2 (+).   **Fig. 22.14.** Models E1 (×) and E2 (+).

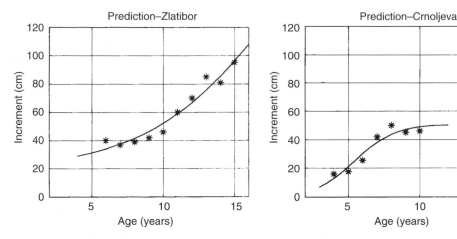

Fig. 22.15. Test of Model A.   **Fig. 22.16.** Test of Model C.

Evaluation of the Proposed Models

In the evaluation of the proposed models, model testing is performed by using two new data sets. Model predictions for suitable sites are tested by the individual curve of height increment that corresponds to the sample plots at the site Zlatibor (Fig. 22.15). It can be seen that Model A (Fig. 22.10, with some standard deviation) includes the curve of the height increment for Zlatibor. We can say that a good agreement has been achieved between theory, Model A, and the new measured data, the curve of height increment of site Zlatibor (Fig. 22.15).

On the other hand, the problem lies in the case of bad (unsuitable) sites or unsuccessful provenances. Model C is tested by the translated individual curve of height increment of the site Crnovljeva Mountain. We have to translate the individual curve of height increment (Fig. 22.16) in both the horizontal and vertical directions. The agreement in this case is very poor. Theoretically, this result is expected (Stamenkovic and Vuckovic, 1988).

Discussion and Conclusions

Evidently, the proposed models cannot be used in any particular cases to obtain precise information for the study sites (locations). However, we can use them to get general information in the statistical sense, for the assessment of more stands with similar conditions for Douglas-fir development. This information is useful for management purposes, for the comparison of different options in decision making. Also, we have seen that NN are very useful and suitable tools for building models of forest processes.

Acknowledgements

This work is part of the project *Structural and productivity characteristics of artificially established coniferous stands and the proposition of optimal forest management.* This research was supported by the Ministry of Science, Technologies and Development of Serbia (BTP.5.06.0516.A).

References

Baldwin, V.C. Jr, Burkhart, E.H., Westfall, A.J. and Peterson, D.K. (2001) Linking growth and yield and process models to estimate impact of environmental changes on growth of loblolly pine. *Forest Science* 47, 77–82.

Haykin, S. (1994) *Neural Networks: a Comprehensive Foundation*. McMillan College Publishing Company, New York, 696 pp.

Johnsen, K., Samuelson, L., Teskey, R., McNulty, S. and Fox, T. (2001) Process models as tools in forestry research and management. *Forest Science* 47, 2–7.

Lavadinovic, V. (1995) Variability of 29 Douglas fir (*Pseudotsuga menziesii*, Mirb. Franco) provenances in test plots in Serbia with the aim of the improvement and introduction of this species. MSc thesis, University of Belgrade, Faculty of Forestry, Belgrade, Yugoslavia [in Serbian].

Lavadinovic, S.V. and Koprivica, J.M. (1996) Development of young Douglas-fir (*Pseudotsuga taxifolia* Britt.) stands of different provenances on beech sites in Serbia. In: *Proceedings of IUFRO Conference 'Modelling Regeneration Success and Early Growth of Forest Stands'*, Copenhagen, Denmark, pp. 390–400.

Lavadinovic, S.V. and Koprivica, J.M. (1997) Development of young Douglas-fir stands of different provenances at oak sites in Serbia. In: Amaro, A. and Tomé, M. (eds) *Empirical and Process-based Models for Forest Tree and Stand Growth Simulation*, 21–27 September, Portugal, pp. 231–241.

NN Toolbox (2000) *Neural Network Toolbox*, Version 6.0.0.88, Release12. MATLAB 6 R12.

Radonja, J.P. (2000) Radial basis function neural networks in tracking and extraction of stochastic process in forestry. In: *Proceedings of the 5th Seminar on Neural Networks Application in Electrical Engineering, NEUREL2000*, 25–27 September. IEEE and Academic Mind, Belgrade, Yugoslavia, pp. 81–86.

Radonja, J.P. (2001) Efficiency procedure of height curve fitting using artificial intelligence method. *Institute of Forestry Collection*, 44–45, Belgrade, 37–50 [in Serbian].

Radonja, J.P., Stankovic, S.S. and Cukanovic, N.R. (2000) Multilayer neural networks in process of height curve fitting. *INFO Science 3/2000*, Savpo, Belgrade, pp. 22–26.

Reed, D.D. (1997) Ecophysiological models of forest growth: uses and limitations. In: Amaro, A. and Tomé, M. (eds) *Empirical and Process-based Models for Forest Tree and Stand Growth Simulation*, 21–27 September, Portugal, pp. 305–311.

Stamenkovic, V. and Vuckovic, M. (1988) *Growth and Productivity of Trees and Forest Stands*. University of Belgrade, Faculty of Forestry, Belgrade, 368 pp [in Serbian].

Vrcelj-Kitic, D. (1982) Plantations of Douglas-fir (*Pseudotsuga menziesii* (Mirb.) Franco) in different site conditions of Serbia. *Monographs, Book 40*, Institute of Forestry and Wood Industry, Belgrade, 151 pp. [in Serbian].

Zeide, B. (1997) What kind of bricks should we use to build growth models: physiological or ecological? In: Amaro, A. and Tomé, M. (eds) *Empirical and Process-based Models for Forest Tree and Stand Growth Simulation*. 21–27 September, Portugal.

Zhang, Q.-B., Hebda, R.J., Zhang, Q.-J. and Alfaro, R.I. (2000) Modeling tree-ring growth responses to climatic variables using artificial neural networks. *Forest Science* 46, 229–239.

23 Simulation and Sustainability of Cork Oak Stands

Nuno de Almeida Ribeiro,[1] Ângelo Carvalho Oliveira,[2] Peter Surovy[1] and Hans Pretzsch[3]

Abstract

Cork oak (*Quercus suber* L.) stands (*montados*) are the most common forest system in southern Portugal. Actual modifications in the management of this agroforestry system have reduced its resilience, compromising sustainability. Simulation results, obtained with a spatial single-tree growth model, were used to test the influence of different strategies of regeneration and management on the sustainability of the system. Crown cover, stand structure and cork production were the variables used to build a sustainability evaluation method.

Introduction

Cork oak (*Quercus suber* L.) stands occupy about 713,000 ha in Portugal (mainly in the south), being one of the most important productive systems due to their ecological and economic outputs.

Actual changes in forest management of cork oak stands, mainly due to the reduction in hand labour and increasing mechanization, combined with installation of new stands, created the need to develop a tool to generate scenarios resulting from these management options (Ribeiro *et al.*, 2001). Spatial tree growth simulators are good tools for creating the necessary scenarios, being able to predict tree growth dependent on site and competition status, permitting the simulation of a large range of management actions (Daniels, 1976; Reynolds *et al.*, 1981; West, 1981; Wensel and Biging, 1988; Holmes and Reed, 1991; Biging and Dobbertin, 1992; Kimmins, 1993; Stage and Wykoff, 1993; Jones and Carberry, 1994; Pukkala *et al.*, 1994; Vanclay, 1994; Bachmann, 1997; Pretzsch, 1997; Bartelink, 1998; Grote and Erhard, 1999).

The cork oak productive system (*montado*) has some peculiarities due to the fact that it is an agro-silvo-pasture system, which implies the existence of conflicting activities. The system is based on the trees, and its sustainability can be jeopardized

[1] Departamento de Fitotecnia da Universidade de Évora, Portugal
Correspondence to: nribeiro@uevora.pt
[2] Departamento de Engenharia Florestal do Instituto Superior de Agronomia, Portugal
[3] Chair of Forest Yield Science, Faculty of Forest Science, Technical University of Munich, Germany

by both intensification of the understorey activities, which leads to a lack of regeneration and consequent disappearance of the crown cover, and extensification, which leads to an invasion of shrubs and other oaks, increasing the competition and the risk of forest fires.

System sustainability is also closely linked to soil loss due to erosion because of the activities related to grazing (soil disking and undercover cultivation). The soil exposure plays an important role in the process of erosion; therefore crown cover is a key factor in controlling the erosive dynamics. The quality of the site for cork oak cultivation is mainly related to the soil depth, structure and nutrient status; therefore erosion has a serious impact on site quality. The loss of crown cover will result in an increasing soil exposure and loss of site quality, leading to an escalating process of stand degradation and consequent decrease in quality and quantity of cork production.

In this chapter simulation results, obtained with a spatial single-tree growth model (CORKFITS 2.1), were used to test the influence of different strategies of regeneration and management on the sustainability of the system.

Materials and Methods

To test system sustainability (ecological and economic) it was decided to analyse the evolution of crown cover, stand structure and cork production (quantity and quality) in 100-year simulation runs using the spatial single-tree growth model CORKFITS 2.1 constructed with sub-growth models that use the potential increment modifier principle (Pretzsch, 1997; Ribeiro *et al.*, 2001).

$$z = zpot \times modifier + \varepsilon$$

where z is the growth variable (height, diameter, bark, etc.), zpot is the potential growth as a function of site, modifier is the reduction factor as a function of spatial competition index and the intensity of debarking, and ε is a random error. CORKFITS 2.1 is shown as a flowchart in Fig. 23.1.

CORKFITS 2.1 is constituted by growth models (cork, stem, tree height and crown), cork production models and mortality models. A structure generator, STRUGEN, based on a filtered Poisson process (Pretzsch, 1992, 1997), and the filters were parameterized for the natural spatial structure of cork oak stands. STRUGEN is used to simulate virtual stands as well as regeneration (Ribeiro *et al.*, 2001). In all the models except potential functions, a random error component is added.

To analyse the results, it was decided to use a cork production index (cpi) that combines production and quality (defined by cork thickness):

$$cpi = dcw \times Q$$

where dcw = is dry cork weight per hectare and Q is a quality index ($Q = \sum_{k=1}^{n} ip_k p_k$ with ip_k = index price for the cork quality k, p_k = proportion of cork weight in cork quality class k, n = number of cork quality classes). The index prices are indexed to the price of the most valuable cork quality class for industry (Table 23.1).

The cpi combined with crown cover is used to evaluate system sustainability in terms of the economic outputs and in terms of soil coverage (protection against erosion).

It was decided to test even-aged and balanced uneven-aged structures as starting points and analyse the evolution of the stands during 100-year simulation runs maintaining constant the site and the intensity of debarking.

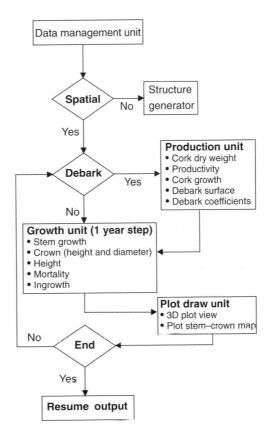

Fig. 23.1. CORKFITS 2.1 flowchart.

Table 23.1. Quality class k and index prices ip_k for the industrial cork quality classes considered.

Calliper class	k	ip_k
14–18 mm	5	22
18–22 mm	4	31
22–27 mm	3	50
27–32 mm	1	100
32–40 mm	1	100
>40 mm	2	66

For the even-aged structures, all spacing combinations were tested between [3×3, 3×4...14×15, 15×15]. For balanced uneven-age structures, all combinations of densities between 60 and 360 trees per hectare distributed by the diameter at breast height (DBH) classes, in the proportions presented in Table 23.2, were tested.

For each structure combination, thirty 100-year simulation runs were made without thinning. This option is related to the restrictions imposed by Portuguese law in relation to the cuttings, as cork oak is a protected species.

The results are presented with the statistics: $\bar{x}\pm95\%$ci where $\bar{x}$ is the estimated mean for 30 repetitions of 1 year of the 100-year simulation and 95%ci is the 95% confidence interval for each mean.

Table 23.2. Proportions of trees in the perimeter classes.

Perimeter class	Proportion
(0,70]	0.40
(70,100]	0.25
(100,130]	0.16
(130,160]	0.10
(160,190]	0.06
(160, + ∞]	0.04

Results

In Figs 23.2 and 23.3 the results are presented for mean cork production index cpi for 100-year simulations of each combination tested. In Figs 23.4 and 23.5 we present the evolution of values and 95%ci of cpi and crown cover percentage for all referred combinations in the 100-year simulation.

In Fig. 23.2 it can be seen that, from an economic point of view, cpi reaches its maximum values in the spacings 5×5, 6×5, 7×5, 8×4 and 8×5, due to the fact that cork quantity (highest in the spacing 5×5) is compensating for cork quality Q (highest in the spacing 8×5). Competition between trees is responsible for reductions in cork growth and therefore its loss of quality.

In Fig. 23.3 it can be observed that for densities above 200 trees/ha the gains in cpi are low, indicting that the higher quantity obtained with higher densities is compensating for lower quality, due to a reduction in cork growth.

If the most usual economic solutions are analysed in terms of the evolution of cpi and crown cover percentage over time (Figs 23.4 and 23.5), information about system feasibility and sustainability can be obtained.

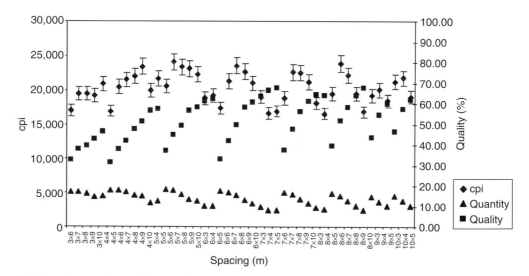

Fig. 23.2. Mean values and 95%ci of cpi, quantity (kg/ha) and quality for each even-aged spacing combination in a 100-year simulation.

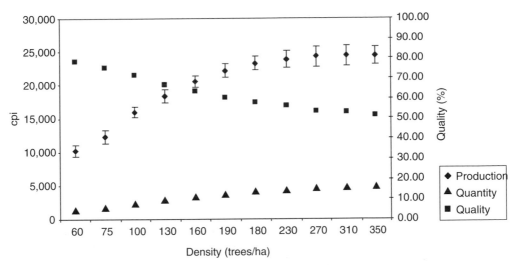

Fig. 23.3. Mean values and 95%ci of cpi, quantity (kg/ha) and quality for each uneven-aged density combination in a 100-year simulation.

Observing Fig. 23.4 for the even-aged structures it can be seen that the more sustainable scenarios (economically and ecologically) are those from the spacings 7×7 (for flat areas, <15% inclination) and 8×5 (for sloping areas, >15% inclination) due to the fact that the crown cover is sufficient for the protection of the soil and the cpi and permits the maintenance of the undercover activities characteristic of the *montado* system. Nevertheless, it can be seen that in these structures all the trees will be decaying at the same time, and therefore the maintenance of a sustained economic output implies that the regeneration (natural or artificial) time must anticipate the disappearance of the older trees.

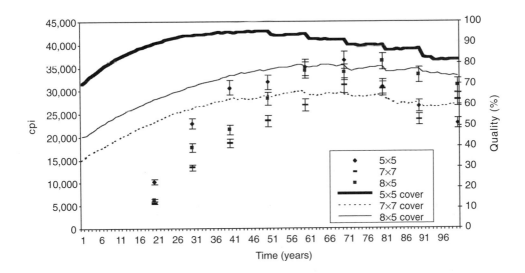

Fig. 23.4. Evolution of values and 95%ci of cpi and crown cover percentage for even-aged structures for 5×5, 7×7, 8×5, spacing combinations in a 100-year simulation.

In Fig. 23.5 the evolution of a balanced uneven-aged structure where the starting density was 160 is the best compromise to maintain the agro-silvo-pasture sustainable system in a flat area, and 200 trees/ha is a good solution for a sloping area. The maintenance of a balanced uneven-aged structure is difficult in this kind of productive system due to grazing management; therefore, although it is more stable than the even-aged structures (Figs 23.4 and 23.5), if no effort is made towards regeneration it will also decay in both crown cover and cpi, compromising the economic and the ecological sustainability.

The solution in order to find the sustainability for this system in time is to force regeneration (natural or artificial) in a periodicity that anticipates crown cover loss as well as production (Fig. 23.6).

As can be seen in Fig. 23.6 it was possible to maintain system stability of cpi and crown cover with only two regeneration moments (at 40 and 60 years) in 100 years. This example shows the importance of correct management in the maintenance of system sustainability with low negative impact on cork production and animal feeding and a positive influence on soil conservation, and therefore on site quality.

Final Remarks

This chapter has demonstrated the flexibility of the spatial tree growth models in the simulation studies of management options concerning the density of the stand.

The combination of the cork production index (cpi) with crown cover and its evolution over time constitutes a good sustainability evaluation method both from an economic and an ecological perspective. The artificial system *montado* is sustainable only if it is economically viable and its ecological stability is maintained.

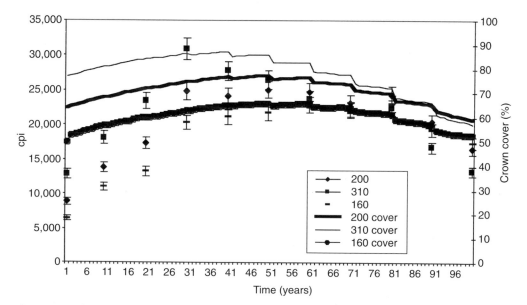

Fig. 23.5. Evolution of values and 95%ci of cpi and crown cover percentage for an uneven-aged density combination in a 100-year simulation.

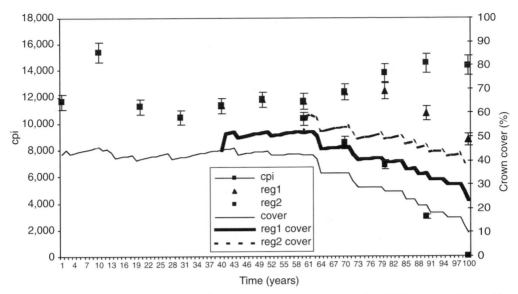

Fig. 23.6. Evolution of values and 95%ci of cpi and crown cover percentage for a 100-year simulation with regeneration at 40 (reg1 cover) and 60 (reg2 cover) years (50 trees/ha).

Both even- and uneven-aged structures are not sustainable without regeneration management. It was shown that it is possible to maintain sustainability with low energy investments and low impact in grazing if regeneration moments anticipate the loss of crown cover.

Finally, we have shown the possibilities of the spatial tree growth simulator CORKFITS 2.1 as a tool to help forest owners make decisions in order to adjust their management to attain system sustainability.

Acknowledgements

This work was supported by ITM QLK5-CT-2000-01349 project.

References

Bachmann, M. (1997) On the effects of competition on individual tree growth in spruce/fir/beech mountain forests. *Allgemeine Forst und Jagdzeitung* 168, 127–130.

Bartelink, H.H. (1998) Simulation of growth and competition in mixed stands of Douglas-fir and beech. PhD thesis, Landbouwuniversiteit Wageningen (Wageningen Agricultural University), Wageningen, The Netherlands.

Biging, G.S. and Dobbertin, M. (1992) A comparison of distance-dependent competition measures for height and basal area growth of individual conifer trees. *Forest Science* 38, 695–720.

Daniels, R.F. (1976) Simple competition indices and their correlation with annual loblolly pine tree growth. *Forest Science* 22, 454–456.

Grote, R. and Erhard, M. (1999) Simulation of tree and stand development under different environmental conditions with a physiologically based model. *Forest Ecology and Management* 120, 59–76.

Holmes, M.J. and Reed, D.D. (1991) Competition indices for mixed species northern hard-woods. *Forest Science* 37, 1338–1349.

Jones, P.N. and Carberry, P.S. (1994) A technique to develop and validate simulation models. *Agricultural Systems* 46, 427–442.

Kimmins, J.P. (1993) *Scientific Foundations for the Simulation of Ecosystem Function and Management in FORCYTE-11.* Report No. NOR-X-328, Edmonton, Alberta, Canada.

Pretzsch, H. (1992) *Konzeption und Konstruktion von Wuchsmodellen für Rein und Mishbestände.* Ludwig-Maximilians-Universität München, Munich, Germany.

Pretzsch, H. (1997) Analysis and modelling of spatial stand structures: methodological considerations based on mixed beech–larch stands in Lower Saxony. *Forest Ecology and Management* 97, 237–253.

Pukkala, T., Kolstrom, T. and Miina, J. (1994) A method for predicting tree dimensions in Scots pine and Norway spruce stands. *Forest Ecology and Management* 65, 123–134.

Reynolds, M.R. Jr, Burkhart, H.E. and Daniels, R.F. (1981) Procedures for statistical validation of stochastic simulation models. *Forest Science* 27, 349–364.

Ribeiro, N.A., Oliveira, A.C. and Pretzsch, H. (2001) Importância da estrutura espacial no crescimento de cortiça em povoamentos de sobreiro (*Quercus suber* L.) na região de Coruche. In: Neves, M.M., Cadima, J., Martins, M.J. and Rosado, F. (eds) *A Estatística em Movimento. Actas do VIII Congresso Anual da Sociedade Portuguesa de Estatística.* SPE, Lisbon, pp. 377–385.

Stage, A.R. and Wykoff, W.R. (1993) Calibrating a model of stochastic effects on diameter increment for individual-tree simulations of stand dynamics. *Forest Science* 39, 692–705.

Vanclay, J.K. (1994) *Modelling Forest Growth and Yield: Applications to Mixed Tropical Forests.* CAB International, Wallingford, UK.

Wensel, L.C. and Biging, G.S. (1988) The CACTOS system for individual-tree growth simulation in the mixed conifer forests of California. In: Ek, A.R., Shifley, S.R. and Burk, T.E. (eds) *Forest Growth Modelling and Prediction.* General Technical Report North Central Forest Experiment Station, USDA Forest Service, Minneapolis, Minnesota, pp. 175–182.

West, P.W. (1981) Simulation of diameter growth and mortality in regrowth eucalypt forest of southern Tasmania. *Forest Science* 27, 603–616.

Part 4

Models, Validation and Decision under Uncertainty

Three broad areas may be identified within this theme: (i) forest models and forest management decision making; (ii) procedures, questions and problems associated with model validation and decision making under uncertainty; and (iii) areas of improvement and future research relevant to model building, model validation and decision making under uncertainty.

Forestry Models and Decision Making

Forestry models play a crucial role in forest management decision making. Over the years, a great many models have been developed. New and improved models are continuing to emerge. The core essence of mimicking or representing reality in an increasingly accurate and precise manner through the scientific modelling process fits particularly well with a decision maker's needs for facilitating the decision process and enhancing the quality of a decision. It appears that no decision maker today could make the right forest management decision without regular recourse to some kind of forest models, although the emphasis and the levels of detail and responsibility for a model builder and a decision maker can be quite different. A sagacious balance between them can sometimes be hard to strike and needs to be continuously pursued.

While recognizing that modelling for understanding is important, many forest models emphasize modelling for prediction instead of modelling for understanding. Forestry modelling as a profession has placed too much emphasis on finding the 'perfect model' which did not serve a decision maker's needs and was rarely useful in the real world. Modellers are called on to develop models that address more of the operational concerns that forest practitioners and decision makers face in day-to-day management of forest resources, and to build models that can really be used to solve real world problems.

The background and the philosophical bent of the modellers influence the models being built. Most decision makers call for simple and practical models instead of complex and idealistic models that modellers wittingly or unwittingly like to build. They also call for models that are relatively easy to use, have reasonably high R^2 values, and are fairly consistent and robust under the wide range of conditions that may be encountered in practice. Simple models also have the merit of closing the communication gaps between decision makers and modellers.

Decision makers care more about the overall performance of a model. Model builders

often focus on the performance of individual components. Since good individual components do not necessarily result in good overall performance (due, in part, to an incomplete
understanding of the interactions among the components), it is very important to assess the
overall performance of a model.

Sensitivity analysis, robustness studies, and risk and uncertainty assessment are important considerations in model building. The core themes of these analyses are the variation of
the input variables to a model to investigate their consequence on the output of the model,
and the need to recognize the importance of the risk of errors and uncertainties in the model.
Decision makers need to look at the potential errors that may be imbedded in the data and
the variables used to develop a model, as well as any structural deficiency of the model and
the modeller's own biases.

Due to the increased use of models in decision making, model credibility is becoming
increasingly important in forest management. This is particularly true when forest managers
and decision makers legitimize their decisions based on models. More recently, reliance on
valid models for making critical decisions about sustainable resource management has
placed model credibility on a more prominent footing. Model validation is one of the most
effective ways to enhance model credibility. Additional efforts should be directed at identifying the important issues relating to model validation. Modellers should devise more precise
and explicit procedures to express and measure the credibility of a model.

There is a large gap between models in theory and models in practice. Decision makers
call for modellers to get in touch with reality more often, and to seek opinions from others;
modellers see models being used for purposes for which they were not designed.

It is important to increase communication among 'non-technical' decision makers, modellers and the public in general. Modellers should make more direct contact with model users
and linguistically-oriented public policy makers. There is a need for modellers to interact and
communicate with people not directly involved in the modelling process.

A shifting paradigm

Traditional models emphasizing a static representation of available data need to be jointed
together with other biological, economic, environmental, social and operational considerations, to allow complex decisions to be made in a more rational manner. Model builders look
more at the technical side of the decision process, and predict what might happen based on
the 'best' available science. Decision makers have responsibilities and accountabilities to
other stakeholders whose livelihood may be dependent on or impacted by the decision.
Expanding modelling to include other relevant considerations allows decision makers to use
a model to assess the full consequence of a decision.

Model Validation – Procedures, Questions and Problems

The achievable goals of model validation are to increase credibility and gain sufficient confidence about a model, and to ensure that model predictions reflect the most likely outcome in
reality. While the idea behind validation may appear simple, it is in fact one of the most convoluted topics associated with model building. There is no set of specific standards or tests
that can easily be applied to determine the 'appropriateness' of a model, nor is there any
established rule to decide what techniques or procedures to use. Every new model seems to
present a unique challenge. There might be some technical and practical difficulties to establishing an agreed-upon 'uniform' validation standard, and the statistical techniques are very
limited in addressing the scope of model validation. However, despite the difficulties, validation remains an integral part of model building and is most effective in establishing the credi-

bility of a model. It is important to have a minimum validation procedure. Such a procedure may not act as 'the standard', but rather as a way to ensure some minimum level of uniformity for validating a great many models in order to separate good models from poor models. Several of the most important elements recommended to be included in such a procedure are proposed and discussed in the contributions to this section, together with some of the relevant questions and unsolved problems:

1. *Independent validation data sets.* An assessment of a model's validity must be done on a separate validation data set that is independent of the model-fitting data. The 'data-splitting' validation approach might still have some utility as a way of cross-validation. However, because of the potential variations involved in data splitting, the dependence between the split data sets, and the easy manipulation of the data-splitting process, it is important to limit the use of this approach unless a more repeatable procedure can be developed and a more consistent reporting standard can be followed.

2. *Dynamic validity and simulations.* There should be some built-in mechanisms to allow the model to be assessed in a dynamic way to better understand the 'black box' (model). The inner-working of a model and the consequence of uncertainty should be evaluated. Sensitivity analysis is an important part of model validation; the systematic investigation of the reaction of model responses to potential scenarios of model input should be considered as an integral part of model validation.

3. *Overall performance and the performance of individual components.* Model builders need to assess the overall performance of a model, as well as the performance of the individual components within the model. Due to the complex nature of the interactions among model components, the understanding of the interactions among model components is very important in model validation.

4. *Model generality.* Good models have greater generality and are expected to show greater usefulness over a wide range of areas. Modellers often put too many unrealistic restrictions on a model, such as 'a fitted model should be used within the range of the original data used to fit the model'. This is easier said than done and is frequently ignored in application. Model users are often unaware of the data range used to fit the model, and an appropriately fit model should be able to be extrapolated beyond the original data to a certain point.

5. *Model simplicity, practicality and operability.* It is important to develop simpler models or simplified versions of models for wide applications. There are many examples in which model users and decision makers have been increasingly frustrated at modellers who build complicated models with no real use in mind and that solve no real world problem.

6. *Theoretical soundness and biological realism.* The biological logic incorporated in a model and the biological interpretation of parameters and model components are very important in building and validating a model.

7. *Independent validation.* Individuals who are not directly involved in the modelling process may have decidedly different objectives and viewpoints, and may not understand the subtleties of a model. Independent validation lends more credibility to a model and is thus desirable to do, and forces a modeller to direct more effort and devise more explicit procedures to convey the modeller's own confidence to a third party not involved in the initial modelling. Independent validation also asks a modeller to be more diligent about the modelling process and places more emphasis on the utility of a model in meeting the objectives of model users and solving real world problems. It provides an added safety measure against spurious results.

8. *Visualization.* Visualization is one of the most important parts in model validation (and model building). The role of visualization needed to be put in a more prominent position; properly laid out graphics can quickly show the goodness of prediction of a model on a single or several graphs. Graphics are also more powerful than statistics in many cases and have the merit of closing the communication gaps between 'technical' and 'non-technical' people.

9. *Validation statistics and statistical tests.* Standard validation statistics such as mean prediction error and percentage error, mean absolute difference, mean square error of prediction and prediction coefficient of determination need to be included in the validation process. As for the use of statistical hypothesis tests, they might be useful for validating or disproving a model. Others maintain that the usefulness of statistical tests in model validation is extremely limited, and that such tests are often misinterpreted. All agree that the role of statistical tests in model validation deserves a more careful evaluation, and the question of whether or not validation needs to involve testing a specific validity hypothesis should be studied further.

10. *Objective documentation of validation results.* The poor and selective use of statistics and statistical tests by some modellers must be avoided and the establishment of a more objective and relatively standard way of reporting validation results promoted.

The contributions that follow address model documentation and validation, as well as decision making under uncertainty. The editors thank Shongming Huang (Canada) for coordination of this section.

24 A Critical Look at Procedures for Validating Growth and Yield Models

S. Huang[1], Y. Yang[1] and Y. Wang[2]

Abstract

This chapter discusses the general procedures and methodologies used for validating growth and yield models. More specifically, it addresses: (i) the optimism principle and model validation; (ii) model validation procedures, problems and potential areas of needed research; (iii) data considerations and data-splitting schemes in model validation; and (iv) operational thresholds for accepting or rejecting a model. The roles of visual or graphical validation, dynamic validation, as well as statistical and biological validations are discussed in more detail. The emphasis in this chapter is placed on the understanding of the validation process rather than the validation of a specific model. The limitations and the pitfalls of model validation procedures, as well as some of the frequent misuses of these procedures are discussed. Several technical and practical recommendations concerning the validation of growth and yield model are made.

Introduction

Once a model is fitted, an assessment of its validity using an independent data set is needed to see if the quality of the fit reflects the quality of predictions. The basic idea behind model validation is to see if a fitted model provides 'acceptable' performance when it is used for prediction. If it provides acceptable performance with a small error and a low variance, the model can be considered appropriate. Otherwise, the model is inappropriate and needs to be adjusted or even re-fitted if it is going to be used at all. The achievable goals of validation are to increase the credibility and gain sufficient confidence about a model, and to ensure that model predictions represent the most likely outcome of the reality.

While the idea behind model validation may appear simple, it in fact is one of the most convoluted and paradoxical topics associated with model building. Numerous research articles have been written on how to conduct appropriate model validation, yet the 'alchemy' of model validation still appears blurred, with no

[1] Forest Management Branch, Land and Forest Division, Alberta Sustainable Resource Development, Canada
Correspondence to: Shongming.Huang@gov.ab.ca
[2] Northern Forestry Centre, Canadian Forest Service, Canada

generic or best solution to the problem. Balci and Sargent (1984) listed a total of 308 references dealing with different methods of model validation. Many more have been published since then (Smith and Rose, 1995; Kleijnen, 1999; Sargent, 1999). A database devoted entirely to model validation is available (manta.cs.vt.edu/biblio). Various approaches have also been suggested in the forestry literature and used in a variety of settings (e.g. Freese, 1960; Ek and Monserud, 1979; Reynolds *et al.*, 1981; Burk, 1988; Marshall and LeMay, 1990; Soares *et al.*, 1995; Nigh and Sit, 1996; Huang *et al.*, 1999).

The validation of a growth and yield model is made more complicated by the fact that the 'model' may consist of many submodels or functions, each independently or simultaneously estimated using different techniques. It is a validation of a system of models rather than a single function. While the behaviour of the individual components within the system plays an important role in determining the final outcome, and it is desirable to have good individual components, the system outcome is usually considered relatively more important in practice when validating a growth and yield model.

Model validation is critical in the development of a growth and yield model and in establishing the credibility of such a model. Unfortunately, there is no set of specific standards or tests that can be easily applied to determine the 'appropriateness' of a model. In addition, there is no established rule to decide what techniques or procedures to use. Every new model presents a new and unique challenge. This is also the same challenge faced by many researchers in other areas (Gass, 1977; Sargent, 1999).

The primary objectives of this chapter are to: (i) review and discuss a number of commonly used validation procedures and associated problems relevant to validating growth and yield models; (ii) assess the role of statistical techniques in validating growth and yield models; (iii) provide a set of guidelines for looking at the validity of growth and yield models; and (iv) discuss some of the misconceptions, limitations and pitfalls associated with model validation. While the graphical, statistical and biological validations are examined in more detail, the emphasis is on the understanding of the validation process rather than on validating a particular model, and on the holistic rather than the segmented approach towards validating growth and yield models. Because of the imprecision, diversity and ambiguity of model validation, several practical recommendations concerning the validation of growth and yield models are made. Due to the contentious nature of model validation, some of the recommendations may not fit into the commonly accepted norms. They are made with the main purpose of stimulating further discussion on model validation, eventually leading to a better procedure for validating forestry models. Six published benchmark data sets were used in this study for illustration (see Appendix). They were chosen from standard regression textbooks and previous validation examples.

The Optimism Principle and Model Validation

Mosteller and Tukey (1977) stated that testing a model on the data that gave birth to it is almost certain to overestimate its performance, because the optimizing process that chose it from among many possible models will have made the greatest use possible of any and all idiosyncrasies of those particular data. As a result, the model will probably work better for these data than for almost any other data that will arise in practice. Picard and Cook (1984) elaborated more on this optimism principle. For the general linear model $Y = X\beta + \varepsilon$, where $\varepsilon \sim (0, \sigma^2 I)$, the ordinary least squares (OLS) estimator for β is obtained by minimizing the sum of squared errors. This estimator (i.e. $b = (X'X)^{-1}X'Y$) is unbiased, consistent and efficient with respect to the class of linear unbiased estimators. It is the best in the sense that it has the minimum variance among all linear unbiased estimators (Judge *et al.*, 1988).

The regression fitted through OLS has a number of properties, including: (i) the sum of the errors is zero, $\Sigma(y_i - \hat{y}_i) = \Sigma e_i = 0$, where y_i is the *i*th observed value and $\hat{y}_i$ is its prediction from the fitted model ($i = 1, 2, ..., n$); and (ii) the sum of the squared errors is a minimum, $\Sigma(y_i - - \hat{y}_i)^2 = \Sigma e_i^2 = \text{min}$. Because Σe_i^2 is a minimum, many statistical measures dependent on Σe_i^2, such as the R^2 (coefficient of determination) and the mean square error (MSE), are the 'best' for the OLS fit. A minimum MSE implies that the model gives the highest precision on the model-fitting data. This will tend to underestimate the inherent variability in making predictions from the fitted model. It is invariably observed that the fitted model does not predict new data as well as it fits existing data (Picard and Cook, 1984; Reynolds and Chung, 1986; Burk, 1988; Neter *et al.*, 1989), and the least squares errors from the model-fitting data are the 'best' (i.e. smallest), and are generally smaller than the prediction errors from the model validation data. In fact, the fitted model from OLS will probably fit the sample data better than the true model would if it were known (Rawlings, 1988), and fit statistics from the model-fitting data give an 'over-optimistic' assessment of the model that is not likely to be achieved in real world applications (Reynolds and Chung, 1986). Picard and Cook (1984) noted that 'a naïve user could easily be misled into believing that predictions are of higher quality than is actually the case'. The understanding of this optimism principle is very important in model validation. What it signifies is that a model can still be acceptable even though it is less accurate and/or less precise on the validation data set.

Given a statistic such as Σe_i or MSE from the model-fitting data set, the optimism principle implies that one should expect that a similar quantity from the validation data set would be larger. However, if it is much larger, the fitted model is likely to be inadequate for prediction because of a large prediction error and/or a large variance. The relevant question becomes: how much larger can the statistic from the validation data set be and still be considered acceptable? Many researchers have considered such a question, but found there was really no simple answer (Berk, 1984; Sargent, 1999).

Model Validation Procedures and Problems

In order to validate a growth and yield model, various procedures have been used. Each of these procedures validates a part of a model, and each has limitations when used in a segmental manner. The calls for a holistic approach are evident in Ek and Monserud (1979), Buchman and Shifley (1983), Soares *et al.* (1995) and Vanclay and Skovsgaard (1997). In general, the following procedures, each with its own shortcomings, need to be considered when validating a growth and yield model.

Visual or graphical validity

Since growth and yield models often consist of many submodels composed of functions describing different components, it is sometimes difficult or time-consuming, if not impossible, to validate these models in an efficient manner. The difficulty is usually caused by the enormous number of potential interactions among the submodels, and the inability to judge the overall goodness of prediction of a model based on the validation of the submodels. Visual or graphical inspection of model predictions provides the most effective method of model validation. It has been used routinely in many fields involving modelling and simulation (Golub and Schechter, 1986; Cook, 1994; Frey and Patil, 2002). Properly laid out graphics can quickly show the goodness of prediction of a growth and yield model on a single or several graphs. These graphs verify not only the individual components of the model, but also the potential interactions among them, and the overall performance of the model. They can also reveal whether the model is behaving within the expected bounds and limits, is in concert with the collected data and agrees with the generally accepted biological theories/consensus of growth patterns.

In practice, visual or graphical inspection of model predictions involves visually examining the fit of the model on the validation data, and then deciding whether the fit appears reasonable. Often, the following graphs are especially valuable in detecting the goodness of prediction: (i) plots showing the fitted curve(s) overlaid on the validation data; (ii) plots showing the observed versus the predicted y-variables; (iii) plots showing the prediction errors versus the predicted y-values; (iv) plots showing the prediction errors versus the observed values of the x-variable(s); and (v) plots showing the trajectories of observed and predicted values over time, and prediction errors over time (e.g. Fig. 24.1). Since many of the growth and yield models involve the time factor and repeated measurements, it is very instructive to plot the trajectory against the time sequence. Such a plot can show whether the prediction variance is changing or if the model is behaving well across the entire range of the time factor, or at various time classes.

The graphs are interconnected. Their main purpose is to show the goodness of predictions in an intuitive, obvious and unfiltered manner. They provide a compelling, yet easy to understand assessment of model predictions without ever going into the more complicated technical details. If the prediction errors look large, skewed and display some obvious undesirable patterns, there is no need to waste additional resources to validate the model further. In addition, visual or graphical validation is extremely powerful in detecting model inadequacies and helping to devise a strategy to correct the error. In fact, it is even more powerful than most of the more intricate procedures and statistical methods. Many of the technologically most advanced models, such as those developed for air battle and missile defence, are verified routinely through graphical methods (Golub and Schechter, 1986).

Visual or graphical validation should be done for the overall performance of a model, as well as for the individual components within the model, although the emphasis and the levels of detail for model builders and model users can be quite

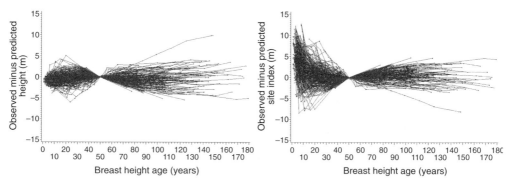

Fig. 24.1. The trajectories of height and site index prediction errors (from Huang, 1997). Each line shows the height or site index prediction errors over time for one stem analysis tree.

different. Model builders need to look at the overall performance and the performance of individual components with equal vigilance, while model users may
choose to put more emphasis on the overall model performance. The caveats are:
(i) good overall performance does not necessarily indicate good individual components; and (ii) good individual components do not necessarily add up to good overall performance. The second point is particularly noteworthy. Often we see
modellers trying very hard to get the 'best' individual components, yet when the
best components are put together, the whole model is almost certain to collapse
(due, at least in part, to an incomplete understanding of the complex nature of the
interactions among the components).

For most practical applications, visually inspecting the fit of the model may be
all that is needed. This is one of the most efficient ways to show the overall picture
of model performance, and to endow the model with credibility and model users
with confidence. To a large extent, the statistical methods, such as those discussed
later, mostly corroborate or reinforce the observations made through a conscientious
examination of the graphics. Many of the statistical methods are in fact less powerful than the graphics.

The disadvantage of the graphical methods is that they can be fairly subjective
at times. Given the same set of validation graphs, different people may reach different conclusions due to the lack of an objective quantitative measure. Because of its
reliance on judgement and opinion, visually inspecting the graphs is considered by
some people to be less definitive and convincing. It is sometimes necessary to validate a model further by combining the graphical approach with statistical methods.

Statistical validity: validation statistics and statistical tests

A large number of validation statistics can be used as quantitative measures for
assessing the goodness of prediction of a fitted model on the validation data. These
assess the size, direction and dispersion of the prediction errors in a quantitative
way (in contrast to the graphical means as discussed earlier). Several of the most frequently used prediction statistics are: mean prediction error ($\bar{e}$) and percentage error
($\bar{e}$%), mean absolute difference (MAD), mean square error of prediction (MSEP), relative error in prediction (RE%) and prediction coefficient of determination (R_p^2):

$$\bar{e} = \sum\nolimits_{i=1}^{k}(y_i - \hat{y}_i)/k = \sum e_i/k \qquad \bar{e}\% = 100 \times \bar{e}/\bar{y} = 100 \times \bar{e}/[\sum\nolimits_{i=1}^{k} y_i/k] \tag{1}$$

$$\text{MAD} = \sum\nolimits_{i=1}^{k}|y_i - \hat{y}_i|/k \tag{2}$$

$$\text{MSEP} = \sum\nolimits_{i=1}^{k}(y_i - \hat{y}_i)^2/k = \sum e_i^2/k \qquad \text{RE}\% = 100\sqrt{\text{MSEP}}/\bar{y} \tag{3}$$

$$R_p^2 = 1 - [\sum\nolimits_{i=1}^{k}(y_i - \hat{y}_i)^2]/\sum\nolimits_{i=1}^{k}(y_i - \bar{y})^2 \tag{4}$$

where k is the number of samples in the validation data, and $\bar{y}$ is the average of the
observed y values. Among the validation statistics, $\bar{e}$ and $\bar{e}$% are the most intuitive
with regard to the size and the direction of the differences between y_i and $\hat{y}_i$. Both
take into account the signs of the prediction errors. The MAD removes the signs of
the prediction errors and gives the absolute differences between y_i and $\hat{y}_i$. The RE%
expresses the MSEP on a relative scale, relative to the $\bar{y}$. The $\bar{e}$, $\bar{e}$% and MAD deal
with 'bias' or 'accuracy', and the MSEP, RE% and R_p^2 with 'precision' of the estimates.

The validation statistics shown in Equations 1–4 provide 'averaged' measures
with respect to the overall model performance. Sometimes, the prediction perfor-

mance at some key data ranges may need to be examined in more detail to determine whether the model performs well at some critical ranges. In these cases, the validation statistics can be computed within classes of the data and by different variables.

Table 24.1 lists the validation statistics obtained for the six validation data sets shown in the Appendix. These statistics can be quite useful in determining the goodness of prediction of a model. However, they have several conspicuous deficiencies. The deficiencies are caused, not by any flaw in the statistic itself, but by the possible misinterpretations of the calculated results. As an example, if $\bar{e}$ is close to zero, one may infer that the model provides an 'unbiased' prediction, and therefore, it is good and can be used for predictions. This may lead to the acceptance of a poor model as illustrated in Huang *et al.* (1999, 2002). In fact, relying on validation statistics alone for determining the adequacy of a model can be misleading. The data in each graph in Fig. 24.2 have the same x and y values, but follow different patterns. The validation statistics calculated using Equations 1–4 would show that the model performed equally well. Clearly this is not true. Graphical validation adds capability that statistics alone do not provide.

Diagnostic tests and their roles in model validation

Many statistical methods, including various tests, are routinely used in determining the goodness of prediction. They include: regression analysis of observed versus predicted values, analysis of variance, paired t-test, regular and non-central χ^2 tests, Theil's inequality test, various F-tests, novel test, sign test, Kolmogorov–Smirnov test, Wilcoxon signed-rank test, Brown–Mood test and analysis of prediction and/or tolerance intervals. Ideally, for a good model, the mean prediction error should not differ from zero so that the prediction is unbiased, and the precision of prediction (i.e. variance) should not exceed some limits so that the variation of the prediction is within some tolerance. Almost all statistical methods are designed to evaluate: (i) if the prediction error is different from zero; and (ii) if the variance of the prediction is larger than some critical value established based on a statistical table or set up specifically for a specific study. Due to the sheer number of the statistical tests available, only some of the most commonly seen are discussed here.

1. The paired t-test, as presented in Snedecor and Cochran (1980) and Rawlings (1988).
2. The χ^2-tests, which can be written in various forms, for example, χ^2_k, χ^2_{k-1} and χ^2_{k-2} (Freese, 1960). Other modified forms of the χ^2-test (Reynolds, 1984; Zucker,

Table 24.1. A summary of the validation statistics for the six validation data sets listed in the Appendix.

Validation data set	$\bar{e}$ (1)	s_e	$\bar{e}\%$ (1)	MAD (2)	MSEP (3)	RE% (3)	R_p^2 (4)
I (MP)	0.027	4.872	0.105	3.845	22.152	18.225	0.896
II (Neter)	0.071	0.047	3.290	0.072	0.007	3.937	0.917
III (Rawl)	289.0	505.5	6.133	329.0	31345	11.882	0.900
IV (West)	−0.191	4.057	−0.340	3.048	15.714	7.079	0.861
V (EM)	8.319	7.371	8.384	8.694	120.14	11.047	0.913
VI (RC)	8.746	55.609	4.872	37.273	3119.8	31.112	0.800

The relevant statistics are defined by the equations in parentheses, and s_e is the standard deviation of the prediction errors calculated according to Snedecor and Cochran (1980).

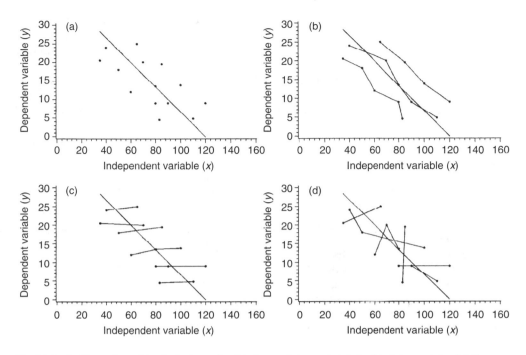

Fig. 24.2. An illustration of the limitations of validation statistics and statistical tests in model validation (from Huang *et al.*, 2002). The same data (dots) appear in each graph, giving the same prediction errors. Validation statistics and tests cannot detect the patterns or trends shown in the data.

1985) have also been conducted. Results are not reported in this study but are available elsewhere (Huang *et al.*, 2002).

3. The *F*-test of the regression $y_i = b_0 + b_1 \hat{y}_i$. If the model is a good one, the regression will be a 45° line through the origin. Thus, the adequacy of the model can be determined by testing if $b_0 = 0$ and $b_1 = 1$, separately (using *t*-tests), or simultaneously using the *F*-test (Montgomery and Peck, 1992).

4. The *F*-test of the regression $e_i = b_0 + b_1 \hat{y}_i$. For a good model, the regression will be a horizontal line running through the origin. The adequacy of the model to be validated can be determined by testing if $b_1 = 0$ (and $b_0 = 0$). The *F*-test from standard regression software can be used for this purpose.

5. The novel *F*-test of the regression $e_i = b_0 + b_1(y_i + \hat{y}_i)$, as described by Kleijnen *et al.* (1998).

6. The Kolmogorov–Smirnov (KS) test and the modified KS test, as described in Stephens (1974).

7. The Wilcoxon signed-rank test, as described in Daniel (1995).

To evaluate the usefulness of the above tests, first, the simple linear regression $y_i = b_0 + b_1 \hat{y}_i$ was fitted for each of the six validation data sets. Two examples of such fits are shown in Fig. 24.3. Table 24.2 lists the results of the tests. For separate *t*-tests, the *t* value for b_0 is read directly from the fit of $y_i = b_0 + b_1 \hat{y}_i$. The *t* value for testing $b_1 = 1$ is computed according to Montgomery and Peck (1992).

The results shown in Table 24.2 are both interesting and puzzling. Each model passed one or more tests, but none of the models passed all tests. In a practical sense, this could imply that one could apply different test statistics, choose the one that satisfies your objective, and accept or reject a model at will. Such a practical argument might have some merits, but it does not necessarily hold true all the time.

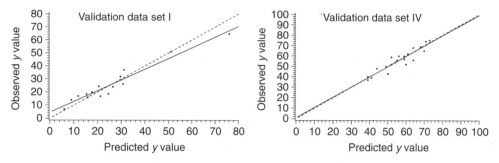

Fig. 24.3. Example regressions (solid lines) between observed and predicted values (dashed lines represent the 45° lines). The validation data sets are listed in the Appendix.

Table 24.2. Results of different tests on six validation data sets.

		Validation data set (listed in the Appendix)					
Test		I (MP)	II (Neter)	III (Rawl)	IV (West)	V (EM)	VI (RC)
Paired t		0.02	11.08*	1.81	−0.22	4.51*	1.25
Chi-squared	χ^2_k	373.75*	62.71	105.79*	78.86*	146.32*	4569.9*
	$\chi^2_{(k-1)}$	373.74*	18.91	77.60*	78.67*	62.04*	4577.8*
	$\chi^2_{(k-2)}$	220.91*	17.90	34.73*	78.53*	55.17*	3762.2*
Separate t	$b_0 = 0$	2.51*	−0.20	−2.24*	0.14	0.39	−2.32*
	$b_1 = 1$	−3.00*	1.72	3.14*	−0.18	1.32	3.36*
Simultaneous F	$b_0 = 0$ and $b_1 = 1$	4.50*	65.16*	8.18*	0.04	11.57*	6.55*
Standard F	$b_1 = 0$	8.99*	2.94	9.87*	0.03	1.74	11.28*
Novel F	$b_0 = 0$ and $b_1 = 0$	2.78	69.23*	11.70*	0.25	13.06*	18.62*
KS		0.17	0.55*	0.41*	0.17	0.49*	0.18*
Modified KS		0.71	4.08*	1.42*	0.80	2.07*	1.47*
Wilcoxon		54	−6.25*	9.50*	94	4.00*	−0.17

The statistics for separate t-tests are obtained from the fit of $y_i = b_0 + b_1 \hat{y}_i$; the statistics for the standard F-test are obtained from the fit of $e_i = b_0 + b_1 \hat{y}_i$, where y_i is the ith observed value, $\hat{y}_i$ is its prediction from the fitted model ($i = 1, 2, ..., k$), k is the number of observations in the validation data, and $e_i = (y_i - \hat{y}_i)$.

In model validation, the inappropriateness of the paired t-test has been discussed by many (Freese, 1960; Ek and Monserud, 1979; Reynolds *et al.*, 1981). It was included in this study because it is still used in practice to evaluate the appropriateness of a model. Freese (1960) pointed out that the paired t-test 'uses one form of accuracy to test for another form, frequently with anomalous results'. Huang *et al.* (1999, 2002) provided other examples in which the paired t-test could result in a misleading outcome. Testing whether or not $\bar{e}$ is zero may be useful sometimes, but using the testing result as a criterion for accepting or rejecting a model is not really relevant in model validation.

The χ^2-tests were used as an alternative for testing the accuracy and determining the applicability of forestry models. Freese (1960) specified an allowable error of ±7.16% within the true (observed) mean. The same allowable error of ±7.16% was used in this study for all data sets. A critical allowable error, $E_{crit.}$, expressed as a percentage of the observed mean, can be computed by re-arranging Freese's (1960) χ^2 statistics, for instance, for validation of data set I,

$$E_{crit.} = E / \bar{y} = \sqrt{\tau^2 \lambda^2} / \bar{y} = \sqrt{\tau^2 \sum (y_i - \hat{y}_i)^2 / \chi^2_{crit.}} / \bar{y} = 0.277 = 27.7\% \qquad (5)$$

If the specified allowable error is within ±27.7% of the observed mean, the χ^2_k test will indicate that the model gives satisfactory predictions; otherwise, it will indicate that the predictions are unacceptable.

The χ^2-tests have some limitations as well. They require a statement of the acceptable accuracy, which is often set in a fairly subjective manner depending on the objective of the study and the specific situation involved. If a more objective accuracy, such as that from the model-fitting data, is used and compared (or tested) against its counterpart from the validation data, the optimism principle dictates that the test result is almost certain to indicate that the model is 'poorer' on the validation data, and this may lead to wrongly concluding that the model is inadequate because it produced less accurate or less precise predictions. Berk (1984) demonstrated that the relative efficiency measure (r_E) defined by r_E = MSEP/MSE could exceed 2 even when a model predicts well. For Montgomery and Peck's (1992) and Neter *et al.*'s (1989) data, the MSEP values are about two and three times, respectively, larger than the MSE values from the modelling data. However, in both cases, they concluded that the MSEPs were not greatly different from the MSEs, and this (together with other criteria), supported the validity of their models. When the χ^2-tests were conducted (Table 24.2), they suggested that, in Montgomery and Peck's case, the model is inaccurate, although $\bar{e}$ is very close to zero; and in Neter *et al.*'s case, the model is accurate, although $\bar{e}$ is far from zero. A more striking phenomenon occurred on data set IV from West (1981). The χ^2-test showed that the predictions are unacceptable, even though all other tests and the graph in Fig. 24.3 suggested that the model predicted the data almost perfectly. All of these results imply that the χ^2-tests produce mixed signals and reject valid models too often. This is evident in Table 24.2.

The separate *t*-tests and the simultaneous *F*-test of the intercept b_0 = 0 and the slope b_1 = 1 also produced mixed signals when compared between themselves or with other tests. For data set II, for instance, separate tests led to the acceptance of b_0 = 0 and b_1 = 1, but the simultaneous *F*-test rejects it. The χ^2-tests accept the model, but the *t*-test and the KS tests reject the model (see Table 24.2).

The simultaneous *F*-test of b_0 = 0 and b_1 = 1 for $y_i = b_0 + b_1 \hat{y}_i$ (see Fig. 24.3) is intuitively the most reasonable test. However, it is a very sensitive test that tends to reject appropriate models much too often. For instance, Montgomery and Peck (1992) indicate that the predictions on data set I are quite reasonable, and Neter *et al.* (1989) conclude that the predictions on data set II are appropriate. In both cases, however, the simultaneous *F*-test leads to the rejection of the predictions (Table 24.2).

A casual look at the model validation literature revealed that the simultaneous *F*-test was one of the most used tests. Presumably, the intuition underlying this test is that if the model is a good one, the regression between y_i and $\hat{y}_i$ should be a 45° line (see Fig. 24.3). However, as early as 1972, Aigner (1972) showed that such an intuition is generally wrong. In more recent studies, Kleijnen *et al.* (1998) and Kleijnen (1999) provided some rather compelling explanations of why such a test is wrong in model validation. They called it the 'naïve test' and showed that even if the model gave excellent predictions, it could still reject b_0 = 0 and b_1 = 1.

The novel *F*-test does not reject a valid model more often than the 'naïve test' does. For example, for data set I, the naïve test rejects the predictions, whereas the novel test does not (see Table 24.2).

Kleijnen *et al.* (1998) also showed other examples in which the naïve test: (i) rejects a truly valid model more often than the novel test does; and (ii) exhibits a 'perverse' behaviour in a certain domain; that is, the worse the fitted model is, the higher the probability of its acceptance. The continued use of the simultaneous *F*-test in validating growth and yield models is not recommended.

The non-parametric tests also produced conflicting results. For instance, for data sets III and VI, the KS and the modified KS tests reject the prediction while the Wilcoxon signed-rank test accepts it (Table 24.2). This again implies that one can selectively choose a test, and accept or reject a model at will.

Many statistical tests and methods are based on some specific distributional assumptions. Even for a non-parametric test such as the KS test, there are still some assumptions required. The assumption of normality is especially common in this regard. For instance, Gregoire and Reynolds (1988) showed that non-normality seriously distorts the power, the interval estimates and the operating characteristic of the χ^2 tests proposed by Freese (1960) and Reynolds (1984). A thorough examination of the assumptions prior to testing may help to decide the 'best' choices of tests and eliminate some of the 'not so desirable' tests. A useful practice in model validation is thus to verify the underlying assumptions. Unfortunately, this is not regularly done in practice, and is exacerbated by the fact that, for instance, the detection of non-normality is not so easy. There are various tests such as Shapiro–Wilk, Anderson–Darling, Shapiro–Francia, Cramér–von Mises, Filliben, and the Watson statistic that can be used, but sometimes with inconsistent outcomes. This was clearly evident when several of the normality tests were conducted on the validation data sets (Huang *et al.* 2002).

The results (Table 24.2) obtained from different tests did show that the usefulness of statistical tests in model validation is very limited. Over the years many statistical tests have been suggested and used in model validation, and new and 'improved' tests are still being proposed. These tests can be useful for some specific purposes, as long as their limitations are adequately realized. However, none of them seems to be generic enough to work in all cases. It appears that testing is a relative concept, relative to a unique situation, a particular set of data, assumptions or constraints. For this reason, some researchers have suggested the use of a number of tests to look at different facets of model behaviour. However, the results obtained in this study showed that this was not really the solution. In fact, it can be the source of confusion. Clearly, this is an area where research is needed. A general strategy for model validation should be to look at how well a model fits new data, rather than use 'a test' to decide whether it is good enough, which may be different depending on the strength of the relationship, the data, model types, study objectives and the 'comfort level' of the individual involved. Model validation is a composite process that may involve use of tests, but test results should not be used as the sole criterion for deciding the validity of a model. In fact, whether or not validation needs to involve testing a specific validity hypothesis is still open to question (see Reynolds, 1984; Burk, 1986).

Biological and theoretical validity

Biological reasonableness is becoming increasingly important in validating growth and yield models. A valid model must 'make sense'. Often, the 'best' model is the one that exhibits some desirable features, such as conformance to some of the basic laws of biological growth, consistency with the understanding of growth theories, and realism and logic in interpretation. Burkhart (1997) pointed out that models that show expected behaviour with respect to biological hypotheses have greater generality and usefulness and inspire confidence on the part of model builders and users.

Other researchers also emphasized the importance of looking at the validity of a model from theoretical and 'bio-logic' points of view (Buchman and Shifley, 1983; Vanclay and Skovsgaard, 1997; Pretzsch *et al.*, 2002).

While inherently the term 'biologically reasonable' appears subjective, due mainly to the lack of a precise definition, there are certain rules and criteria that can be followed. For instance, based on the numerous growth and yield curves developed over time, there is some general agreement on what these curves should look like. All stand growth curves have the same general shape, with increasing growth rates initially, approaching an asymptote or reaching a maximum, then gradually slowing down or steadily declining to zero. This can be illustrated on the growth of the spruce–aspen stands in Alberta (Fig. 24.4). Clearly, the exact shape of an aspen curve will be different from that of a spruce curve. The aspen curve exhibits more rapid early growth with an earlier peak and a much more profound decline following this peak. Based on numerous yield curves developed for Alberta, we have a reasonably good idea of when these peaks occur and what the maximum growth rates and volumes should be for different species and sites at these peaks. Although we cannot say exactly what these values may be (which will be dependent on the actual data), we do have a fairly good sense of whether they are biologically reasonable, as illustrated in Fig. 24.4. A total of 11 aspects, which are by no means exhaustive, may need to be looked at when evaluating the biological reasonableness of growth and yield models. They are listed in Huang *et al.* (2002).

While the biological correctness of growth and yield models is very important, modellers should be cautioned against overemphasis on the 'biological basis' of these models. Occasionally, modellers try too hard to attach 'biological interpretations' to various model parameters. They often take little notice of the fact that many of these so-called biological models are purely statistical or empirical. In this regard, the auto-catalytic logistic model is a particularly 'unfortunate' model, because it was derived purely empirically by Verhulst in 1838 (Seber and Wild, 1989), but in some recent forestry applications it has been treated as if it was derived from a grand theory of the growth of trees. In some researchers' minds, it has acquired a status far beyond that accorded to a model that was empirically derived. In fact, as pointed out by Sandland and McGilchrist (1979) and Seber and Wild (1989), the biological bases of growth models have sometimes been taken too seriously and have led to the development of ill-fitted models. Schnute (1981) showed that the assumptions that constitute the biological bases of biologically derived models could easily be made differently in different situations. Thompson (1989) indicated that the distinction between biological and appropriately built empirical models might not be that succinct.

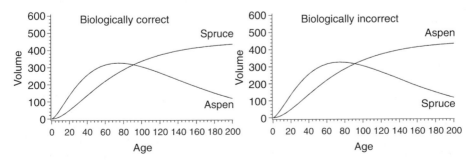

Fig. 24.4. An illustration of biologically correct and incorrect spruce–aspen growth patterns in boreal mixed-wood stands. The curves were generated based on $y = ax^b \exp(-cx)$.

Dynamic and structural validity

As defined by Gass (1977), dynamic validity is concerned with the determination of how the model will be maintained and modified so that it will represent, in an acceptable fashion, the real dynamic world system. Gass (1977) developed a number of principles and criteria to help understand dynamic validity. Several of the relevant ones are: (i) are there procedures for collecting, analysing or incorporating new data? (ii) are there mechanisms built-in for detecting and incorporating the changes into the model? (iii) is there a procedure available to discuss and resolve what to do about divergences between the model solution and the actual outcomes? and (iv) are there procedures available for determining, on a continuing basis, whether the model is still valid in a changing environment?

Dynamic validity recognizes that a model is a dynamic entity, and model validation is not a 'one-shot' process, but, rather, an iterative process (Buchman and Shifley, 1983; Burk, 1986; Soares *et al.*, 1995). There should be some built-in mechanisms to allow the model to be assessed in a dynamic way. The works of Mowrer and Frayer (1986), Gadbury *et al.* (1998) and Kangas (1999) belong to dynamic validity. Dynamic validity can also deal with the structural flexibility and/or compatibility of a model to see if it can be integrated or disintegrated, and what may happen when this is done (Ritchie and Hann, 1997), or if it can be refined to incorporate the most up-to-date data, and the most recent technological advances and knowledge. Model localization procedures (Huang *et al.*, 2002) can also be considered as a part of dynamic validation, as can sensitivity analysis and model extrapolation.

Sensitivity analysis and risk assessment

Sensitivity analysis (SA) and risk assessment of a model is an important consideration in model validation (Gass, 1977; Sargent, 1999; Pretzsch *et al.*, 2002), and the literature on SA is burgeoning (e.g. French, 2001; Frey and Patil, 2002). SA is sometimes called 'what-if' analysis. It can be defined as the systematic investigation of the reaction of model responses to some potential scenarios of the model's input, or to some changes in the model's components, structure and make-up. The core theme of SA is the varying of input variables in a model to investigate their consequence on the output of the model. In growth and yield modelling, SA often relates to questions such as what happens to the growth and yield 'if site index doubles'; 'if an extreme density is used'; 'if one or more of the critical variables involved in model predictions are measured with uncertainty or a 20% error'; 'if one or several of the model components is modified or changed'; and 'if an alternative decision/action is taken'. French (2001) listed many different purposes for which SA may be needed. The most important of these is the development of consequence models to pull together statistical, biological, economic, environmental and operational considerations, and inject these into the analysis to make a more rational decision about a model.

SA requires a set of model runs with the values of the variables used in the model defined by the potential scenarios. The output from the runs needs to be evaluated carefully, especially when it varies for input variations that are within the bounds of realism. One of the primary goals of SA is to decipher the 'black-box' (model) and understand its behaviours better under varying conditions.

Saltelli *et al.* (2000), French (2001) and Frey and Patil (2002) provided good syntheses of various SA methods. Pannell (1997) summarized the utility of these methods from a more practical point of view. Many of these methods are highly mathematical and quite theoretical in nature. They are beyond the scope of this chapter. Interested readers can find details in the cited references. Two simple and practical methods of conducting SA in growth and yield modelling are the computation of a sensitivity index and the graphical techniques of SA. These are described in more detail in Pannell (1997) and Frey and Patil (2002), and summarized in Huang *et al.* (2002).

In spite of the countless applications of SA in other areas of modelling, there are not many cases in which the sensitivity of a growth and yield model has been assessed. Most have dealt with limited entities rather than an entire projection system (Gertner, 1989; Mowrer, 1991; Gadbury *et al.*, 1998). This is surprising considering the usefulness and the wide applications of SA in other fields. It appears that forest modellers have made limited contributions in SA, and that this is an area of much-needed research.

Extrapolation behaviours

Many model builders suggested that a fitted model should be applied within the range of the data used to fit the model (to achieve the best outcomes), and some are totally against the extrapolation of the model outside the range of the original data. This can be rather restrictive and is frequently ignored in practice. Model users are often unaware of the data range used to fit the model, and an appropriately built model should allow extrapolation beyond the original data to some degree.

An assessment of extrapolation behaviours can be done in two ways (Frey and Patil, 2002). First, the input variables can be stretched beyond the original data range. Secondly, specific combinations of values that have not appeared in the modelling can be used as model inputs. If the model still behaves well and makes biological sense, then the model is a reasonable one. Otherwise, it needs to be modified or changed. Regression analysis based on 'best-fit' functions (e.g. polynomials) may approximate the data well. However, since there is no biological basis for choosing these functions, they often lead to unrealistic predictions when extrapolated outside the range of the modelling data. Typically when a base model is chosen based on sound biological assumptions, it will be likely to perform well when extrapolated outside the original data to some degree (not necessarily for all values of the input variables). SA can be used to show how well a model reacts when it is extrapolated (Frey and Patil, 2002). It will also reveal the desirable data ranges to which the model can reasonably be extrapolated.

Operational validity

Operational validity is concerned with how well a model performs in an operational environment (Gass, 1977). Many of the criteria shown in Buchman and Shifley (1983) relate to operational validity. Several specific aspects that operational validity addresses are given in Huang *et al.* (2002). Too often, forest modellers are concerned too much with the technicalities of 'the model', but not enough with the usefulness and operability of the model in describing and coping with real world situations that are the main focus of most model users. Operational validity calls for the model to be checked in a practical, operational environment. As the model is being used in the real world, both the modellers and model users will develop an intuitive feeling about how good and useful the model is, and how operable and viable the model is (in addition to how accurate and precise).

Operational validity plays an important part in deciding the final fate of a model. Due to the lack of operability, some good models are seldom used in practice, yet other models that might be poorly fitted are widely used. Monserud (2001) cautioned that sometimes it does not follow that the best model will serve the user's needs, or will even be implemented. In addition, one person's 'best' model is almost guaranteed not to be the best for another. These cautionary notes, however, should not be taken as an excuse for those modellers whose models are rarely used in practice. Instead, modellers should try harder to address the operational concerns that forest practitioners must face, and to find models that can solve real world problems. This is not contrary to the main objective of many modellers, who attempt to build models that mimic reality, although the emphasis and the level of understanding may be different and a sagacious balance between the 'best model' and the 'best operable model' among various stakeholders can be difficult to strike. This is an old challenge that modellers and model users must face.

Third party validity

There were numerous cases in which modellers validated their own models and showed that their models were valid and better when compared with other models. This was not surprising because each modeller who built a model had a special insight into the operation of the model, and had developed some special attachments to the model. A modeller's own validation may be useful, but it often causes some preconceived preference (or bias) towards the modeller's own model. The fact that almost all models were shown to be valid or better than other models by modellers themselves, yet good models are not often seen in practice, attests to this bias by modellers. While self-validations invariably lead modellers to pat themselves on the back, model users and decision makers are becoming increasingly frustrated that different models built on similar data, each claimed to be scientifically defendable and shown to be better than others by its builders, gave quite different answers to the same problem (Richels, 1981).

The third party validity of a model is thus a very important aspect in model validation (Gass, 1977; Holmes, 1983). It reduces the weight of the opinions (or hidden praises) a modeller showers on his or her own model, and removes the modeller's own biases. Individuals who are not directly involved in the modelling process often have a more sceptical attitude about the model, and their viewpoints can be more objective, pertinent and decidedly different from those of the modellers. The requirement for third party validity also forces a modeller to direct more efforts and devise more explicit procedures to convey the modeller's own confidence to a third party not directly involved in the modelling process. Holmes (1983) presented such a procedure of five elements in an articulate and persuasive manner. The procedure is well suited to the validation of growth and yield models.

Data Validity and Data Splitting

Because the quality of fit does not necessarily reflect the quality of prediction, an assessment of a model's validity on a separate data set is needed. The most critical element relevant to validation data is that such data must be independent of the model-fitting data. Due to the scarcity of such data, some statisticians have proposed splitting the data into two portions of various percentages, such as 50–50%, 75–25% or 80–20%, and use one subset for fitting and the other for an 'independent' validation. Many forest modellers have adopted such an approach. It is interesting to note that the reported outcomes from such an approach always appear the same: the model is good on both the fitting and validation data sets. There is hardly any reporting of a failed model from such an approach although, in reality, one tends to find that many models behave undesirably in application.

What is wrong here? The 'trick' is the 'data-splitting scheme'. Because the split data sets are not independent of each other, the data-splitting scheme used in model validation is not validating a fitted model; instead, it validates the sampling technique used to partition the data. Since there is no standard procedure on how the two portions of the data should be partitioned, various alternatives can be used:

1. The data can be split by 50–50%, 75–25%, 80–20% or any other proportion as the modeller sees fit.
2. Various sampling techniques can also be used to derive two representative sample portions.
3. Assuming that the data were to be split 50–50% randomly, the modeller could repeatedly split the data in endless ways to derive the 'correct' 50–50% split. There is a strong possibility that, in practice, this kind of approach can be easily misused due to its lack of consistency and repeatability. In fact, it may also be 'manipulated', as one can keep sampling until the 'right' sample comes up.
4. The sampling proportion is large. Even for an 80–20% split, 20% of the population is sampled for model validation. This is a huge percentage considering that most of the forest inventories in Alberta and elsewhere sample much less than 1% of the population. Sampling 50% of the population in a 50–50% split is half way to 'census' instead of 'sampling'. It provides the favourable proportions for mirroring the population either way, leading to the best possible illusion for model validation.

In addition to data splitting, other procedures might also be used to evaluate the goodness of model prediction. These include: the conditional mean squared error of prediction (C_p), the PRESS statistic, Amemiya's statistic, various resampling methods with the funny names of 'bootstrap' and 'jackknife', and Monte Carlo simulations (Judge *et al.*, 1988; Dividson and Mackinnon, 1993). All these procedures are correct in their own right. For instance, the C_p and Amemiya's statistic are similar to other goodness of fit measures. The PRESS statistic is similar to data splitting. The resampling methods are used when appropriate sampling results are not available and one needs a non-parametric method of estimating measures of precision (Judge *et al.*, 1988). Through resampling the estimated errors after a model has been fitted to the data, some 'pseudo sample data' are generated to emulate the modelling data, which permit the re-fit of the model. Monte Carlo simulations involving the pseudo sample data are used to approximate the unobservable sampling distributions and provide information on simulated sampling variability, confidence intervals, biases, etc. All these procedures can provide some informative statistics and can be of use for looking at a model from different angles, but their utility in model validation is quite dubious and not clearly understood, for they are heavily dependent on the model-fitting data. This dependence is not consistent with the prerequisite of validating a model on independent data set(s).

While recognizing that data splitting is useful for other purposes (Picard and Berk, 1990), it was felt that because of the variations and the potential subjectivity related to data splitting (re-creation), the practice of splitting the data into two parts should not be used further in validating forestry models, for the reserved data are not independent of the modelling data and there are numerous ways in which the data can be chosen to substantiate a modeller's own objectives and, sometimes, bias. In some ways, the fact that there is hardly any reporting of a failed model from this approach attests to its insoluble flexibilities (and subjectivity). Forest modellers should look for a more steady, consistent and repeatable approach towards the data used for validation. If one cannot gather more data, perhaps, as Davis and Johnson (1987) pointed out unambiguously, one should wait a number of years to obtain the data before such a validation is made. Hastily validating a model on split data or 'pseudo sample data' may produce some nice statistics, but they prove little about the performance of the model on independent validation data.

Thresholds for Accepting or Rejecting a Model

Once the various aspects of a model have been carefully considered, a decision on whether the model is acceptable or not must be made. This sounds simple, but is often the most difficult and tricky task, and the process itself requires a number of fairly subjective pre-decisions on several fronts.

One of the essential elements for determining whether the predictions are acceptable or not is the need for a statement on the required accuracy. A yield table (model) that gave a volume prediction differing by 150 m^3 from the observed volume of 300 m^3 would be considered unacceptable by most people. A yield table that gave a volume prediction within 15 m^3 of the observed volume of 300 m^3 would probably be acceptable to most people, but what about if a yield table gave 20, 30, 50, 60 or 75 m^3 error?

Freese (1960) used an allowable error of ±7.16%. Mowrer and Frayer (1986) suggested that the coefficient of variation should not exceed 20%. Gertner *et al.* (1995) used a guideline of within ±20% of the mean. The same guideline was adopted and recommended by McRoberts (1996). Burk (1986) listed many cases in which the prediction errors from various models are about 5–20% of the mean, with 10% most common. Pillsbury *et al.* (1995) used an allowable error of ±12%. In Alberta, a ±10% prediction error at 95% confidence level was used in several studies as the allowable error.

Obviously, the question of how good is good enough could be answered differently. Any statement on the required accuracy is inherently subjective, and is dependent on various factors (the uniqueness of the data, model, study objective, and the 'comfort level' of the individual involved). While the exact number that determines the acceptability of a model could be argued interminably, for the most part, such argument has very little practical significance. Berk (1984) and Sargent (1999) illustrated this well. As evident from the references cited, a general consensus within the forestry modelling community appears to be that a number somewhere around 10%–20% is realistic and reasonable. Based on the experience in dealing with different growth and yield models and data, the following thresholds are provided here as practical guidelines for accepting or rejecting a model, whenever a quantitative statistic needs to be used in decision making.

1. If the mean prediction error ($\bar{e}$) is within ±10% of the observed mean at 95% confidence intervals, the model should be acceptable, provided the graphical validation shows no adverse pattern.
2. If $\bar{e}$ is greater than ±10%, but less than or equal to ±20%, acceptance or rejection of the model is dependent on the specific circumstances involved. This is the region of uncertainty, and additional data and validation may be needed to help reach a more definitive answer.
3. If $\bar{e}$ is more than ±20% of the observed mean, the model is rejected regardless of the theory.

Although it can be said that models that produce ±10% or more prediction errors are less desirable, the actual choice of the acceptance and rejection levels is somewhat subjective. The practice of relying on ē% as the barometer for accepting or rejecting a model is itself an over-simplification of the immensely complex and ambiguous processes of model validation. In any case, if one chooses to use adjustment techniques (Huang *et al.*, 2002), it is always possible to achieve a zero mean prediction error.

There are other more involved procedures that can be used to determine the thresholds, such as making use of prediction intervals and tolerance intervals (Patel, 1989; Odeh, 1990). Reynolds (1984), Reynolds and Chung (1986) and Nigh and Sit (1996) provided interval-based validation examples in the forestry context. The computations of the intervals, such as the critical error in Equation 5, can provide some useful statistics and assist modellers evaluating model behaviour and establishing thresholds. However, several researchers have recommended caution in the use of an interval-based approach (see Huang *et al.*, 2002). The danger one needs to be aware of in this approach is that a modeller may sometimes set up the thresholds within or outside the computed critical values or intervals subjectively to satisfy some specific purposes.

Towards a Unified Approach to Model Validation

While there is a large and diverse literature on model validation, including a considerable number of reviews, the gap between the theories and the reality of model validation in actual practice has taken on 'canyon sized proportions' (Holmes, 1983). There are numerous unorthodox ways of validating a model (Balci and Sargent, 1984), and no one validation method enjoys widespread application. The fact that each modeller has shown that his/her model was valid or better than others using similar data and for the same question has made the decision process on the validity of a model extremely intricate and discordant.

The need for a unified approach towards model validation is obvious, even though the question of whether such an approach exists remains unsolved, and a universally agreed upon approach may not be easy to find. However, there are certain general rules and criteria that can be followed more consistently so that model validation can be judged upon some standard baselines or benchmarks that apply across the spectrum. This is another area of needed research. As an initial attempt towards such an approach, the following options are presented with the main purpose of inviting further discussion on such a topic, eventually leading to some general consensus or, at best, a solution for validating forestry models.

The 'do-nothing' approach

If the data used in modelling are collected through a well-designed sampling process, and the model is built on some sound theoretical basis (e.g. it 'best' fits the data and is biologically meaningful), the quality of fit will probably reflect the quality of predictions within the same population. Thus, as a rule of thumb, 'don't worry about model validation, so long as the data gathering and the modelling processes are sound, and the model is to be used within the defined peripheries'.

Another consideration for recommending the 'do-nothing' approach is explained by the variations in validation procedures and the lack of a consistent way to use and interpret these procedures. The statistical procedures shown in this study attest to such variations and inconsistencies. Thus, given the inadequacy of the current validation techniques, 'don't worry about model validation' becomes a viable option, for 'a model possesses some inherent invalidities' (Burk, 1986), 'its predictive ability is always open to question' (Soares *et al.*, 1995) and 'it is almost impossible to validate any model' (Monserud, 2001). Leave the problem to the experts and let them figure it out first.

The 'quick and clean' approach

The 'quick and clean' approach involves: (i) examining various prediction graphs; (ii) calculating the prediction statistics as discussed above; and (iii) assessing the size of the prediction statistics in relation to the thresholds and deciding the validity of the model. A prerequisite of the 'quick and clean' approach is that independent validation data must be used.

The quick and clean approach does not involve statistical tests. The validation approach shown by Neter *et al.* (1989) and Montgomery and Peck (1992) can be considered as examples of this approach. For most practical applications, the quick and clean approach is the most efficient and reliable approach that both the modellers and model users can apply without great difficulty.

The conventional approach (the 'data-splitting' approach)

The conventional approach involves splitting the data into two parts, one part for model fitting and the other part for model validation. Sometimes, this approach has also been called 'cross-validation'. It has some utility, but because of: (i) the potential variations involved in data splitting; (ii) the dependence between the split data sets; (iii) the easy manipulability of the data-splitting process; and (iv) the fact that there is hardly any failure from this approach, the continued use of this approach is not recommended, unless a more consistent, repeatable and verifiable procedure can be established. The existing 'data-splitting' approach mostly validates the sampling scheme, not the fitted model.

The 'statistical test' approach

A diverse array of statistical tests has been used in model validation. However, as emphasized in this and some other articles, no one test can incorporate all the elements of model validation and provide a composite and consistent method for accepting or rejecting a model. It is a bit naïve to rely on some kind of test to determine whether a model is adequate and should be accepted, although such a practice is still in use in model validation. Over the years, statisticians have made breakthroughs and discovered many tests that decided the outcome of debates, but it appears that they have not found any reliable test that can decide the fate of a model. It has been suggested that one should look at the results from a number of tests, but such suggestions deserve more scrutiny, because the test results can frequently be conflicting, contradictory and at odds with each other. This is clearly evident in Table 24.2. Statisticians might rightly suggest that forest modellers and model users check the assumptions before each test is conducted but, during the process, other assumptions are introduced. These 'assumptions on assumptions' are difficult to verify or even identify, and are often violated in practice in some way. This is the dilemma that faces statisticians and forest modellers alike.

The 'cynical' approach

Because there is a diverse array of statistical tests that can potentially be used in model validation, and each test may produce different results for the same question and on the same data, a 'smart' method of model validation is to look at a number of statistical tests and selectively choose one or more tests that 'best' suit one's purpose. This type of 'cynical' approach, which has its adherents in practical and non-practical model validation, takes advantage of the variations and difficulties inherent in model validation. Such an approach, however, rarely stands up to the scrutiny of experts and model users, and should be avoided whenever possible. It is sometimes tempting, especially for a decision maker, to choose a statistical test to prop up a decision, and to hope that others are convinced of the test results and unaware of the risks and drawbacks associated with such a test.

The comprehensive approach

The comprehensive approach usually involves: (i) the use of an independent validation data set; (ii) the visual or graphical validation; (iii) the calculation of validation statistics on the entire validation data set or on some predefined classes of the data for the variable(s) of interest; (iv) the examination of the biological and theoretical validity; (v) the understanding of the dynamic behaviour of the model under the changing conditions or operating environment; (vi) considerations on the practicality and operability of a model; and (vii) the validation from a third party not directly involved in the modelling process. The comprehensive approach may involve the use of hypothesis testing, but it advocates a general strategy of looking at how well a model fits the new data from various aspects, rather than using a test to decide whether or not it is good enough, which can be quite different depending on various factors. Gass and Thompson (1980), Holmes (1983) and Sargent (1999) provided ample and succinct descriptions on how such an approach can be conducted in other areas. They are well suited for the validation of growth and yield models in forestry.

Acknowledgements

The reviews and constructive suggestions from Dr Jack Heidt, Dave Morgan, Grant Klappstein and Dr Geoff Wang are greatly appreciated.

References

Aigner, D.J. (1972) A note on verification of computer simulation models. *Management Science* 18, 615–619.

Balci, O. and Sargent, R.G. (1984) A bibliography on the credibility assessment and validation of simulation and mathematical models. *Simuletter* 15(3), 15–27.

Berk, K.N. (1984) Validating regression procedures with new data. *Technometrics* 26, 331–338.

Buchman, R.G. and Shifley, S.R. (1983) Guide to evaluating forest growth projection systems. *Journal of Forestry* 81, 231–234.

Burk, T.E. (1986) Growth and yield model validation: have you ever met one you liked? In: Allen, S. and Cooney, T.M. (eds) *Data Management Issues in Forestry: Proceedings of a Computer Conference and 3rd Annual Meeting of the Forest Resources Systems Institute*, 7–9 April, Atlanta, Georgia, pp. 35–39.

Burk, T.E. (1988) Prediction error evaluation: preliminary results. In: Wensel, L.C. and Biging, G.S. (eds) *Forest Simulation Systems*. Proceedings of IUFRO Conference, 2–5 November, UC Berkeley, California, 81–88.

Burkhart, H.E. (1997) Development of empirical growth and yield models. In: Amaro, A. and Tomé, M. (eds) *Empirical and Process-based Models for Forest Tree and Stand Growth Simulation*. 21–27 September, Portugal, pp. 53–60.

Cook, R.D. (1994) *An Introduction to Regression Graphics*. John Wiley & Sons, New York, 253 pp.

Daniel, W.W. (1995) *Biostatistics: a Foundation for Analysis in the Health Sciences*, 6th edn. John Wiley & Sons, New York, 780 pp.

Davis, L.S. and Johnson, K.N. (1987) *Forest Management*, 3rd edn. McGraw-Hill, New York, 790 pp.

Dividson, R. and Mackinnon, J.G. (1993) *Estimation and Inference in Econometrics*. Oxford University Press, New York, 875 pp.

Ek, A. and Monserud, R.A. (1979) Performance and comparison of stand growth models based on individual tree and diameter-class growth. *Canadian Journal of Forest Research* 9, 231–244.

Freese, F. (1960) Testing accuracy. *Forest Science* 6, 139–145.

French, S. (2001) Modelling, making inferences and making decisions: the roles of sensitivity analysis. In: *Proceedings of the Third International Symposium on Sensitivity Analysis of Model Output*. 18–20 June, Madrid, Spain, p. 75.

Frey, H.C. and Patil, S.R. (2002) Identification and review of sensitivity analysis methods. *Risk Analysis* 22, 553–578.

Gadbury, G.L., Iyer, H.K., Schreuder, H.T. and Ueng, C.Y. (1998) *A Nonparametric Analysis of Plot Basal Area Growth Using Tree Based Models*. USDA Forest Service Research Paper RMRS-RP-2, 14 pp.

Gass, S.I. (1977) A procedure for the validation of complex models. In: Avula, X.J.R. (ed.) *Proceedings of the 1st International Conference on Mathematical Modeling*, 29 August–1 September, pp. 247–258.

Gass, S.I. and Thompson, B.W. (1980) Guidelines for model evaluation. *Operations Research* 28, 431–479.

Gertner, G.Z. (1989) The sensitivity of measurement error in stand volume estimation. *Canadian Journal of Forest Research* 20, 800–804.

Gertner, G.Z., Cao, X. and Zhu, H. (1995) A quality assessment of a Weibull based growth projection system. *Forest Ecology and Management* 71, 235–250.

Golub, J. and Schechter, B. (1986) Graphical verification revisited. In: Crosbie, R. and Luker, P. (eds) *Proceedings of the 1986 Summer Computer Simulation Conference*. Reno, Nevada, 28–30 July, pp. 182–186.

Gregoire, T.G. and Reynolds, M.R. (1988) Accuracy testing and estimation alternatives. *Forest Science* 34, 302–320.

Holmes, W.M. (1983) Confidence building in simulation models as a practical process. In: *Proceedings of the 1983 Summer Computer Simulation Conference*, Vancouver, BC, pp. 195–199.

Huang, S. (1997) Development of compatible height and site index models for young and mature stands within an ecosystem-based management framework. In: Amaro, A. and Tomé, M. (eds) *Empirical and Process-based Models for Forest Tree and Stand Growth Simulation*. 21–27 September, 1997, Portugal, pp. 61–98.

Huang, S., Titus, S.J., Price, D. and Morgan, D.J. (1999) Validation of ecoregion-based taper equations for white spruce in Alberta. *Forestry Chronicle* 75(2), 281–292.

Huang, S., Yang, Y. and Wang, Y. (2002) *Validating and Localizing Growth and Yield Models: Procedures, Problems and Prospects*. Internal Report, Publication No. FMB-06547-G01-2002-06-28, Land and Forest Division, Government of Alberta, Edmonton, Alberta, Canada, 38 pp.

Judge, G.G., Hill, R.C., Griffiths, W.E., Lütkepohl, H. and Lee, T.C. (1988) *Introduction to the Theory and Practice of Economics*. John Wiley & Sons, New York, 1024 pp.

Kangas, A. (1999) Methods for assessing uncertainty of growth and yield predictions. *Canadian Journal of Forest Research* 29, 1357–1364.

Kleijnen, J.P.C. (1999) Validation of models: statistical techniques and availability. In: Farrington, P.A., Nembhard, H.B., Sturrock, D.T. and Evans, G.W. (eds) *Proceedings of the 1999 Winter Simulation Conference*, pp. 647–654.

Kleijnen, J.P.C., Bettonvil, B. and Groenendaal, W.V. (1998) Validation of trace-driven simulation models: a novel regression test. *Management Science* 44, 812–819.

Marshall, P. and LeMay, V. (1990) *Testing Prediction Equations for Application to Other Populations*. School of Forestry, Wildlife Resources, Virginia Tech, Publication No. FWS-3-90, pp. 166–173.

McRoberts, R.E. (1996) Estimating variation in field crew estimates of site index. *Canadian Journal of Forest Research* 26, 560–565.

Monserud, R.A. (2001) The role of models in answering questions of forest sustainability. Presented at the IUFRO Forestry Modelling Conference, 12–17 August, Vancouver, BC, Canada.

Montgomery, D.C. and Peck, E.A. (1992) *Introduction to Linear Regression Analysis*. John Wiley & Sons, New York, 527 pp.

Mosteller, F. and Tukey, J.W. (1977) *Data Analysis and Regression*. Addison-Wesley, 588 pp.

Mowrer, H.T. (1991) Estimating components of propagated variance in growth simulation model projections. *Canadian Journal of Forest Research* 21, 379–386.

Mowrer, H.T. and Frayer, W.E. (1986) Variance propagation in growth and yield projection. *Canadian Journal of Forest Research* 16, 1196–1200.

Neter, J., Wasserman, W. and Kutner, M.H. (1989) *Applied Linear Regression Models*, 2nd edn, Irwin, 667 pp.

Nigh, G.D. and Sit, V. (1996) Validation of forest height–age models. *Canadian Journal of Forest Research* 26, 810–818.

Odeh, R.E. (1990) Two-sided prediction intervals to contain at least *k* out of *m* future observations from a normal distribution. *Technometrics* 32(2), 203–216.

Pannell, D.J. (1997) Sensitivity analysis of normative economic models: theoretical framework and practical strategies. *Agricultural Economics* 16, 139–152.

Patel, J.K. (1989) Prediction intervals: a review. *Communications in Statistics: Theory and Methods* 18, 2393–2465.

Picard, R.R. and Berk, K.N. (1990) Data splitting. *American Statistician* 44, 140–147.

Picard, R.R. and Cook, R.D. (1984) Cross-validation of regression models. *Journal of the American Statistical Association* 79, 575–583.

Pillsbury, N.H., McDonald, P.M. and Simon, V. (1995) Reliability of tanoak volume equations when applied to different areas. *Western Journal of Applied Forestry* 10(2), 72–78.

Pretzsch, H., Biber, P., Ďurský, J., von Gadow, K., Hasenauer, H., Kändler, G., Kenk, G., Kublin, E., Nagel, J., Pukkala, T., Skovsgaard, J.P., Sodtke, R. and Sterba, H. (2002) Recommendations for standardized documentation and further development of forest growth simulators. *Forstwissenschaftliches Centralblatt* 121, 138–151.

Rawlings, J.O. (1988) *Applied Regression Analysis: a Research Tool*. Wadsworth, Belmont, California, 553 pp.

Reynolds, M.R. (1984) Estimating the error in model predictions. *Forest Science* 30, 454–468.

Reynolds, M.R. and Chung, J. (1986) Regression methodology for estimating model prediction error. *Canadian Journal of Forest Research* 16, 931–938.

Reynolds, M.R. Jr, Burkhart, H.E. and Daniels, R.F. (1981) Procedures for statistical validation of stochastic simulation models. *Forest Science* 27, 349–364.

Richels, R. (1981) Building good models is not enough. *Interfaces* 11(4), 48–54.

Ritchie, M.W. and Hann, D.W. (1997) Implications of disaggregation in forest growth and yield modelling. *Forest Science* 43, 223–233.

Saltelli, A., Chan, K. and Scott, E.M. (2000) *Sensitivity Analysis*. John Wiley & Sons, Chichester, UK, 475 pp.

Sandland, R.L. and McGilchrist, C.A. (1979) Stochastic growth curve analysis. *Biometrics* 35, 255–271.

Sargent, R.G. (1999) Validation and verification of simulation models. In: Farrington, P.A., Nembhard, H.B., Sturrock, D.T. and Evans, G.W. (eds) *Proceedings of the 1999 Winter Simulation Conference*, pp. 39–48.

Schnute, J. (1981) A versatile growth model with statistically stable parameters. *Canadian Journal of Fisheries and Aquatic Sciences* 38, 1128–1140.

Seber, G.A.F. and Wild, C.J. (1989) *Nonlinear Regression*. John Wiley & Sons, New York, 768 pp.

Smith, E.P. and Rose, K.A. (1995) Model goodness-of-fit analysis using regression and related techniques. *Ecological Modelling* 77, 49–64.

Snedecor, G.W. and Cochran, W.G. (1980) *Statistical Methods*, 7th edn, Iowa State University Press, Ames, Iowa, 507 pp.

Soares, P., Tomé, M., Skovsgaard, J.P. and Vanclay, J.K. (1995) Evaluating a growth model for forest management using continuous forest inventory data. *Forest Ecology and Management* 71, 251–265.

Stephens, M.A. (1974) EDF statistics for goodness-of-fit and some comparisons. *Journal of the American Statistical Association* 69, 730–737.

Thompson, J.R. (1989) *Empirical Model Building.* John Wiley & Sons, New York, 242 pp.

Vanclay, J.K. and Skovsgaard, J.P. (1997) Evaluating forest growth models. *Ecological Modelling* 98, 1–12.

West, P.W. (1981) Simulation of diameter growth and mortality in regrowth eucalypt forest of southern Tasmania. *Forest Science* 27, 603–616.

Zucker, P.A. (1985) Combined evaluation of predicted model means and variances. In: *Proceedings of the 1985 Summer Computer Simulation Conference,* Chicago, Illinois, pp 164–167.

Appendix

Table A.1. Model validation data sets (part 1).

Case	Data set I		Data set III		Data set IV		Data set V	
	y_i	$\hat{y}_i$	y_i	$\hat{y}_i$	y_i	$\hat{y}_i$	y_i	$\hat{y}_i$
1	51.00	50.9230	2380	2320	62	61	95.4	85.1
2	16.80	21.1405	3190	3300	40	39	88.4	83.6
3	26.16	30.7557	3270	3290	60	56	146.2	137.0
4	19.90	17.6207	3530	3460	52	52	167.1	138.2
5	24.00	26.4366	3980	3770	75	73	52.7	50.4
6	18.55	15.2766	4390	4210	39	41	67.4	70.4
7	31.93	29.6602	5400	5470	50	49	87.9	85.8
8	16.95	11.8576	5770	5510	54	54	106.0	98.0
9	7.00	6.0307	6890	6120	68	64	95.4	86.9
10	14.00	9.0033	8320	6780	37	39	79.4	71.2
11	37.03	31.1640			56	64	125.1	119.6
12	18.62	24.5482			69	68	130.7	118.1
13	16.10	15.8125			59	52	136.1	135.9
14	24.38	20.4524			48	44	63.0	50.6
15	64.75	76.0820			74	71	126.9	110.8
16					60	59	19.8	12.8
17					56	55		
18					65	70		
19					43	50		
20					52	60		
21					57	59		

Data set I (MP) from Montgomery and Peck (1992); Data set II (Neter) from Neter *et al.* (1989); Data set III (Rawl) from Rawlings (1988); Data set IV (West) from West (1981); Data set V (EM) from Ek and Monserud (1979); Data set VI (RC) from Reynolds and Chung (1986); y_i observed 'true' value; and $\hat{y}_i$, predicted value from the model to be validated.

Table A.2. Model validation data sets (part 2).

Case	Data set II y_i	Data set II $\hat{y}_i$	Data set VI y_i	Data set VI $\hat{y}_i$
1	2.0326	1.9316	62.1	44.2
2	2.4086	2.3025	49.2	75.2
3	2.2177	2.0962	55.8	59.3
4	1.9078	1.8700	41.1	56.6
5	2.0035	1.9532	91.3	72.6
6	2.0945	2.0786	61.9	70.9
7	1.7652	1.7559	63.5	82.1
8	1.7925	1.7372	61.9	70.2
9	2.1292	2.0429	62.3	92.0
10	2.2295	2.2075	33.0	54.9
11	2.1524	2.1175	78.0	113.7
12	2.3188	2.2615	88.4	91.3
13	1.9039	1.7725	50.9	92.5
14	2.0508	1.9805	60.4	92.9
15	2.6525	2.6454	118.1	113.6
16	2.2053	2.1170	87.7	101.7
17	1.9246	1.9334	70.3	103.0
18	2.1541	2.0647	103.8	108.0
19	2.4970	2.3941	84.7	83.2
20	1.7237	1.6916	55.1	78.7
21	2.8339	2.7049	40.2	79.6

Case	Data set II y_i	Data set II $\hat{y}_i$	Data set VI y_i	Data set VI $\hat{y}_i$
22	2.1282	2.0570	50.7	82.9
23	2.6884	2.5488	106.4	90.3
24	2.4284	2.3381	122.6	97.2
25	2.0261	2.0317	139.9	120.5
26	2.0843	1.9674	85.6	103.9
27	2.2826	2.1668	115.4	104.4
28	2.2073	2.1295	32.3	61.6
29	2.0443	1.9470	100.8	123.4
30	2.4863	2.3866	90.1	114.3
31	1.9037	1.7530	110.7	130.3
32	2.6647	2.4914	165.6	190.6
33	1.9071	1.9007	154.8	200.9
34	1.9093	1.9030	182.6	179.4
35	2.4389	2.3870	265.2	225.2
36	2.3343	2.2420	169.1	158.5
37	1.3379	1.2946	217.3	194.1
38	2.1996	2.0904	168.9	184.3
39	1.8795	1.8608	254.4	198.5
40	2.1504	2.0920	154.1	168.2
41	1.4330	1.3888	196.2	215.6
42	2.4381	2.3488	187.1	195.6

Case	Data set II y_i	Data set II $\hat{y}_i$	Data set VI y_i	Data set VI $\hat{y}_i$
43	2.1075	2.0231	179.3	184.7
44	2.2843	2.2833	269.5	266.0
45	2.1615	2.1097	275.1	267.3
46	2.0558	1.8935	204.2	225.4
47	2.7249	2.6428	202.2	315.7
48	2.0520	2.0697	318.2	213.7
49	2.6810	2.5889	227.3	223.0
50	2.2604	2.1449	323.5	281.1
51	2.2553	2.2234	492.3	316.2
52	2.1745	2.1117	511.8	312.9
53	2.0224	1.9723	320.3	235.9
54	2.1413	1.9983	362.7	289.3
55			262.3	275.6
56			291.4	222.4
57			434.8	284.0
58			333.1	361.7
59			322.9	255.5
60			359.0	334.6
61			423.8	362.6
62			290.7	373.5
63			416.3	282.2

Data set I (MP) from Montgomery and Peck (1992); Data set II (Neter) from Neter *et al.* (1989); Data set III (Rawl) from Rawlings (1988); Data set IV (West) from West (1981); Data set V (EM) from Ek and Monserud (1979); Data set VI (RC) from Reynolds and Chung (1986); y_i, observed 'true' value; and $\hat{y}_i$, predicted value from the model to be validated.

25 Model Testing by Means of Cost-plus-loss Analyses

Tron Eid[1]

Abstract

Tests of models on independent data as a part of the model development often end when the errors are calculated and evaluated, and compared with corresponding results from previously developed models. The error calculations, however, provide only limited information with respect to the application of a model. Valuable information may also be obtained if a link between the errors, the consequential incorrect treatment decisions and the corresponding economic losses can be established. A case study, based on models for prediction of basal area mean diameter and number of trees/ha in forest stands, where the economic losses related to incorrect timing of final harvests when decisions were based on erroneous predicted information, is presented.

The case study demonstrates that calculation of the economic losses provides information that is useful as a supplement to considerations based on errors only when it comes to application of the models. The quantification of the economic losses, expressed in monetary terms, provides information that enables the modeller to relate effects of model errors directly to inventory costs. The variations in losses under different forest conditions also demonstrated a potential for reduced costs where there is a choice between use of the models when the economic losses are low and use of field measurements when the economic losses are high.

Introduction

Testing on independent data is an important part of model validation. Such tests are important irrespective of whether it is a simple model describing height over age or a model describing ecologically complicated growth variations over time under different conditions and treatments. Quite often these tests end when the errors are calculated and compared with corresponding results from similar previously developed models (see, for example, Næsset and Tveite, 1999).

However, a quantification of the errors provides only limited information when it comes to the application of a model. Since the objective of developing a model is quite often to provide information that can be used for decision making with respect

[1] Agricultural University of Norway, Department of Forest Sciences, Norway
Correspondence to: tron.eid@isf.nlh.no

to silvicultural treatment, an additional step may be taken if a link between the errors, the consequential incorrect treatment decisions and the corresponding economic losses is established. Such an approach can probably provide valuable information that can be used as a supplement to considerations based on errors only.

Hamilton (1978) suggested cost-plus-loss analyses as a possible way to create a link between errors and economic losses. In cost-plus-loss analyses, the *total* costs, i.e. the costs of the inventory and the expected economic losses as a result of future incorrect decisions due to errors in measurements, are minimized. This approach has been used several in evaluations related to the design and intensity of inventories (Burkhart *et al.*, 1978; Larsson, 1994; Ståhl, 1994; Ståhl *et al.*, 1994; Eid, 2000).

This chapter investigates how cost-plus-loss analyses can be applied in model testing based on independent data. A key issue for discussion is to what extent a quantification of economic losses (i.e. economic losses as a result of future incorrect decisions due to errors in models) provides information that can be used as a supplement to considerations founded solely on the errors. A case study based on models for prediction of basal area mean diameter (D_{ba}) and number of trees/ha (N) in forest stands (Eid, 2001) is presented. A test data set of 85 observations was applied.

Materials and Methods

Background

Information about D_{ba} and N in a forest stand is generally important in forest management planning in Norway. For analyses with large-scale forestry scenario models such as AVVIRK-2000 (Eid and Hobbelstad, 2000) and GAYA-JLP (Lappi, 1992; Hoen and Gobakken, 1997), these variables are used as input in submodels for diameter growth (Blingsmo, 1984), height development (Braastad, 1977; Tveite, 1977) and mortality (Braastad, 1982; Eid and Øyen, 2003), as well as in models predicting timber values (Blingsmo and Veidahl, 1992) and logging costs (e.g. Eid, 1998).

Eid (2001) developed models for prediction of D_{ba} and N in forest stands in order to use them in practical management planning in Norway as an alternative to field measurements. The models were adapted to variables available from two different types of stand inventories, i.e. 'relascope inventories' where the mean height by basal area (H_L) is computed from height measurements of sample trees selected by means of the relascope, and the basal area/ha (BA) is measured by the relascope at subjectively selected sample plots, and 'visual inventories' with a direct and subjective determination of volume/ha (V) in the field. For both types of inventories, stand age (A) and stand site quality (S) (i.e. dominant height in metres at breast height, age 40 years) are determined through subjectively selected samples in stands.

Questions related to model type, to statistical properties of the fitted models (e.g. precision, correlation and collinearity), to logical consistency of the models and to tests on independent data were all assessed when the models were developed (Eid, 2001). In the final models adapted to relascope inventories (Model 1), BA, H_L and A were selected as independent variables, while V, A and S were selected in the models adapted to visual inventories (Model 2).

The main conclusions of Eid (2001) were that Model 1 could be applied at an aggregated level for use in large-scale forestry scenario analyses because there was no evidence of systematic errors in tests on independent data. The substantial level of the random errors, however, indicated that one should be cautious in exclusively relying on the model with respect to decisions at the stand level. Model 2 was not recommended for extensive use because it might produce systematic errors, in addition to substantial random errors. For scattered stands with poor site quality and/or low standing volume (i.e. for small parts of a total area and for the least valuable stands), the model was still recommended for use, however, because the consequences of errors in such cases were considered to be small.

Economic losses

The uncertainty related to applications of the models is attached not only to the initial description of D_{ba} and N, but also to projections made by the large-scale forestry scenario models applied in practical management planning, since these decision tools use D_{ba} and N in several submodels for prediction of future biological and economic conditions. A quantification of to what extent the treatment decisions are changed as a result of erroneous predictions and, if they are changed, how large the economic losses will be is therefore considered important.

Figure 25.1 illustrates how losses in net present value (NPV), i.e. economic losses, may appear due to errors, when timing of final harvest is considered. The figure shows the true NPV from final harvests at time T. If it is assumed that observed data provide information that enables the decision maker to take 'correct decisions' (i.e. final harvest at time *Tobs*), the result is a maximum NPV (i.e. NPV*obs*). If the decisions are based on information from predicted data with uncertainty involved (errors), the final harvests will be carried out at, for example, time *Tpred*$_1$ or *Tpred*$_2$, instead of *Tobs*. The resulting economic losses may then be found as the difference between NPV*obs* and NPV*pred*$_1$ and NPV*pred*$_2$, respectively.

Calculations as illustrated in Fig. 25.1 were made for a test data set of 85 observations (Table 25.1). Let NPV*pred*$_i$ be the NPV/ha of observation i, i = 1, 2, ..., n in a test data set when the initial description of D_{ba} and N was based on prediction made by the models, and NPV*obs*$_i$ be the NPV/ha of observation i, i = 1, 2, ..., n, when the initial description was based on observed values for D_{ba} and N. The economic loss per hectare, when the models were applied, for observation number i, was then calculated as: $Eloss_i = NPVobs_i - NPVpred_i$.

The calculations related to the timing of final harvests were done with GAYA-JLP, which is a large-scale forestry scenario model based on simulation of treatments in each stand (Hoen and Gobakken, 1997) and linear programming for solving management problems at the forest level (Lappi, 1992). A management strategy maxi-

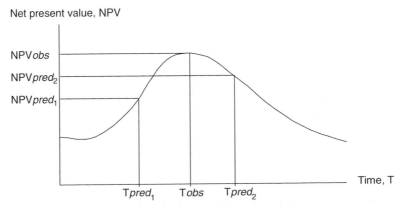

Fig. 25.1. An illustration of how economic losses may arise due to erroneous predicted data.

Table 25.1. Test data set.

Symbol	Definition	Mean	Min.	Max.
V	Volume (m³/ha)	224	57	553
BA	Basal area (m²/ha)	24.5	9.4	47.6
N	Number of trees (per ha)	670	200	1925
D_{ba}	Basal area mean diameter (cm)	22.2	15.8	30.5
H_L	Mean height by basal area (m)	18.8	11.5	26.1
S	Site quality (m)	14.1	6.0	24.6
A	Age (years)	103	40	170
RA	Relative age (actual age over optimum economic rotation age)	1.21	0.72	1.88

mizing the NPV without any constraints was analysed. An annual rate of discount of 3% was applied. Projections were done for 5-year periods.

Results

Mean differences between predicted and observed D_{ba} and N as a percentage of the observed values ($\bar{D}$), and the corresponding standard deviations for the differences (SD), were calculated for the 85 observations in the test data set. The mean differences between predicted and observed D_{ba} were 1.8 and 2.6%, respectively, for Model 1 and 2 and, correspondingly, −1.2 and 6.0% between predicted and observed N (Table 25.2). No differences were significantly different from zero. The mean standard deviations for the differences between predicted and observed D_{ba} were 11.9 and 16.3%, respectively, for Model 1 and 2. The corresponding standard deviations for N were 22.3 and 45.3%.

The mean economic loss was €5.83/ha for Model 1 and €17.22/ha for Model 2. For Model 1, economic losses (i.e. losses due to incorrect timing of final harvests), appeared for 15 out of 85 observations, while the corresponding number for Model 2 was 21 (Table 25.2.).

When Model 1 was used for prediction, D_{ba} was significantly overestimated for high values of basal area/ha (BA) ($P < 0.01$), mean height by basal area (H_L) ($P < 0.001$) and volume/ha (V) ($P < 0.01$) (Table 25.3). N was significantly underestimated for high values of H_L ($P < 0.01$) and significantly overestimated for low values of BA ($P < 0.05$). No obvious patterns were seen for the standard deviations of the differences over the different parts of the data. Large economic losses were found for low values of BA, H_L and V.

Table 25.2. Differences ($\bar{D}$) and standard deviations for the differences (SD) between predicted and observed D_{ba} and N, as percentages of the observed mean D_{ba} and N, and the corresponding economic losses, in the test data set.

	No. of observations	Diameter (D_{ba})			Number of trees (N)			No. of observations with economic losses	Economic losses (€/ha)
		Observed (cm)	$\bar{D}$ (%)	SD (%)	Observed (N/ha)	$\bar{D}$ (%)	SD (%)		
Model 1	85	22.2	1.8ns	11.9	670	21.2ns	22.3	15	5.83
Model 2	85	22.2	2.6ns	16.3	670	6.0ns	45.3	21	17.22

Significance levels (Bonferroni: Miller (1981)): ns, not significant ($P>0.05$); *$P<0.05$; **$P<0.01$; ***$P<0.001$

Table 25.3. Differences ($\bar{D}$) and standard deviations (SD) between predicted and observed D_{ba} and N, as percentages of the observed mean D_{ba} and N, and the corresponding economic losses, over different parts of the test data set. Model 1.

Part of the data	No. of observations	Diameter (D_{ba})			Number of trees (N)			Economic losses (€/ha)
		Observed (cm)	$\bar{D}$ (%)	SD (%)	Observed (per ha)	$\bar{D}$ (%)	SD (%)	
BA (m²/ha)								
≤16.0	16	22.5	−5.5ns	8.3	378	16.3*	20.5	17.97
16.1–22.0	18	22.6	2.9ns	15.7	524	−1.2ns	27.8	4.67
22.1–28.0	25	22.2	0.0ns	9.3	673	1.9ns	18.5	2.51
≥28.1	26	21.9	7.5**	10.5	949	−7.7ns	19.5	2.27
H_L (m)								
≤15.0	10	19.6	−4.9ns	10.0	530	11.4ns	26.5	24.40
15.1–18.0	27	22.1	−3.7ns	11.1	623	7.0ns	22.2	5.79
18.1–21.0	28	22.2	1.4ns	7.5	736	0.2ns	15.8	1.08
≥21.1	20	23.7	12.1***	11.6	713	−17.7**	21.4	3.14
V(m³/ha)								
≤150	19	22.0	−4.4ns	9.3	425	10.4ns	25.8	15.75
151–200	21	22.1	0.1ns	14.2	587	2.5ns	25.3	3.20
201–250	17	22.4	2.8ns	11.2	665	1.1ns	19.3	4.90
≥251	28	22.4	6.6**	10.3	903	−7.8ns	18.9	1.56
A (years)								
≤80	19	21.8	0.7ns	10.9	779	3.8ns	19.9	5.44
81–100	16	22.4	4.4ns	11.2	669	−5.0ns	18.8	5.55
101–120	13	21.8	3.6ns	14.3	663	−5.1ns	27.3	13.12
≥121	37	22.7	−0.5ns	11.7	599	0.4ns	23.9	1.66
S (m)								
≤9.50	16	20.1	0.1ns	12.9	598	−0.9ns	27.5	15.25
9.51–12.50	22	23.3	−2.9ns	10.6	581	6.4ns	21.3	4.13
12.51–15.50	14	22.3	6.9ns	13.2	632	−8.0ns	23.4	3.93
15.51–18.50	12	23.0	2.1ns	9.7	675	−1.9ns	17.5	2.10
≥18.51	21	22.2	4.7ns	11.8	842	−3.2ns	21.2	3.72
RA								
≤1.00	23	21.2	−1.1ns	9.6	672	7.9ns	15.9	18.72
1.01–1.25	20	21.8	6.0ns	13.7	718	−10.8ns	25.3	1.35
1.26–1.50	29	22.2	3.0ns	11.1	709	−3.7ns	19.8	0.88
≥1.51	13	24.9	−1.9ns	12.8	508	5.8ns	28.1	0.78
All	85	22.2	1.8ns	11.9	670	−1.2ns	22.3	5.83

Significance levels (Bonferroni: Miller (1981)) ns, not significant ($P>0.05$); $*P<0.05$; $**P<0.01$; $***P<0.001$.

When Model 2 was used for prediction, D_{ba} was significantly overestimated for high values of BA ($P < 0.01$) (Table 25.4). No other significant differences were found for D_{ba} or N over any part of the data when the observations were sorted according to BA, H_L and V. Large economic losses were found for low values of BA and H_L.

For Model 1, no significant differences between predicted and observed values were found when the data were sorted according to age (A), site quality (S) or relative age (RA) (Table 25.3). Large economic losses were found for the poorest site quality classes and for the observations with the lowest RA. The large losses for the observations with low values of RA can also be seen in Fig. 25.2, where the economic losses were plotted over RA. Out of 15 observations where economic losses occurred, seven had an RA lower than 1.0. The mean in-optimality-loss for observations with RA lower than 1.0 was €18.72/ha.

For Model 2, significant differences between predicted and observed values were found for the observations with the lowest and the highest values of A (Table 25.4). Significant differences were found for low values of RA in predicting D_{ba} as well as N. For the observations with RA lower than 1.0, the overestimation of N was >50%. The mean in-optimality-loss for these observations was €61.47/ha, and 14 out of 21 observations where economic losses occurred belonged to this group of observations (Fig. 25.3).

Discussion and Conclusions

Tests on independent data by means of differences and standard deviations of the differences between predicted and observed values for D_{ba} and N were performed when the models were developed (Eid, 2001). Such tests were also done for this chapter. The results of these calculations (Table 25.2) confirmed the high level of uncertainty related to application of the models. A standard deviation of 45% (prediction of N with Model 2) is a substantial error level according to all standards. As an example, an investigation of practical Norwegian forest inventories from 14 different sites, where the accuracy of volume determination was evaluated, showed standard deviations between measured volume and reference volume varying between 15 and 31% (Eid and Næsset, 1998). In addition, the tests on the 85 observations gave evidence of biases in certain parts of the data, e.g. when predicting D_{ba} for high values of BA, H_L and V (Tables 25.3 and 25.4). These biases were less obvious in the tests performed by Eid (2001).

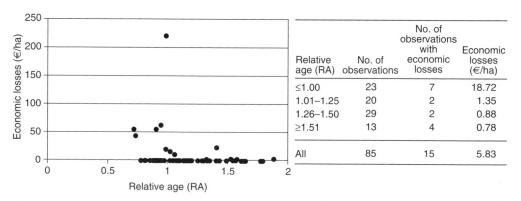

Relative age (RA)	No. of observations	No. of observations with economic losses	Economic losses (€/ha)
≤1.00	23	7	18.72
1.01–1.25	20	2	1.35
1.26–1.50	29	2	0.88
≥1.51	13	4	0.78
All	85	15	5.83

Fig. 25.2. Economic losses over relative age (actual age over optimum economic rotation age). Model 1.

Table 25.4. Differences ($\bar{D}$) and standard deviations (SD) between predicted and observed D_{ba} and N, as percentages of the observed mean D_{ba} and N, and the corresponding economic losses, over different parts of the test data set. Model 2.

Part of the data	No. of observations	Diameter (D_{ba})			Number of trees (N)			Economic losses (€/ha)
		Observed (cm)	$\bar{D}$ (%)	SD (%)	Observed (per ha)	$\bar{D}$ (%)	SD (%)	
BA (m²/ha)								
≤16.0	16	22.5	-3.7ns	11.2	378	27.0ns	38.8	43.38
16.1–22.0	18	22.6	-3.2ns	17.1	524	30.2ns	57.8	19.85
22.1–28.0	25	22.2	1.8ns	17.1	673	10.3ns	44.2	11.76
>28.1	26	21.9	11.6**	14.5	949	-11.2ns	35.4	4.56
H_L (m)								
≤15.0	10	19.6	6.6ns	15.8	530	-5.9ns	38.4	51.20
15.1–18.0	27	22.1	5.7ns	17.4	623	-10.6ns	47.1	8.60
18.1–21.0	28	22.2	-1.5ns	15.8	736	17.9ns	44.7	19.25
≥21.1	20	23.7	2.5ns	15.7	713	12.8ns	40.2	9.05
V(m³/ha)								
≤150	19	22.0	0.0ns	14.0	425	11.4ns	45.2	37.15
151–200	21	22.1	1.4ns	18.0	587	9.8ns	50.0	7.96
201–250	17	22.4	-1.5ns	15.7	665	17.8ns	46.1	27.80
≥251	28	22.4	7.8ns	16.3	903	-2.8ns	40.5	4.23
A (years)								
≤80	19	21.8	-11.0***	10.8	779	46.4***	26.8	31.44
81–100	16	22.4	-2.1ns	16.5	669	14.8ns	45.1	22.58
101–120	13	21.8	8.7ns	14.6	663	-13.4ns	26.4	18.96
≥121	37	22.7	11.9***	12.8	599	-25.4***	36.2	1.76
S (m)								
≤9.50	16	20.1	7.5ns	16.0	598	-14.9ns	38.8	32.00
9.51–12.50	22	23.3	6.7ns	14.8	581	-12.4ns	42.7	9.38
12.51–15.50	14	22.3	0.6ns	17.2	632	13.1ns	47.1	12.81
15.51–18.50	12	23.0	-1.6ns	16.6	675	20.3ns	50.1	18.90
≥18.51	21	22.2	-1.4ns	17.0	842	20.6ns	39.0	16.17
RA								
≤1.00	23	21.2	-13.4***	8.5	672	53.7***	26.2	61.47
1.01–1.25	20	21.8	1.1	16.8	718	5.5ns	0.58	0.58
1.26–1.50	29	22.2	11.3***	12.0	709	-21.1ns	0.98	0.98
≥1.5	13	24.9	11.6*	13.8	508	-19.9ns	0.78	0.78
All	85	22.2	2.6ns	16.3	670	6.0ns	45.3	17.22

Significance levels (Bonferroni: Miller (1981)) ns, not significant ($P>0.05$); *$P<0.05$; **$P<0.01$; ***$P<0.001$

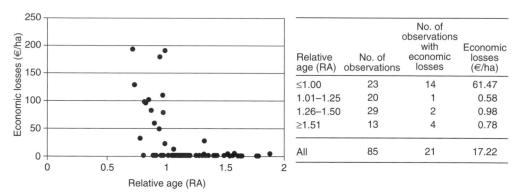

Relative age (RA)	No. of observations	No. of observations with economic losses	Economic losses (€/ha)
≤1.00	23	14	61.47
1.01–1.25	20	1	0.58
1.26–1.50	29	2	0.98
≥1.51	13	4	0.78
All	85	21	17.22

Fig. 25.3. Economic losses over relative age (actual age over optimum economic rotation age). Model 2.

The approach of the present chapter, where economic losses due to incorrect timing of final harvest as a result of erroneous data were calculated, accentuated the uncertainty of the models even more compared with the conclusions of Eid (2001), and further strengthened solutions in favour of field measurements instead of applying the models. For Model 2, where the losses were €17.22/ha, there is little doubt about such a conclusion in favour of field measurements. Even for Model 1 it would probably be possible to determine D_{ba} and N by means of field measurements for less than €5.83/ha, and still end up with a lower error level than those seen in Table 25.2.

When the mean figures Table 25.2 were studied, the close connection between the error levels and the economic losses was quite obvious, i.e. the high error level of Model 2 corresponded to high economic losses. However, when different parts of the data were studied, this picture changed. As an example, when Model 1 was applied for low values of basal area (BA= 16.0 m^2/ha), the economic losses were almost €18/ha, while the losses were about €2/ha for high values of basal area (BA >28 m^2/ha) (Table 25.3). However, the error levels for these two parts of the data were still quite similar. The same tendencies were also seen in different parts of the data over H_L and V, and also when Model 2 was tested (Table 25.4).

As long as a certain error level seems to produce economic losses that are quite different under varying forest conditions, one may consider a 'cost-minimizing strategy' by switching between the two possibilities; field measurements when the economic losses are high and use of the models when the economic losses are low. To be able to carry out such a strategy in practice, however, the forest conditions producing the large economic losses have to be identified.

The fact that large economic losses were found for low values of BA, H_L, V and S was quite confusing. Intuitively one would expect the large losses to appear on sites with a high growing stock and a high productivity. However, when the data were sorted according to RA, i.e. actual age over optimum economic rotation age (Figs 25.2 and 25.3), the explanation of how the economic losses were distributed became quite clear. For the observations where RA is near 1.0, i.e. when age is near optimum economic rotation age, it is important to have precise information in order to make a correct decision with respect to the timing of the final harvest. The more that RA exceeds 1.0, i.e. the more economically mature the forest is, the less important the accuracy of the information becomes (the decision still will be to harvest as soon as possible). Generally this means that less attention may be paid to the uncertainty of the models when the decision is 'obvious', as for the observations with the highest RA, and more attention should be paid to the uncertainty when the decision is 'less obvious', as for the observations where the RA is near 1.0. According to this, a cost-minimizing strategy would be to apply the models in the oldest forest, and carry out field measurements when the age is close to optimum economic rotation age.

Any quantification of the losses related to the timing of final harvests only is, of course, a simplification. A more complex situation will arise if the decision maker must respond to, for example, temporarily fluctuating timber prices or rapidly decreasing tree vitality. The selected annual rate of discount of 3% also has a great impact on the results. By, for example, changing the rate from 3% to 4%, fewer of the oldest stands would have large economic losses. On the other hand, more old stands would be involved if the rate was changed from 3% to 2%. Decisions with respect to thinnings may also have been included in the analyses, since such decisions depend heavily on D_{ba} and N. A strategic decision that involves postponed harvests (e.g. non-declining harvests over time) may also change the results quite radically. In a situation with postponed harvests, accurate information about the oldest stands becomes more important, compared with the present case, because it would then be not only a question of timing of the harvest, but also a question of prioritizing between a large number of over-mature stands.

The example of the case study included only those considerations related to model errors. A complete cost-plus-loss approach should in principle include two alternatives: (i) applying the models and (ii) performing a field inventory to determine D_{ba} and N. The models should only be used if the total costs of applying them are lower than the total costs of performing a field inventory.

Figure 25.4 illustrates the basic features of the cost structure for the alternative where the models are applied. The losses due to model errors are fixed and not influenced by the intensity of the field work. The losses due to errors in the independent variables, however, depend on the intensity of the field work, i.e. the more effort put into measurements, the lower error level in independent variables and, accordingly, also losses due to errors. The total costs of applying the models are determined by summarizing the losses due to model errors, the losses due to errors in independent variables and the cost related to the field work. A similar approach should also be applied for field measurements of D_{ba} and N, i.e. cost-plus-loss analyses based on empirical assumptions for field costs, intensity of field work and the resultant error levels.

In addition to cost-plus-loss analyses, more general considerations with respect to the application of the present models should be done. Since the models aim at providing input to large-scale forestry scenario models and analyses, it is important to consider the generally numerous sources of uncertainty in such analyses (see, for example,

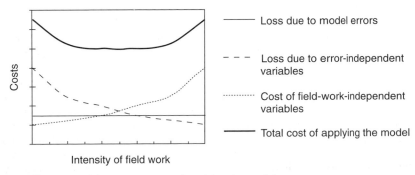

Fig. 25.4. An illustration of the cost structure of applying the models.

Eid, 2000). One may therefore in general argue as follows: as long as the models for D_{ba} and N are unbiased, they will in most cases not introduce any substantial change with respect to the final uncertainty of the large-scale forestry scenario analyses.

Although the uncertainty of the models was accentuated even more, the main conclusions with respect to application of the models were not significantly changed as a result of studying the economic losses. Cost-plus-loss analyses in model development can never replace considerations related to precision, correlation, collinearity and logical consistency of the models, and a quantification of error-based independent data. However, cost-plus-loss analyses may undoubtedly provide information that is useful as a supplement to such considerations. First of all, a quantification, in monetary terms, of the simultaneous effects of systematic as well as random errors provides information that enables the modeller to relate effects of model errors directly to inventory costs. The results with respect to variations in losses under different forest conditions also demonstrate a potential for reduced costs by alternating between use of the models when the losses are low, and use of field measurements when the losses are high. The approach may possibly be enhanced by implementing and comparing cost-plus-loss analyses based on empirical assumptions for both application of the models and performing a field inventory to determine D_{ba} and N.

Acknowledgements

The present study was supported by the Forest Trust Fund, Norwegian Ministry of Agriculture.

References

Blingsmo, K. (1984) *Diametertilvekstfunksjoner for bjørk-, furu og granbestand*. Research Paper of Norsk Institutt for Skogforskning 7/84, 22 pp. [in Norwegian].

Blingsmo, K. and Veidahl, A. (1992) *Funksjoner for bruttopris av gran-og furubestand*. Research Paper of Skogforsk 8/92, 23 pp. [in Norwegian].

Braastad, H. (1977) *Tilvekstmodellprogram for bjørk*. Research Paper of Norsk Institutt for skogforskning 1/77, 17 pp. [in Norwegian].

Braastad, H. (1982) *Naturlig avgang i granbestand*. Research Paper of Norsk Institutt for Skogforskning 12/82, 46 pp. [in Norwegian].

Burkhart, H., Stuck, R. and Reynolds, M. (1978) Allocating inventory resources for multiple-use planning. *Canadian Journal of Forest Research* 8, 100–110.

Eid, T. (1998) *Long Range Prognosis and Use of Production Functions to Estimate Costs of Mechanized Harvesting*. Research Paper of Skogforsk 7/98, 31 pp. [in Norwegian with English summary].

Eid, T. (2000) Use of uncertain inventory data in forestry scenario models and consequential incorrect harvest decisions. *Silva Fennica* 34, 89–100.

Eid, T. (2001) Models for prediction of basal area mean diameter and number of trees for forest stands in South-eastern Norway. *Scandinavian Journal of Forest Research* 16, 467–479.

Eid, T. and Hobbelstad, K. (2000) AVVIRK-2000 – a large scale forestry scenario model for long-term investment, income and harvest analyses. *Scandinavian Journal of Forest Research* 15, 472–482.

Eid, T. and Næsset, E. (1998) Determination of stand volume in practical forest inventories based on field measurements and photo-interpretation: the Norwegian experience. *Scandinavian Journal of Forest Research* 13, 246–254.

Eid, T. and Øyen, B.H. (2003) Models for prediction of mortality in even-aged forest. *Scandinavian Journal of Forest Research* 18, 64–77.

Hamilton, D.A. (1978) *Specifying Precision in Natural Resource Inventories: Proceedings of the Integrated Inventories of Renewable Resources*. USDA Forest Service, General Technical Report RM-55:276–281.

Hoen, H.F. and Gobakken, T. (1997) *Brukermanual for Bestandssimulatoren GAYA v1.20*. Internal Report, Department of Forest Sciences, Agricultural University of Norway, 59 pp. [in Norwegian].

Lappi, J. (1992) *JLP: a Linear Programming Package for Management Planning*. Research Paper 414, The Finnish Forestry Research Institute, Suonenjoky, 134 pp.

Larsson, M. (1994) *The Significance of Data Quality in Compartmental Forest Registers in Estimating Growth and Non-optimal Losses: a Study of final Felling Compartments in Northern Sweden*. Department of Biometry and Forest Management, Swedish University of Agricultural Sciences, Umeå, Report 26, 73 pp. [in Swedish with English summary].

Miller, R.G. (1981) *Simultaneous Statistical Inference*, 2nd edn. Springer-Verlag, New York, 299 pp.

Næsset, E. and Tveite, B. (1999) Stand volume functions for *Picea abies* in Eastern, Central and Northern Norway. *Scandinavian Journal of Forest Research* 14, 164–174.

Ståhl, G. (1994) Optimal stand level forest inventory intensities under deterministic and stochastic stumpage value assumptions. *Scandinavian Journal of Forest Research* 9, 405–412.

Ståhl, G., Carlson, D. and Bondesson, L. (1994) A method to determine optimal stand data acquisition policies. *Forest Science* 40, 630–649.

Tveite, B. (1977) Bonitetskurver for gran. *Communications of Norsk Institutt for skogforskning* 33, 1–84. [in Norwegian].

26 Regulating the Yield of Goods and Services from Forests: Developing Tools to Support Management Decisions and Policy Development for Multiple Objective Forest Management

P.R. van Gardingen[1]

Abstract

Growth and yield models have been developed for many tropical countries with the intention of supporting management decisions relating to the yield of timber from forests. These tools have, however, largely failed to deliver significant improvements in management practice. Forest managers and policy makers increasingly have to consider the supply of a much wider range of goods and services, including non-timber forest products and a range of environmental services. It is obvious that existing approaches for yield regulation, which have failed to deliver management improvements for the regulation of timber, will be even less suitable for application in systems of multiple objective forest management supporting the livelihoods of a wide range of stakeholders. New tools and approaches are required to support decision making for management and policy development.

The multiple objective forest management (MOFORM) cluster of research (www.moform.org) aims to support the development of new knowledge and tools for the regulation of the yield of goods and services from tropical forests. This will involve the modification and extension of existing growth and yield prediction tools (models) to develop a yield regulation toolbox. This will include tools for the prediction, allocation, monitoring and reporting of yield from forests linked with appropriate economic and financial tools and instruments supporting multiple objective forest management.

This chapter reports on the implementation of yield regulation pilot studies in Indonesia and Guyana utilizing two growth and yield models MYRLIN and SYMFOR to support forest management decisions and the development of policy. The lessons learned from these studies are discussed in terms of how to design tools and approaches that better meet the needs of forest managers and policy makers for multiple objective forest management.

[1] Centre for the Study of Environmental Change and Sustainability, The University of Edinburgh, UK
Correspondence to: P.vangardingen@ed.ac.uk

Introduction

Growth and yield models have been developed to support the management of tropical forests. The available models range in complexity from relatively simple size
class-based models to much more complex spatially dependent individual-based
models. Such models have been described and reviewed elsewhere (Vanclay, 1995;
Porte and Bartelink, 2002).

The availability of sophisticated growth and yield tools supporting the regulation of yield of timber from tropical forests seems to have done little to improve forest management, the livelihoods generated from forests, or conservation of tropical
forest resources. For this reason, forest managers and policy makers in tropical
countries often question the relevance of such tools. The current study aims to
describe processes by which growth and yield models may be applied more effectively to support the generation of livelihoods from tropical forests and demonstrate
benefits of such approaches to forest managers and policy makers.

This chapter is based on the results of two separate approaches. The first
involved consultation with stakeholders in a range of countries to determine the
need for new research supporting the application of growth and yield management
tools in tropical countries. The second involved the application of existing growth
and yield tools for yield regulation planning for pilot studies in Indonesia and
Guyana

Demand for New Tools and Approaches

A process of consultation to define the needs for new tools and approaches supporting the regulations of yield of goods and services from tropical forests commenced
in 2001. A multi-stakeholder workshop supported by the Forestry Research
Programme (FRP) of the UK Department for International Development (DFID) was
held in Edinburgh, Scotland, in June 2001 to assess and document demand for new
research and dissemination activities.

The workshop identified a series of priority constraints. A common theme was
the need to make existing knowledge more readily available to those responsible for
making policy and management decisions. This, combined with the lack of effective
financial and economic instruments, was considered to be the main reason for the
failure of existing growth and yield tools to improve management practice in the
field. The participants judged that there was little need to invest in further technical
development of existing growth and yield tools.

It was recognized that major challenges remain in many tropical countries
because of the lack of incentives for sustainable management of forests and the
prevalence of illegal logging activities. These factors combine to reduce the impact
of application of traditional approaches to yield regulation in the tropics for reasons
including:

- alternative land uses are potentially more profitable;
- there are high opportunity costs associated with the adoption of improved (sustainable) forest management practices; and
- some options for sustainable forest management may maintain or increase the
  marginalization of significant stakeholder groups, leading to increased conflict
  and the potential for illegal harvesting or destruction of forest resources.

The findings of the June 2001 workshop were used to develop a concept for a
multidisciplinary research programme to support multiple objective forest manage-

ment (MOFORM, www.moform.org). Three main topics of new research were suggested (van Gardingen, 2002).

State of knowledge review on yield regulation

This review should consider methods for the prediction, allocation, control, monitoring and reporting of the yield of goods and services from forests. The review should document the needs for new approaches for yield regulation resulting from the involvement of a wider range of stakeholders in forest management and the production of a wider range of goods and services from forests. The review should also consider how experience from temperate forests and developed countries can be adapted and transferred to meet needs for tropical forest systems and developing countries.

Yield regulation toolbox

The need to adapt existing yield regulation tools to meet the needs of policy and decision makers has been clearly identified. These include methods for the prediction, allocation, control, monitoring and reporting of the yield of goods and services from forests and their linkage to other approaches used to support forest management decisions, including financial and economic instruments.

Financial and economic instruments

There have been a limited number of studies that integrate methods of yield regulation with financial analysis (McLeish and Susanty, 2000; Ashton *et al.*, 2001; Huth and Ditzer, 2001; McLeish and van Gardingen, 2001). These studies considered the return on investment for the production of timber using simple financial cost–benefit analysis. There is a need to extend this work to consider a wider range of goods and services as well as the economic benefits of good forest management. The consultation suggested a need for research to produce guidelines to which economic and financial tools and instruments are most appropriate under specific circumstances. This may require the development of methods that permit rational debate between stakeholders who only use monetarized methods of valuation with those who also consider social and cultural values. Suitable approaches that combine multiple criteria decision analysis and participatory methods have been developed for use with other land-use systems and watershed catchment management (Proctor, 2001, 2002).

Designing Yield Regulation Pilot Studies

A series of yield regulation pilot studies have been implemented in Indonesia and Guyana during 2001 and 2002. The locations for the pilot studies were chosen to illustrate that each has characteristics that require site-specific approaches determined by factors including:

- the size and management objectives of the forest enterprise;
- availability of data describing the forest enterprise, including:
  - *static inventory,*

- ○ *growth estimates from repeated growth measurements (permanent sample plots),*
- ○ *spatial data (GIS) describing management compartments,*
- ○ *financial data;*
- • availability of suitably trained staff and local technical advisors.

The pilot studies were implemented in Indonesia and Guyana, representing contrasting forest types and management regimes. Natural forests in Indonesia are dominated by commercial dipterocarp species characterized by relatively high growth rates and resulting potential productivity. In contrast, the majority of forests in Guyana are on poor soils with low growth rates and commercial productivity. Both countries have implemented systems of large industrial forest concessions for many years and have recently started to encourage alternatives, including community forest management.

Data availability varied between locations. In each country, there was one location where a large industrial concession held information describing management compartments, static inventory and permanent sample plots. These data permitted the application of a range of growth and yield tools. One Indonesian site also had detailed financial information integrated as a model of the concession. In Guyana, one community forest enterprise had limited information available, but in Indonesia no suitable location could be found.

In both countries it was necessary to implement an extensive programme of training for local partners to support the pilot studies. This training involved the principles of yield regulation, application of tools and presentation of results to other local stakeholders. Local researchers were trained to apply yield regulation tools and then asked to design the studies for Indonesia (Ridwan and Redhahari, 2001) and Guyana (Khan and Singh, 2001).

Four types of pilot studies are described here. These are:

1. Analysis of policy options for industrial forest management in Indonesia.
2. Linkage of growth and yield models with financial tools in Indonesia.
3. Yield regulation for large-scale industrial forest management in Indonesia and Guyana.
4. Yield regulation for community forest management in Guyana.

The pilot studies were implemented to demonstrate how existing yield regulation tools could support policy and management decisions and to identify how these tools and their application could be further improved. Two contrasting yield regulation tools that could be applied in Indonesia and Guyana were available from earlier research projects. These are, MYRLIN (methods of yield regulation with limited information, www.myrlin.org) and SYMFOR (silvilculture and yield management for tropical forests, www.symfor.org). The MYRLIN toolbox was designed to produce a simple growth and yield model to be applied at concession level, based on limited data, specifically a single static inventory measurement. In contrast, application of the SYMFOR framework requires long-term data from permanent sample plots to calibrate an individual-based ecological model and is typically applied to describe results of management interventions at plot or compartment level.

The MYRLIN toolbox

The MYRLIN toolbox is based around a simple diameter-class projection model (Wright and Alder, 2000). The current version of the software and documentation is available electronically (Alder *et al.*, 2002). MYRLIN is implemented as three simple spreadsheet tools designed to assist the process of yield regulation in natural tropical forest. These are: (i) the production of stand tables; (ii) estimation of growth and mortality rates; and (iii) stand projection and harvesting model. These tools are directed at the needs of organizations and enterprises that have only a limited research base of information on forest dynamics for their locality. MYRLIN is designed to run with data describing a series of forest compartments (or management units) describing part or all of a forest enterprise (typically a concession).

The SYMFOR framework

The SYMFOR framework was initially developed for mixed dipterocarp forests in Indonesia (Phillips *et al.*, 2002a) and includes spatially explicit, individual-based models of ecological processes and forest management practice. New ecological models have been developed for Guyana (Phillips *et al.*, 2002b) and work is ongoing with versions for forests in Brazil, Ecuador and Bolivia. The development of an ecological model for the SYMFOR framework requires data obtained from repeated measurements of permanent sample plots. The model simulates the growth, recruitment and mortality of each individual tree in a plot, typically 1 ha in size.

The model simulates a number of stochastic processes and thus should be run repeatedly for each of several replicate plots in one forest type. A typical SYMFOR run may involve 20 replicate simulations for each of up to 20 plots. Such an analysis will normally take at least several hours, but this will vary depending on the complexity of the task, the number and replication of plots and the length of the simulation, as well as the speed of the computer. Data output by the SYMFOR framework then needs to be subjected to further statistical analysis before being made available for use by forest managers and policy makers.

Pilot Study Results

Study 1: policy analysis in Indonesia

The SYMFOR framework was applied in conjunction with statistical analysis of forest management data in order to evaluate the standard silvicultural prescriptions for the dipterocarp forests in Indonesia. The model was programmed to simulate the yield regulation system applied through the Indonesian selective logging and replanting system (TPTI), which utilizes a diameter limit of either 50 or 60 cm for all commercial species with a cutting cycle of 35 years (McLeish and Susanty, 2000). The study showed the yield from the TPTI system was not sustainable and compared it with a suggested alternative system of a maximum eight stems per hectare (Sist *et al.*, 1998; van Gardingen *et al.*, 1998). This research concluded that a system based on a maximum yield specified as either a maximum of eight stems per hectare or a felled volume of 50 m^3/ha was sustainable when combined with a 35–45 year cutting cycle. These results were summarized in a short briefing paper to make the results more readily available to policy makers (Purnama *et al.*, 2001).

Study 2: linkage of growth and yield models with financial analysis

A number of studies have described the linkage of the SYMFOR framework with a concession-scale model of financial performance (Fadilah, 1997) for the PT Inhutani I concession in Berau, East Kalimantan, Indonesia. A general guide for the linkage of the models (McLeish and van Gardingen, 2001) showed that a relatively simple approach is adequate, which only requires estimates of the predicted yield per hectare and number of stems from SYMFOR and estimates of the costs of harvesting and silvicultural treatments for the financial model. This approach was applied in an extensive study of yield regulation options for forests in the concession that compared the TPTI system based on a 50 cm diameter limit with alternatives of eight stems per hectare and maximum yields of either 50 m^3/ha or 60 m^3/ha (van Gardingen *et al.*, 2003). This study repeated the results of the previous policy study (McLeish and Susanty, 2000; Purnama *et al.*, 2001) showing that the sustained yields could only be obtained when regulated using either maximum volume or number of stems. The financial analysis suggested that the best management option was likely to be yield regulated to 50 m^3/ha combined with a 35-year cutting cycle or 60 m^3/ha and a 45-year cutting cycle. Both of these management regimes gave an internal rate of return that was predicted to exceed 10% over three cutting cycles.

Study 3: yield regulation for large-scale industrial forest management

The studies described above all involved the application of SYMFOR to the analysis of forest management scenarios for industrial forest management in Indonesia, represented by a limited number of sample plots from a single management unit and forest type. MYRLIN was applied to a study of yield regulation at concession scale for the PT Intracawood concession in Indonesia and the Barama Company Ltd concession in Guyana. In both studies, data were derived from a GIS (geographic information system) linked to stratified inventory data. These data were combined to produce a description of the compartments to be managed. In each study, users calibrated the MYRLIN growth and yield model using the spreadsheets provided in the toolbox before running management scenarios to evaluate the sustainable yield. The users found the toolbox easy to apply but noted that the management options were too simplistic to allow the simulation of standard harvesting practice in either country. The studies both recommended that the model required more flexibility in management scenarios and that the presentation of results required some indication of uncertainty.

The application of MYRLIN in Guyana was done in parallel to an application of SYMFOR calibrated using sample plots from the same area. The two models produced very similar results at low harvest intensities. MYRLIN predicted lower yields than SYMFOR when harvesting intensity was increased. This was attributed to differences in the growth models since SYMFOR models competition and increases the growth rate of trees in the residual stand following harvesting, whereas MYRLIN has constant growth rates within a species group in all conditions.

Study 4: yield regulation for community forestry management

MYRLIN was applied to community forest management with the ITUNI small loggers association (ISLA) in Guyana. This group had very limited information about their forest managed under a woodcutting lease for an area of 33,300 ha. The ISLA are required to produce a management plan for this area in order to secure long-term access to the forest resources. Researchers from the Guyana Forestry Commission assisted the ISLA to apply MYRLIN to calculate estimates of sustainable yield for the concession area. Data describing the forest were limited to basic vegetation maps. These were improved through field surveys, interpolation from soils maps, and a participatory mapping exercise. The resulting vegetation map was combined with inventory data from an adjacent area of forest to provide data for the compartments to be managed. The data were then used to calculate expected yield using MYRLIN.

The results from this study predicted a sustained annual yield of around 14,000 m³ for the concession based on a cutting cycle of 20 years and an annual coupe of 940 ha. This represented the first estimate of likely production and will form a major component of the management plan for the area. The study was implemented as the first step in an iterative process to improve yield estimates. The users of the results identified a number of issues that need to be improved, including the collection of inventory data from the concession and the implementation of more realistic harvesting options in MYRLIN. It was noted by all those involved that the participation of forest managers in the study ensured that the results were more relevant to management decisions and that the managers were more likely to act on the outcomes of the study. One result of this process was the decision of the ILSA to implement their own inventory for use in the next phase of planning.

Discussion

The pilot studies in Indonesia and Guyana have demonstrated how existing growth and yield tools can be used to support management and policy making and suggested ways to improve the effectiveness of this process. The studies were implemented as an example of action research where, if possible, potential users of the results were involved in the planning and implementation of the research. This process has generated a number of lessons for future developments to build tools to support decision making and policy development for tropical forests. These are summarized here.

Adapt existing growth and yield models.

All of the studies suggested that the structure and design of the existing growth and yield models were adequate for the applications supporting management and policy development. The users did not identify any requirement for further development of the biological and ecological content of the models they had used. Their comments suggested that it was far more important for the models to be adapted to improve the management options available and the method than that results were presented to users.

The comments about management options mainly related to the available methods of yield regulation that could be simulated. In MYRLIN this is limited to harvesting a proportion of the standing volume above a user-specified diameter limit. It was noted that this system could not simulate the current logging code of practice in Guyana, nor could it simulate the suggested modification to TPTI in Indonesia based on a maximum number of harvested stems or volume. SYMFOR was recognized as having more flexible harvesting options, but users sometimes found these difficult to implement. Users also expressed a need to extend the analysis of both models to represent a wider range of goods and services provided by forests, initially to non-timber forest products and then to include environmental services.

Users assessed the method of presentation of results from simulations as being an extremely important part of the design of the tools. The graphical representation of results by MYRLIN was considered to be very helpful, but all users suggested that this needed to be combined with some indication of reliability or uncertainty and standard definition of terminology (in particular 'annual allowable cut'). Users found SYMFOR more difficult to use, as it required subsequent statistical analysis of results before they could be used for management or policy decisions. It was suggested that there is a need for more guidance to be provided on how to apply models to support management and policy (as opposed to users' manuals) and on how best to present results from simulation modelling.

In both countries it was possible to compare the two models. The simulated yield estimates were similar, but MYRLIN tended to produce lower estimates when competition decreased after heavy disturbance. This was considered a favourable outcome, as MYRLIN was designed to work with limited information and would produce conservative estimates of yield. It was noted that this did provide an incentive for managers to collect more extensive data that could be used in more complex growth and yield models. Of the models compared, MYRLIN was well suited to operational concession-scale management, while SYMFOR was best suited to more strategic policy analysis. This emphasized that no one tool should be expected to meet the needs of all potential users and that the choice of appropriate tools would depend on the nature of the application and availability of data. The users expressed the need for assistance in choosing appropriate tools for their applications.

Availability of data to apply growth and yield tools

The availability of data to develop or calibrate models and run simulations was identified as an important limiting factor in all of the pilot studies. The choice of locations for the studies was determined by the presence of existing inventory or permanent sample plots. The ISLA study demonstrated the importance of having at least adequate inventory data for the area to be managed, and this was reinforced by the inability to include a community forest management group for Indonesia. Suitable data may not be available in many developing countries, especially with the increasing trend for small-scale community involvement in forest management. The ISLA study demonstrated how such groups could become involved in yield scheduling planning using existing sources of information and encourage further data collection.

Integration with other forest management tools

All of the studies demonstrated the benefits of integrating growth and yield models with other standard forest management tools used for planning purposes. The models need to interact with results from the static forest or compartment inventory and this is often done via a GIS for the site. More sophisticated forestry operations may also have a management information system (MIS) that contains information about productivity and operating expenses.

The financial analysis carried out in Indonesia relied on the availability of information for the concession studied. The guidelines and user interface developed for this study (McLeish and van Gardingen, 2001) demonstrated that integration of tools requires the definition of protocols and standards for transfer of information between systems. In the study reported here, the link was implemented through manual data entry, but more sophisticated approaches may involve real-time integration of software. One example is the integration of a growth and yield model with a GIS system for a forest concession as part of the Iwokrama project in Guyana (www.iwokrama.org).

Financial and economic instruments

It is essential to link growth and yield models with financial and economics analysis in order to be able to influence management and policy decisions. In nearly all situations, such decisions will be made in an environment comparing the management of a production system with either an alternative production system or, more frequently, alternative resource or land-use options. The process of decision making should normally involve the consideration of at least financial, economic, ecological and social factors. The Indonesian pilot study demonstrated that it was possible to design a forest management system that produced a reasonable return on initial investment. This alone was, however, insufficient to encourage changes in management practice, as the study showed that the adoption of improved forest management practice was associated with high initial opportunity costs and returned profits that were much less than alternative (albeit often illegal) management activities.

This indicates that there is a need for new research to guide which economic and financial tools and instruments are most appropriate under specific circumstances. One suggestion has been to explore the development of methods that allow rational debate between stakeholders who only use monetarized methods of valuation and those who also consider social and cultural values. Theses approaches could then be applied to the aim of reaching a consensus on decisions that produce the greatest net social benefit. Research is also needed to consider compensation mechanisms that have the potential to transfer value captured by downstream users and consumers of goods and services, to benefit those forest managers who adopt good land and forest management practices upstream. This must include the development of equitable mechanisms for the capture and distribution of benefits (value) resulting from good forest management.

Involve end-users in the studies

The pilot studies demonstrated the very real benefits of involving end-users of research results in the design and implementation of the studies. Their involvement ensured that the research was more focused on the outcomes required to support management and policy decisions. It also meant that feedback on the performance of the tools reflected their needs rather than the academic interests of the researchers. It is notable that none of the end-users requested improvements in the biological content of the models used in the studies. Their emphasis was on making the results more relevant to management and policy, through the provision of realistic management options and effective (simple) presentation of results.

Present results to wider range of stakeholders

Forest management in most developing countries now involves a wide range of stakeholder groups. Management decisions and policy development should now normally involve dialogue between these groups in order to reduce potential conflict and develop a shared vision for the forest resources.[2] Results from the pilot studies in Indonesia and Guyana were presented in multi-stakeholder fora that included forest managers, policy makers, national and local government, civil society groups, educational institutions and the media. The presentations made at each forum stimulated debate between these groups, each of which had an interest in the forest resource but had previously not engaged in debate about its management.

[2] Accepting that it may not be possible to get all groups to agree to this vision.

This approach generated demand in each country for additional studies and to use the resulting knowledge to contribute to the process of negotiation on forest management and policy issues. The feedback at these sessions indicated that new approaches were required to present results to meet the needs of different stakeholder groups.

Iterative approach

Each pilot study needed to involve an iterative approach to link growth and yield tools to management and policy. This was driven by a number of issues, including data availability, experience of the users and the refinement of the research question as the study progressed. It is recognized that such an approach should be the norm. The ISLA study in Guyana provides a good example, where the simple initial study was used to demonstrate the value of yield scheduling for the association's management plan and is now being used to justify future investment in the collection of static inventory. It is expected that the yield-scheduling exercise will be repeated when the new inventory data are available and the group has further refined the purpose of the study.

Training

The provision of training to users of the tools and resulting knowledge was given a consistently high priority in all of the pilot studies. Users in the partner countries expressed a need for training in a range of topics to support the application of the tools. In addition to training in running the tools, they assessed needs for training in:

- basic forest management;
- statistics and data processing;
- effective presentation of results to support management and policy decisions; and
- interpretation of results from scientific and financial analysis and the concepts of risk and uncertainty (to be provided to forest managers and policy makers).

The training needs assessment associated with the pilot studies emphasized that the impact of tools supporting forest management will only be increased when the application of these tools is integrated with a general improvement in the skill-base used for forest management and policy development.

Conclusions

The results from the pilot study showed that existing management tools could be used to support management decisions and policy making for tropical forests in developing countries. The changing nature of these activities means that the successful implementation of these tools requires integration within a wider process of dialogue between stakeholders. Some of the most important conclusions of the studies are summarized here.

- Existing growth and yield models for timber production are adequate for management and policy applications.

- The management options simulated by these models need to be expanded and improved.
- The available tools need to be expanded to cover other goods and services provided by forests.
- Users require guidelines on how to apply tools to support management and policy and the presentation of results.
- More than one type of tool may be required to meet the range of potential applications. The choice of a tool will depend on the application and availability of data. Users need assistance in choosing the best tool for their own circumstances.
- Data availability may limit some application tools, but much can be done now with existing data sources.
- There needs to be better integration of yield regulation tools with other standard forest management tools, including GIS and financial analysis.
- Financial and economic tools need to be expanded to provide new instruments supporting the sustainable and equitable management of forest resources.
- Involve end-users in a process of iterative action 'research'.
- Results should be made available to a wide range of stakeholders in a form that they can understand and use.
- Training is essential to increase the effectiveness of tools supporting multiple objective forest management.

The yield regulation pilot studies described in this chapter have identified a number of issues that limit the application of growth and yield tools to support forest management in developing countries. The results illustrate that stakeholders need to be encouraged to overcome these through a variety of approaches to link growth and yield tools within a framework for forest and land-use management and policy. It can be argued that some form of forest inventory is an essential prerequisite for any type of forest management. The studies have demonstrated that such data can be utilized through the application of appropriate yield prediction tools to provide projections of future productivity. The remaining challenge is how to integrate these tools into the decision-making and policy process for forests in developing countries such that they deliver against accepted management and development objectives.

Acknowledgements

This chapter draws upon the work of many colleagues in institutions working in partnership to develop the approaches described. Their assistance and support is gratefully acknowledged. This document is an output from a project funded by the UK Department for International Development (DFID) for the benefit of developing countries. The views expressed are not necessarily those of DFID R6915 Forestry Research Programme.

References

Alder, D., Wright, H.L. and Baker, N. (2002) *MYRLIN: Methods of Yield Regulation with Limited Information*. Oxford Forestry Institute, University of Oxford. Available at www.myrlin.org
Ashton, M.S., Mendelsohn, R., Singhakumara, B.M.P., Gunatilleke, C.V.S., Gunatilleke, I.A.U.N. and Evans, A. (2001) A financial analysis of rain forest silviculture in southwestern Sri Lanka. *Forest Ecology and Management* 154, 431–441.

Fadilah, D. (1997) *Minimum Area Model: a Financial Approach to Determine Minimum Area of Forest Management Unit. Report and User Guide.* Berau Forest Management Project, Jakarta, 36 pp.

Huth, A. and Ditzer, T. (2001) Long-term impacts of logging in a tropical rain forest: a simulation study. *Forest Ecology and Management* 142, 33–51.

Khan, M.T. and Singh, J. (2001) *Methods of Yield Regulation in Moist Tropical Forest with Minimal Data (MYRLIN), The University of Oxford, and Silviculture and Yield Modelling for Tropical Forests (SYMFOR), University of Edinburgh, 10–28 September.* Guyana Forestry Commission, Georgetown, Guyana, 16 pp.

McLeish, M.J. and Susanty, F.H. (2000) *Yield Regulation Options for Labanan: a Financial and Economic Analysis of Yield Regulation Options for Logged-over Forest at PT Inhutani I, Labanan Concession.* Berau Forest Management Project, Tanjung Redeb, 45 pp.

McLeish, M.J. and van Gardingen, P.R. (2001) *Linking SYMFOR with Financial Models for Forest Concessions: SYMFOR Technical Note.* University of Edinburgh, 21 pp.

Phillips, P.D., Brash, T.E., Yasman, I., Subagyo, S. and van Gardingen, P.R. (2002a) An individual-based spatially explicit tree growth model for forests in East Kalimantan (Indonesian Borneo). *Ecological Modelling* 159, 1–26.

Phillips, P.D., van der Hout, P., Arets, E.J.M.M., Zagt, R. and van Gardingen, P.R. (2002b) *An Ecological Model for the Management of Natural Forests Derived from the Tropenbos Permanent Sample Plots at Pibiri, Guyana.* University of Edinburgh, 21 pp.

Porte, A. and Bartelink, H.H. (2002) Modelling mixed forest growth: a review of models for forest management. *Ecological Modelling* 150, 141–188.

Proctor, W. (2001) *Valuing Australia's Ecosystem Services Using a Delibrative Multi-criteria Approach.* European Society for Ecological Economics. Frontiers 1 conference, Fundamental issues of environmental economics, 15 pp.

Proctor, W. (2002) *Assessing Ecosystem Services in Australia: 7th Biennial Conference of the International Society for Ecological Economics,* Sousse, Tunisia, 6–9 March, 16 pp.

Purnama, B., Rulsi, Y., McLeish, M.J. and van Gardingen, P.R. (2001) *Policy Brief: Yield Regulation for Logged-over Forest.* Ministry of Forestry, Jakarta, 2 pp.

Ridwan and Redhahari (2001) *Methods of Yield Regulation in Moist Tropical Forest with Limited Information (MYRLIN), University of Oxford, and Silviculture and Yield Management for Tropical Forests (SYMFOR), University of Edinburgh, 10–28 September 2001.* Universitas Mulawarman, Samarinda, Indonesia, 21 pp.

Sist, P., Nolan, T., Bertault, J.G. and Dykstra, D.P. (1998) Harvesting intensity versus sustainability in Indonesia. *Forest Ecology and Management* 108, 251–260.

van Gardingen, P.R. (2002) Yield regulation and multiple objective forest management. *ETFRN News*, 51–53.

van Gardingen, P.R., Clearwater, M.J., Nifinluri, T., Effendi, R., Rusmantoro, W., Noor, M., Mason, P.A., Ingleby, K. and Munro, R. (1998) Impacts of logging on the regeneration of lowland dipterocarp forest in Indonesia. *Commonwealth Forestry Review* 77, 71–82.

van Gardingen, P.R., McLeish, M.J., Phillips, P.D., Fadilah, D., Tyrie, G. and Yasman, I. (2003) Timber yield, financial and ecological analysis for the sustainable management of logged-over Dipterocarp forests in Indonesian Borneo. *Forest Ecology and Management* (in press).

Vanclay, J.K. (1995) Growth-models for tropical forests: a synthesis of models and methods. *Forest Science* 41, 7–42.

Wright, H.L. and Alder, D. (2000) *Proceedings of a Workshop on Humid and Semi-humid Tropical Forest Yield Regulation with Minimal Data.* OFI occasional papers, No. 52, Oxford Forestry Institute, Oxford, 95 pp.

27

CAPSIS: Computer-aided Projection for Strategies in Silviculture: Advantages of a Shared Forest-modelling Platform

François de Coligny,[1] Philippe Ancelin,[1] Guillaume Cornu,[2] Benoît Courbaud,[3] Philippe Dreyfus,[4] François Goreaud,[5] Sylvie Gourlet-Fleury,[2] Céline Meredieu[6] and Laurent Saint-André[2]

Abstract

Forest scientists build models to study, understand and represent stand growth and dynamics. *Ad hoc* computer tools are often developed to implement these models, to test and to explore the consequences of the underlying hypotheses. Some teams have also developed more integrated tools, often based on a parameterizable growth model.

The aim of this chapter is to present CAPSIS, a more generic forest simulator integrating various models, and to discuss the advantages for modellers and foresters of using such a shared forest-modelling platform for research, management and education.

CAPSIS is free portable software, with a kernel providing the overall structure for simulations and generic data descriptions for stands, trees, soil, etc. These descriptions can be completed in modules – one for each model – which implement a proper data structure and specific evolution functions. A plug-in architecture allows the user to add tools for management, data exportation and representation, or connections with other software. At the present time, CAPSIS hosts six modules and two specific libraries.

This shared forest-modelling platform has already shown many advantages: allowing modellers to re-use submodels and share specific tools; and providing to foresters and students a single easy-to-use software platform to compare different management strategies.

[1] UMR AMAP, France
Correspondence to: coligny@cirad.fr
[2] Cirad Forêt, France
[3] Cemagref, EPM, France
[4] INRA, UR Forestières Méditerranéennes, France
[5] Cemagref, LISC, France
[6] INRA, Equipe Croissance et Production, France

Introduction

In order to manage forest stands, foresters need specific tools to predict stand growth and wood quality. Many models have been developed that simulate the dynamics of a forest stand at various scales (e.g. Pretzsch *et al.*, 2002).

Modellers often have to implement these models in computer programs, to ease their calibration, their evaluation on concrete cases and their dissemination. Most models are implemented in *ad hoc* software (e.g. CEP, FAGACÉES, FORMIND, SELVA, SEXI, SIMCOP) which let their authors have complete control over them and choose freely their data structures, their algorithms and their user interface. However, these isolated developments produce specific programs which are sometimes not easy to use by others and may cause difficulties in communicating with each other because their architectures are too different. To benefit from re-usable tools in various situations, some teams have chosen to build more generic simulators implementing a parameterizable model (e.g. BWIN, CO2FIX, FVS, JABOWA, MOSES, PROGNAUS, SILVA, SORTIE, TASS). Generally, a common data structure and a set of functionalities are proposed (growth, competition, environment, management, analysis). The simulator can then be used for several species or forest types. However, only a few software packages (e.g. LMS, MUSE, SYMFOR) propose a more generic approach allowing us to use, compare or compose different growth models with a single interface.

The aim of this chapter is to present such a generic forest simulator – CAPSIS – and to discuss the advantages for modellers and foresters of using a shared forest-modelling platform for research, management and education.

The CAPSIS Forest-modelling Platform

CAPSIS is a generic forest simulator, developed since 1994 (Dreyfus and Bonnet, 1997). The objective of the project is to build a perennial, open and shared modelling platform: (i) to contribute to the development and evaluation of models; (ii) to share tools and methods; (iii) to compare results of different models; (iv) to transfer models to forest managers; and (v) to serve as teaching material. The CAPSIS4 release is generic enough to integrate very different kinds of models: stand models, distance-independent or spatially explicit tree models, wood quality models, seed dispersal models, etc. It provides forest management tools to establish and compare different silvicultural scenarios. CAPSIS is free software (membres.lycos.fr/coligny).

General architecture

The architecture of CAPSIS clearly identifies a *kernel* and some *modules*. The kernel offers common services (detecting and loading models, running simulations, offering import/export facilities, etc.) and generic data structures (for stand, trees, soil, etc.) facilitating the integration of new models. Growth models – in a broad sense – are integrated as independent modules. These modules contain their own data structures, extending some templates chosen from the kernel. They also contain the algorithms used to model stand evolution along time. Management can be automated in the modules or processed externally, in platform *extensions* (plug-ins), such as: thinning, pruning, storms, etc. Each simulation owns a *root step* which is associated with an initial stand. A simulation consists of computing successive states of this stand, by delegating *evolution* phases to the module, and by processing *interventions* with extensions.

CAPSIS extensions also include *viewers* and *data extractors* that provide representations of the stand at a certain time or along time (maps, curves, histograms, tables), and allow the user to observe the simulation results (Fig. 27.1). CAPSIS also provides *libraries* of data structures and processes which can be used by every modeller. Finally, simulations are driven through a *pilot* corresponding to a usage context. The bilingual (French, English) *interactive* pilot proposes a graphical user interface with menus and dialogues, which can easily be adapted to other languages. The *script* pilot can be used to run long or repetitive simulations with no user actions required.

Very different models are already developed with CAPSIS4

MOUNTAIN is a distance-dependent model of uneven-aged monospecific conifer forest. A detailed light interception submodel calculates the energy intercepted by each tree and on the soil, which is then used to determine tree growth, probability of death and regeneration. MOUNTAIN has been used until now to simulate long-term dynamic patterns in monospecies forests, to compare silvicultural strategies (Courbaud *et al.*, 2001) and to train students and forest managers in tree marking.

SELVA is a distance-dependent tree model, designed for the natural tropical rainforests of French Guyana (Gourlet-Fleury and Houllier, 2000). The three fundamental processes of forest dynamics are described: growth, mortality (split into standing dead and two types of windthrow) and recruitment. For particular species, the whole regeneration cycle from seed dispersal to ingrowth is modelled.

The PNN module is a distance-independent tree growth model for pure even-aged stands of black pine (*Pinus nigra nigricans*) in France and Hungary. Thinning

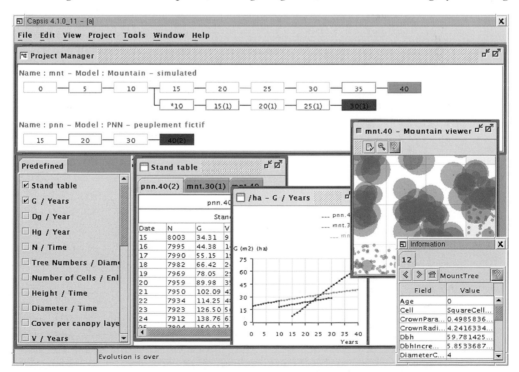

Fig. 27.1. A glance at CAPSIS4: Project manager, stand viewers with tree inspector, data extractors.

design considers target stocking (N, G, V, relative spacing, CCF after thinning) and ratio of tree average diameter (quadratic mean) after thinning:before thinning.

PP3 is a distance-independent tree growth model with a whole-stand growth regulation for pure even-aged stands of maritime pine (*Pinus pinaster*) in Landes of Gascogne (south-west France). The branching model and stem profile relationship are to be implemented soon.

EUCALYPT is a distance-independent tree model developed for eucalyptus clonal plantations in Congo (Saint-André et al., 2002). Three specific features are developed in this module: (i) the management of coppice (several stems per stump); (ii) the integration of a great number of models (sometimes one model per clone); and (iii) the link between CAPSIS and a geographic information system (GIS) (arcInfo/arcView).

VENTOUX is a semi-spatialized model dealing with the evolution of aged artificial pine stands towards mixed stands in the French Mediterranean hinterland mountains. Seedling occurrence of invasive native species is driven by the distance to potential seed sources (beech and silver fir stands). For a multi-level spatialization, the forest is divided into homogeneous units (site, overstorey species, age), then in grids of small cells (10 m × 10 m) containing non-spatialized trees and seedlings.

Two libraries already offer generic tools for forest modellers

The SPATIAL library contains specific tools concerning the spatial structure of forest stands. The RIPLEY data extractor describes the spatial pattern of a stand, using Ripley's K-function (Ripley, 1977). Moreover, we have implemented specific classes to simulate virtual stands of various structures, using Poisson, Gibbs or Neyman–Scott processes. These tools are available for all individual tree models (distance-dependent or independent tree models).

The BIOMECHANICS library proposes classes to describe the internal structure of trunks (stem profile, succession of growth units) and simulate their biomechanical behaviour under external stress (Ancelin, 2001). Associated with the WINDSTORM extension, which is compatible with both distance-dependent and independent tree models, it simulates the effect of wind in forest stands. WINDSTORM should soon become an intervention mechanism to study the impact of silviculture on damage caused by wind as well as their consequences on the stand dynamics.

Discussion: Advantages of a Shared Forest-modelling Platform

After 6 years of activity, the success of the CAPSIS project confirms the advantages of such shared forest-modelling platforms. Firstly, a shared modelling platform is a good opportunity to build a forest modellers' community. The CAPSIS project brings together the main French institutions concerned with forest modelling, and encourages a real partnership between more than 70 actors, both modellers and foresters, who contribute to its specifications and coordination. It is also currently opening to colleagues from other countries.

Secondly, such a shared platform can really allow the participants to share tools and models. Within the CAPSIS project, the organization of regular training sessions facilitates the interactions between computer scientists, in charge of the kernel and common technical aspects, and modellers, in charge of the integration of their model or library. As CAPSIS is free software (general public licence), it is thus very easy to re-use submodels from other modules or from the kernel within the CAPSIS community. For instance, many models have been easily integrated into new modules by simply modifying the evolution law of an already existing model of the same type. Libraries also provide an efficient opportunity to re-use modelling tools. For instance, the spatial tools currently used in the MOUNTAIN module to simulate virtual stands of various structures will soon be used in the SELVA module too. This is also obviously true with interfaces and graphics, which are developed by computer scientist only once and re-used for each module.

Thirdly, a shared modelling platform is a very practical tool to compare models and silvicultural scenarios, because the different models are integrated in a single platform, because data (such as the initial state) are compatible, and because there is a unique scenario manager. Therefore is it easy to compare the results of a simulation of two models using the same initial state, or to compare three different scenarios on the same model. These facilities are of great use for model evaluation by researchers, but also for decision making by foresters and for teaching.

Finally, we think these advantages really justify a change in our programming habits, and recommend broadening the use of such shared forest-modelling platforms.

References

Ancelin, P. (2001) Modélisation du comportement biomécanique de l'arbre dans son environnement forestier: application au pin maritime. PhD thesis, University of Bordeaux I, France.

Courbaud, B., Goreaud, F., Dreyfus, P. and Bonnet, F.R. (2001) Evaluating thinning strategies using a Tree Distance Dependent Growth Model: some examples based on the CAPSIS software 'Uneven-Aged Spruce Forests' module. *Forest Ecology and Management* 145, 15–28.

Dreyfus, P. and Bonnet, F.R. (1997) CAPSIS (Computer-Aided Projection of Strategies In Silviculture): an interactive simulation and comparison tool for tree and stand growth, silvicultural treatments and timber assortment. In: Nepveu, G. (ed.) *Proceedings of IUFRO WP S5.01-04 Workshop 'Connection between Silviculture and Wood Quality Through Modelling Approaches and Simulation Software'*, 26–31 August 1996, Berg-en-Dal, South-Africa. ERQB-INRA Nancy, France, pp. 194–202.

Gourlet-Fleury, S. and Houllier, F. (2000) Modelling diameter increment in a lowland evergreen rain forest in French Guiana. *Forest Ecology and Management* 131, 269–289.

Pretzsch, H., Biber, P., Dursky, J., von Gadow, K., Hasenauer, H., Kändler, G., Kenk, G., Kublin, E., Nagel, J., Pukkala, T., Skovsgaard, J.P., Sodtke, R. and Sterba, H. (2002) Recommendations for standardized documentation and further development of forest growth simulators. *Forstwissenschaftliches Centralblatt* 121, 138–151.

Ripley, B.D. (1977) Modelling spatial patterns. *Journal of the Royal Statistical Society* B39, 172–212.

Saint-André, L., Laclau, J.-P., Deleporte, P., Ranger, J., Gouma, R., Saya, A. and Joffre, R. (2002) A generic model for the dynamics of nutrient concentration within the stem wood over the whole stand rotation. *Annals of Botany* 90, 1–12.

28

Expected Volume and Value of Structural-dimension Lumber From 25-, 30-, 35-, 40- and 50-year-old Loblolly Pine Plantation Timber

Honorio F. Carino[1] and Evangelos J. Biblis[1]

Abstract

The expected volume and value of structural-dimension lumber derived from plantation-grown timber of 25-, 30-, 35-, 40- and 50-year-old loblolly pine (*Pinus taeda* L.) were estimated and used to determine stumpage values through regression analysis. Also, incremental analysis in conjunction with discounted-cash-flow techniques were used to determine the economic desirability of deferring the harvesting of a loblolly pine plantation stand of a given age for purposes of producing structural-dimension lumber. The results of the study provide compelling evidence that the volume and quality of sawlogs from loblolly pine plantations do indeed generally increase with stand age. It was found that for the 25-, 30-, 35-, 40- and 50-year-old loblolly pine stands investigated, the expected total volume yield of stress-graded dimension lumber could be as much as 94.4, 122.4, 151.0, 179.0 and 235.0 m^3/ha, respectively. The expected annual increase in volume yield could be about 5.61 m^3/ha. The increased quality of older loblolly pine plantation timber was clearly reflected by the percentages (based on volume yield) of stress-graded dimension lumber graded No. 1 and better, which were 16.0, 36.2, 46.7, 49.3 and 83.0%, respectively, for 25-, 30-, 35-, 40- and 50-year-old stands. This was also reflected by the expected unit value yield which was $106, $123, $140, $157 and $192/m^3, respectively. On average, the expected annual increase in unit value yield could be about $3.42/m^3. Not surprisingly, the stumpage value of loblolly pine plantations based solely on the value of structural-dimension lumber produced also increases with stand age. For 25-, 30-, 35-, 40- and 50-year-old stands like those investigated, the expected total value yield of stress-graded dimension lumber could be as much as $10,006, $15,055, $21,140, $28,103 and $45,120/ha, respectively. On average, the expected annual increase in value yield could be about $1399/ha. Also, it was determined that it would be economically desirable to defer the harvesting of a 25-year-old plantation stand of loblolly pine like the one investigated by 5, 10, 15 or 25 years if the minimum acceptable rate of return on investment (**MARR**) is less than 8.5, 7.7, 7.1 or 6.2%, respectively.

[1] School of Forestry and Wildlife Sciences, Auburn University, USA
Correspondence to: carinhf@auburn.edu

Introduction

Loblolly pine (*Pinus taeda* L.) is the most important species of timber growing in the southern region of the USA. It is considered an ideal tree for site restoration and forest management, largely because of its hardiness and versatility in terms of its ability to reproduce and grow rapidly on diverse sites (Schultz, 1997). More than half of the southern yellow pine timber now available to sawmillers in the region comes from plantations consisting mostly of loblolly pine. However, dimension lumber producers are noticeably avoiding, whenever possible, the use of timber from intensely managed loblolly pine plantations, apparently because of the fact that fast-grown pine trees produce a high proportion of juvenile wood at young ages. Juvenile wood, which in loblolly pine could occupy from 10 to 20 annual rings (Zobel and Kirk, 1972), is generally unsuitable for solid wood product production. It is known to be weaker and less stiff than mature wood, due to its lower specific gravity, shorter tracheids, larger fibril angles, thinner cell walls, larger lumen diameter, lower percentage of summerwood, more compression wood and larger longitudinal shrinkage (Pearson and Gilmore, 1971; Zobel and Kirk, 1972; Zobel and Blair, 1976).

It has been determined that the structural quality of dimension lumber from loblolly pine plantation sawtimber increases with stand age and stand density (Biblis *et al.*, 1993, 1995, 1997; Biblis and Carino, 1999). This confirms the widespread belief among wood industry people that quality, and not just volume, of loblolly pine plantation timber increases with stand age. Therefore, the stumpage value also increases with stand age, and a stratified stumpage pricing reflecting the quality differences of sawlogs (Carino and Biblis, 2000) or sawtimber from loblolly plantation stands of different age groups should be considered.

Also, a recently conducted study (Biblis *et al.*, 1998) presents compelling evidence that sawtimber regimes of loblolly pine plantations with longer rotations (e.g. 50 years) could provide competitive returns to landowners relative to pulpwood regimes with shorter rotations (e.g. 20 years). The expected economic advantage might even be greater if loblolly pine plantation stands were managed to produce sawtimber purposely for structural-dimension lumber production, as shown in this chapter. For instance, it was estimated, based on Summer 2000 prices quoted from Alabama sawmillers, that a 25-year-old loblolly pine plantation on an average site (e.g. stand top height or mean dominant height of 27.4 m at 50 years old) and with a 2.4 × 2.4 m initial spacing could yield about $10,006/ha of stress-graded dimension lumber from the harvested sawtimber. In contrast, the same stand, if managed to produce pulpwood, would yield only up to $2471/ha (based on a rather optimistic timber harvest of 257 t/ha at $9.62/t).

There is no doubt that older stands of loblolly pine plantation timber will yield higher expected income for the owners. However, there is a dearth of information about the volume and value of structural-dimension lumber yields from loblolly pine plantation timber of various ages. More importantly, the economic harvest rotation of loblolly pine plantations for structural-dimension lumber production is not known. Not much is known, for instance, about whether it would be economically advantageous and desirable for a sawmiller/landowner to defer for another 5, 10, 15 or 25 years the harvesting of a 25-year-old plantation stand of loblolly pine sawtimber for structural-dimension lumber production. This chapter provides invaluable insights into this important issue.

Methods

Five loblolly pine plantation stands, representing 25-, 30-, 35-, 40- and 50-year age classes, were used for this study. These were selected primarily due to their very similar characteristics, including site indices, original spacing and thinning regimes as listed in Table 28.1. The diameter at breast height (DBH) distribution was established by measuring the DBH of all trees in four 0.10 ha sample plots. Fifty trees representing the predetermined DBH distribution in each of the 25-, 30- and 35-year-old stands were selected for sawing. Also selected in a similar fashion were 30 trees from the 40-year-old stand and 12 trees from the 50-year-old stand. All selected sample trees were identified, harvested, measured and segregated prior to bucking them into sawlogs. Each sawlog was measured, recorded and spray painted on each end with different colours – one to indicate the stand age and the other the log location within the tree.

For each batch of sawlogs of a given type (i.e. based on stand age and tree section – butt, middle or top), the sawn pieces of lumber were tallied and sorted by width and length. About 12.9, 11.0 and 24.7 m^3 of lumber was recovered from each group of 50 trees harvested and sawed, representing the 25-, 30- and 35-year-old stands, respectively. Moreover, 14.3 and 9.5 m^3 of lumber was recovered from the 30 and 12 trees harvested from the 40- and 50-year-old stands, respectively. The lumber was then kiln-dried to approximately 15% moisture content and subsequently dressed to the mill's specifications for dimension lumber.

The lumber produced from each plantation stand investigated was subsequently tested mechanically using the destructive method of determining flexural properties (moduli of rupture (MOR) and elasticity (MOE)). For this purpose, 338, 374 and 474 lumber samples were randomly selected from the 25-, 30- and 35-year-old stands, respectively. On the other hand, all the lumber pieces (468 and 356) produced from sawlogs originating from the sample trees (30 and 12 trees, respectively) representing the 40- and 50-year-old stands were tested. These were stored indoors at the Auburn University Forest Products Laboratory for a minimum of 3 weeks prior to testing. Every piece was tested edgewise in flexure to failure with two-point loading according to ASTM D198-84 (American Society for Testing and Materials, 1999), using a Tinius–Olsen hydraulic testing machine with a capacity of 54,545 kg. It should be noted that only two length categories (2.4 and 3.7 m) of lumber were tested. The 5.08 × 15.24 cm and 5.08 cm × 20.32 cm pieces were tested over a span of 3.35 m, and the 5.08 cm × 10.16 cm pieces were tested over a 2.3-m span. For flexure tests, the machine was equipped with an extended base made from a steel double I-beam 6.25 m long and a steel loading head 2.13 m long. Load and corresponding deflection-to-failure data were obtained with a Hewlett-Packard (H-P) data acquisition system connected

Table 28.1. Characteristics of 25-, 30-, 35-, 40- and 50-year-old loblolly pine (*Pinus taeda* L.) stands considered in this study. Site index at base age 50 years.

Stand age (years)	Site index (m)	Original spacing (m × m)	Thinned at age (years)	No. of trees per ha	Average diameter at breast height (cm)	Average merchantable length (m)	Basal area (m²/ha)
25	29	2.4 × 2.4	18	381	27.2	13.4	9.1
30	28	2.4 × 2.4	20	430	24.6	13.2	8.6
35	30	2.4 × 2.4	18 + 27	321	27.4	19.1	13.6
40	27	2.4 × 2.4	25	450	35.6	23.3	18.1
50	27	2.4 × 2.4	18 + 40	272	37.3	25.1	12.0

to an H-P desk computer for processing. From the obtained data, the MOR and MOE for each tested piece of lumber were calculated. To obtain the extreme fibre value in bending 'F_b' for every tested piece, its MOR value was divided by the 2.1 combine safety factor recommended in ASTM D245 (American Society for Testing and Materials, 1999). This 'F_b' value in combination with the corresponding appropriate value 'E' provides the stress grade for each lumber size listed in Tables 1a, 1b and 1c of the SPIB standard grading rules (Southern Pine Inspection Bureau, 1994).

Analysis

The volume and value of the test sample boards for each combination of size and stress-grade classes originating from a stand of a given age were then determined. The valuation of lumber was based on lumber prices (in $/m³) quoted directly from southern yellow pine lumber producers for the summer months of 2000. From the test sample data, it was then possible to estimate the probability of occurrence (or percentage distribution) of stress-graded dimension lumber coming from each sample stand of a given age based on volume (see Table 28.2). Also, it was possible to estimate the weighted average unit value (in $/m) of the tested dimension lumber from each sample stand of a given age, using the following relationship:

$$V_k = \sum_i \sum_j P_{ij} R_{ij} \tag{1}$$

where: V_k is the weighted (by size and grade) average unit value ($/m³) of tested (i.e. stress-graded) dimension lumber from a loblolly pine plantation of a given age

Table 28.2. Percentage distribution of stress-graded dimension lumber from 25-, 30-, 35-, 40- and 50-year-old loblolly pine plantation stands considered in this study[a].

Stand age (years)	Lumber size (cm × cm)	Percent distribution based on volume yield			
		No. 1 and better	No. 2	No. 3 and below	Total
25	5.08 × 10.16	12.0	9.4	49.9	71.3
	5.08 × 15.24	4.0	3.5	21.2	28.7
		16.0	12.9	71.1	100.0
30	5.08 × 10.16	28.6	12.3	42.2	83.1
	5.08 × 15.24	7.6	2.4	6.9	16.9
		36.2	14.7	49.1	100.0
35	5.08 × 10.16	17.0	10.1	16.0	43.1
	5.08 × 15.24	24.4	10.4	11.9	46.7
	5.08 × 20.32	5.3	0.7	4.2	10.2
		46.7	21.2	32.1	100.0
40	5.08 × 10.16	29.6	12.8	14.1	56.5
	5.08 × 15.24	9.6	6.2	12.6	28.4
	5.08 × 20.32	10.1	2.0	3.0	15.1
		49.3	21.0	29.7	100.0
50	5.08 × 10.16	12.8	1.0	0.6	14.4
	5.08 × 15.24	62.2	6.3	8.3	76.8
	5.08 × 20.32	8.0	0.4	0.4	8.8
	Total	83.0	7.7	9.3	100.0

[a]Stress grades are based on SPIB allowable design values of 'F_b' and 'E' as listed in Tables 1a, 1b and 1c, pages A3, A4 and A5 of Standard Grading Rules for Southern Pine Lumber (Southern Pine Inspection Bureau, 1994).

k; P_{ij} is the probability of occurrence of tested dimension lumber of a given size i and grade j from a loblolly pine plantation of a given age; and R_{ij} is the unit price ($/m^3) of tested dimension lumber of a given size i and grade j.

The estimated weighted average unit values (in $/m^3) of the tested dimension lumber from the 25-, 30-, 35-, 40 and 50-year-old stands are $100, $127, $150, $149 and $192/m^3, respectively.

The total volume of dimension lumber per hectare of loblolly pine plantation of a given age was estimated using the data on the number of sample trees harvested for the study, the volume of dimension lumber produced from such sample trees, and number of trees at harvest time. For instance, in the case of the 25-year-old stand, the total volume of dimension lumber is about 98 m^3/ha (i.e. given the number of trees at harvest time = 380/ha; number of sample trees harvested = 50; volume yield of 50 trees = 12.9 m^3). Calculated in a similar fashion, the estimated total volume of dimension lumber from the 30-, 35-, 40- and 50-year-old stands was 94.4, 158.6, 215.1 and 215.7 m^3/ha, respectively.

Given the data on weighted average unit value (in $/m^3) and the estimated total volume of dimension lumber per hectare, it was then possible to estimate the total value of dimension lumber per hectare of loblolly pine plantation of a given age. The estimated total value of dimension lumber of the 25-, 30-, 35-, 40- and 50-year-old stands was $9808, $12,054, $23,836, $32,094 and $41,296/ha, respectively.

For the purposes of predicting expected values (see Table 28.3) as well as observing trends and rates of change of the unit value, total volume and value of stress-graded dimension lumber produced per hectare of loblolly pine plantation of various ages from 25 to 50 years, the following regression equations were developed:

$$Y_1 = -45.80 + 5.61X \qquad r^2 = 0.82 \qquad\qquad (2)$$

where: Y_1 is the expected total volume of stress-graded dimension lumber per unit area (m^3/ha) of loblolly pine plantation of a given age and X is the stand age (years).

$$Y_2 = 20.59 + 3.42X \qquad r^2 = 0.94 \qquad\qquad (3)$$

where: Y_2 is the expected unit value ($/m^3) of stress-graded dimension lumber from a loblolly pine plantation of a given age and X is the stand age (years).

The expected value of stress-graded dimension lumber produced per hectare of loblolly pine plantation of a given age was estimated by multiplying the values given by Equations 2 and 3. In this study, a modified version of stumpage valuation was used in the analysis. Stumpage was estimated by the value of structural-dimen-

Table 28.3. Expected volume and value yields of stress-graded dimension lumber from 25-, 30-, 35-, 40- and 50-year-old loblolly pine plantation stands.[a]

Stand age (years)	Expected total volume yield (m^3/ha)	Expected unit value ($/m^3)	Expected total value yield ($/ha)
25	94.4	106	10,006
30	122.4	123	15,055
35	151.0	140	21,140
40	179.0	157	28,103
50	235.0	192	45,120

[a]The expected unit value and total volume were calculated using Equations 2 and 3, respectively, and the expected value yield per hectare is the product of the two.

sion lumber yield based on actual volume recovery (instead of scaled volume as conventionally done in practice). With such a modified valuation of stumpage notwithstanding, the economic desirability of deferring the harvesting sawtimber for structural-dimension lumber production from a loblolly pine plantation of a given age was determined using incremental analysis in conjunction with discounted-cash-flow techniques. For this analysis, the following relationship proved to be very useful:

$$R = [(F/P)^{1/n} - 1] \times 100 \qquad (4)$$

where: R is the expected incremental rate of return (%); P is the expected present value ($) of stress-graded dimension lumber produced per hectare of loblolly pine plantation of a given age; F is the expected future value ($) of stress-graded dimension lumber produced per hectare of loblolly pine plantation after n years from a base age; and n is the stand age (years).

A decision to harvest the stand at a certain age (e.g. 25 years), and not later (e.g. after 5 years or 30 years of age), would be implemented if R is less than a specified minimum acceptable rate of return on investment (**MARR**). Otherwise, it would be more economically desirable to defer the harvesting of the stand until more favourable economic conditions exist or whenever an R equal to or higher than the **MARR** could be achieved (see Table 28.4).

Results and Discussion

From Table 28.3, it is apparent that the value of stumpage in loblolly pine plantations increases with the age of the stand. The estimated total value yields of stress-graded dimension lumber of the 25-, 30-, 35-, 40- and 50-year-old stands are $10,020,

Table 28.4. Estimated time value (or incremental rate of return) of stumpage of loblolly pine plantations of various ages managed for sawtimber production when harvesting is deferred.[a]

Stand age (years)		Incremental rate of return (%)
Current or base	Harvest time	
25	30	8.5
	35	7.7
	40	7.1
	50	6.2
30	35	6.9
	40	6.4
	50	5.6
35	40	5.9
	50	4.8
40	50	4.8

[a]In this case, stumpage is based on the value of stress-graded dimension lumber product yield. Stress grades are based on SPIB allowable design values of 'F_b' and 'E' as listed in Tables 1a, 1b and 1c, pages A3, A4 and A5 of Standard Grading Rules for Southern Pine Lumber (Southern Pine Inspection Bureau, 1994).

$15,073, $21,081, $28,037 and $44,808/ha, respectively. These represent an annual average increase of about $1399/ha, as determined by regression analysis. In terms of the unit value of dimension lumber yield, the 25-, 30-, 35-, 40- and 50-year-old stands yielded about $106, $123, $140, $157 and $192/m^3, respectively. This represents an annual rate of increase in unit value yield of about $3.42/m^3, as reflected by the coefficient of X in Equation 3. Such age-related increases in value yield could be attributed to increases or improvement in both the volume and quality of dimension lumber yields.

Table 28.3 shows that for the 25-, 30-, 35-, 40- and 50-year-old stands, the estimated volume of dimension lumber produced is about 94.4, 122.4, 151.0, 179.0 and 235.0 m^3/ha, respectively, of which (see Table 28.2) approximately 16.0, 36.2, 46.7, 49.3 and 83.0% graded No. 1 and better, respectively. The annual rate of increase in volume yield is about 5.61 m^3, as reflected by the coefficient of X in Equation 2. Using Equation 4 and the expected total value yields per hectare shown in Table 28.3, the time value (expressed in terms of percentage incremental rate of return or R) of stumpage of loblolly pine plantations managed primarily for the production of sawtimber converted into structural-dimension lumber was determined. For instance, if the harvesting of the 25-year-old stand considered in this study was deferred by 5, 10, 15 or 25 years, the R value could be about 8.5, 7.7, 7.1 or 6.2%, respectively. Similarly, if the harvesting of the 30-year-old stand was deferred by 5, 10 or 20 years, the R value could be about 6.9, 6.4 or 5.6%, respectively. In the case of the 35-year-old stand, if its harvesting was deferred by 5 or 10 years, the R value could be about 5.9 or 4.8%, respectively. If the harvesting of the 40-year-old stand was deferred by 10 years, the R value could be about 4.8%. All these estimated R values are shown in Table 28.4. It should be noted that these are conservative estimates because, as pointed out earlier, the value of structural-dimension lumber derived from the sawtimber produced was the sole basis for the stumpage valuation. The value of 1-inch boards and pulpwood from the top portion of the tree was not included in estimating the stumpage value.

Evidently, the harvesting of a loblolly pine plantation stand of a given age should be deferred if the estimated R is equal to or higher than the minimum acceptable rate of return on investment or **MARR**, whatever that may be. For example, if the **MARR** is 7.5%, it would probably be prudent on the part of the owner of a 25-year-old loblolly pine plantation stand to defer harvesting it for another 10 years, because the R value then or at 35 years of age would be about 7.7%. On the other hand, it would seem more economically desirable for owners of 30-, 35-, 40- and 50-year-old stands to do the harvesting now and not defer it to a later date, because the estimated R values (see Table 28.4) are less than the 7.5% **MARR**. Certainly, harvesting decisions have to be reviewed whenever a change in the hurdle rate or **MARR** occurs. The bottom line is that harvesting of the stand should be implemented only when favourable economic conditions exist, or deferred to a later date whenever the expected R is equal to or higher than the **MARR**.

Summary and Conclusion

Stumpage values of 25-, 30-, 35-, 40- and 50-year-old plantation stands of loblolly pine based solely on the volume and value (based on summer 2000 prices) of stress-graded dimension lumber produced therefrom were determined. Regression equations for predicting the expected total volume and value of structural-dimension lumber yield per hectare of loblolly pine plantation stands were developed. The expected yield values were used in an incremental analysis in conjunction with discounted-cash-flow techniques to determine the economic desirability of deferring the harvesting of a loblolly pine plantation stand of a given age for the purpose of producing structural-dimension lumber. The following can be inferred from the results of the analysis:

1. Structural-dimension lumber volume yield of loblolly pine timber from managed plantations increases with stand age. For 25-, 30-, 35-, 40- and 50-year-old stands like those investigated, the expected total volume yield of stress-graded dimension lumber could be as much as 94.4 , 122.4 , 151.0, 179.0 and 235.0 m^3/ha, respectively. On average, the expected annual increase in volume yield could be about 5.61 m^3/ha.
2. The quality of loblolly pine plantation timber increases with stand age. This was clearly reflected by the percentages (based on volume yield) of structural-dimension lumber graded No. 1 and better, which are 16.0, 36.2, 46.7, 49.3 and 83.0%, respectively, for 25-, 30-, 35-, 40- and 50-year-old stands. This was also reflected by the expected unit value yield of structural-dimension lumber from the 25-, 30-, 35-, 40- and 50-year-old stands, which are $106, $123, $140, $157 and $192/$m^3$, respectively. On average, the expected annual increase in unit value yield could be about $3.42/$m^3$.
3. Stumpage value of loblolly pine plantations based solely on the value of dimension lumber produced increases with stand age. For 25-, 30-, 35-, 40- and 50-year-old stands like those investigated, the expected total value yield of stress-graded dimension lumber could be as much as $10,006, $15,055, $21,140, $28,103 and $45,120/ha, respectively. On average, the expected annual increase in value yield could be about $1399/ha.
4. It would be economically desirable to defer the harvesting of a 25-year-old plantation stand of loblolly pine like the one investigated by 5, 10, 15 or 25 years if the minimum acceptable rate of return on investment (**MARR**) is less than 8.5, 7.7, 7.1 or 6.2%, respectively. For a 30-year-old stand, it would be economically desirable to defer harvesting it by 5, 10 or 20 years if the **MARR** is less than 6.9, 6.4 or 5.6%, respectively. For a 35-year-old stand, it would be economically desirable to defer harvesting it by 5 or 10 years if the **MARR** is less than 5.9 or 4.8%, respectively. And for a 40-year-old stand, it would be economically desirable to defer harvesting it by 10 years if the **MARR** is less than 4.8%.

References

American Society for Testing and Materials (ASTM) (1999) Standard test methods of static tests of lumber in structural sizes D 198–98. *Standard Practice for Properties of Visually Graded Lumber.* D 245–99. Annual Book of ASTM Standards, Section 4, Vol. 04.10. West Conshohocken, Pennsylvania.

Biblis, E.J. and Carino, H.F. (1999) Flexural properties of lumber from a 50-year-old loblolly pine plantation. *Wood and Fiber Science* 31, 200–203.

Biblis, E.J., Carino, H.F. and Teeter, L. (1998) Comparative economic analysis of two management options for loblolly pine timber plantations. *Forest Products Journal* 48(4), 29–33.

Biblis, E.J., Carino, H.F. and Brinker, R. (1997) Flexural properties of lumber from two 40-year-old loblolly pine plantations with different stand densities. *Wood and Fiber Science* 29, 375–380.

Biblis, E.J., Carino, H.F., Brinker, R. and Mckee, C.W. (1995) Effect of stand density on flexural properties of lumber from two 35-year-old loblolly pine plantations. *Wood and Fiber Science* 27, 25–33.

Biblis, E.J., Brinker, R., Carino, H.F. and McKee, C.W. (1993) Effect of stand age on flexural properties and grade compliance of lumber from loblolly pine plantation timber. *Forest Products Journal* 43(2), 23–28.

Carino, H.F. and Biblis, E.J. (2000) Comparative analysis of the quality of sawlogs from 35-, 40-, and 50-year-old loblolly pine plantation stands. *Forest Products Journal* 50(11/12), 48–52.

Pearson, R.G. and Gilmore, R.C. (1971) Characterization of the strength of juvenile wood of loblolly pine *(Pinus taeda* L.). *Forest Products Journal* 21(1), 23–30.

Southern Pine Inspection Bureau (1994) *Grading Rules*. SPIB, Pensacola, Florida, 134 pp.

Schultz, R.P. (1997) *Loblolly Pine: the Ecology and Culture of Loblolly Pine* (Pinus taeda *L.*). Agriculture Handbook 713. USDA Forest Service, Washington, DC, 3 pp.

Zobel, B.J. and Blair, R. (1976) Wood and pulp properties of juvenile wood and topwood of the southern pines. In: *Proceedings of the 8th Cellulose Conference and Applied Polymer Symposium*, Vol. 28, pp. 421–433.

Zobel, B.J. and Kirk, D.G. (1972) Wood properties of young loblolly and slash pines. In: *Proceedings of a Symposium on the Effect of Growth Acceleration on the Properties of Wood*. USDA Forest Service, Forest Products Laboratory, Madison, Wisconsin.

29 Comparing Models for Growth and Management of Forest Tracts

J.J. Colbert,[1] Michael Schuckers[2] and
Desta Fekedulegn[2]

Abstract

The Stand Damage Model (SDM) is a PC-based model that is easily installed, calibrated and initialized for use in exploring the future growth and management of forest stands or small wood lots. We compare the basic individual tree growth model incorporated in this model with alternative models that predict the basal area growth of trees. The SDM is a gap-type simulator. It is a non-spatial, individual sample-tree diameter growth model, following the work of Botkin, Shugart and others. Within SDM, the basic growth model is adjusted to account for shading, competition and climate. Here we make those adjustments by calibrating the growth model to historical data for individual sample trees. We fit alternative sigmoid growth models to the same historical data and compare these models' ability to predict short-term (5-year) and longer term growth of trees. Accuracy and potential effects of bias are discussed relative to the age and source locations of sample trees used in this study.

Introduction

The Stand Damage Model (SDM), a component of the Gypsy Moth Life System Model, calculates tree diameter growth, current diameter and height, and tree mortality for each year of a simulation. The model is a distance-independent, tree-growth simulator, a gap model based on the work of Botkin (1993), Shugart (1984) and others. Each year, the model calculates the diameter growth of trees as a function of relative stocking (a measure of tree crowding), shading, heat and defoliation. The model can be used to predict future growth of a forest stand or tract that the user wants to consider as a homogeneous unit. The user can design tree removals to simulate various kinds of silvicultural regimes at any year of a simulation. The user provides data for management actions, including the year of entry, selection criteria for removals and species-specific targets. The model has the ability to simulate predefined defoliation scenarios over a 20-year period from the initial stand conditions or the user can decide to have defoliation occur at any time and at intensities as input through the

[1] USDA Forest Service, USA
Correspondence to: jcolbert@fs.fed.us
[2] Department of Statistics, West Virginia University, USA

user interface. With the predefined scenarios come a large number of graphs and tables automatically generated to assist the user in assessing possible future impacts of gypsy moth defoliation.

Weather drives photosynthesis and tree growth. Within the model, a cumulative heat unit measure, degree-days above a single threshold (4.4°C), is used for all tree species as a primary driver of diameter growth. Default constant weather data can be overridden by entering weather variation for each year simulated. Tree mortality comprises three factors: a base rate, stress-induced mortality and gypsy moth defoliation-induced mortality. The base rate is increased as diameter growth is reduced. Following mortality calculations, tree growth is updated for the residual stems, and new stems are recruited to the smallest classes.

For details on model formulation or computational algorithms, see Colbert and Sheehan (1995). Colbert and Racin (1995, 2001) provide users with directions for use of the model, and Racin and Colbert (1995) give a complete description of the program structure for the user interface system. The model interface now includes parameters for 71 tree species or species groups. The user can add up to 10 additional species for a stand simulation; 20 species can be included in a single simulation. Seven additional categories for including non-commercial or other unidentified species (grouped by susceptibility to defoliation as conifers or deciduous trees) allow users to provide a more complete and accurate set of initial conditions. In addition to tree counts by species and diameter class, you can assign one of three soil-moisture categories for the stand. Users supply information on defoliation history (in broad categories by species for the 2 years prior to the simulation) and describe defoliation scenarios for each species each year as a percentage for the overstorey and for the understorey.

State of the Art

As a stand-alone model, the SDM provides the user with the ability to assess the effects of defoliation on a tree and stand level. Through multiple simulations, the model can be used to assess the possible future impacts of a gypsy moth outbreak on a large forested area (Colbert *et al.*, 1997). The model has also been used for a much broader assessment of possible forest impacts of the gypsy moth (Gottschalk *et al.*, forthcoming; Guldin *et al.*, forthcoming). The model can also be used as a stand growth and forest management simulator, disregarding the gypsy moth and associated impacts. Through the user interface, one can describe a forest stand, set growth parameters, design future management entries, and project an initial inventory into the future to assist in viewing the effects of various scenarios. The current model can carry a stand through cutting cycles, but the current regeneration algorithms are limited and research is under way to enhance the models for predicting regeneration success and growth of young stands.

General Technical Report NE-211 (Liebhold *et al.*, 1995) was the source for species feeding preference of the gypsy moth; data from Botkin (1993) and Shugart (1984) contributed significantly to the formulations used in the model. To better understand and interpret the current SDM, review the documentation for Version 1.0 (Colbert and Racin, 1995; Colbert and Sheehan, 1995; Racin and Colbert, 1995). The 'How-To' document (Colbert and Racin, 2001) provides an installation diskette and a complete description of the updates and enhancements for Version 2.0 of the model and a tutorial for use of the new data entry system. Plot data from any design of fixed area and prism point samples are converted automatically into both stand-level summary tables and an initial-conditions file for further use in simulations.

The user interface allows data to be stored and retrieved through the expanded and enhanced menu system. Documentation and software downloads can be found at the US Forest Service website: www.fs.fed.us/ne/morgantown/4557/gypsymth. Available at this website are descriptions of the models and related publications as well as links for downloading this and other models that have been developed.

Included with the 'How-To' document is a PC/Windows installation diskette. Once installed, the user is provided with two example data sets. One is from a series of variable-radius plots from a stand in Ohio. The second is from a fixed area three-plot cluster in Pennsylvania. These example data are used in the tutorial section for learning plot data entry. In addition, there is an Excel file that contains a complete 15-year history of the three-plot cluster for use as a validation and test data set.

Platforms supported by the Stand Damage Model

The model currently runs under MS-DOS and within a DOS window or emulator under Windows 3.1/95/98/NT on Intel or compatible processors.[3] It operates as a DOS package of programs, using a graphical interface for management of input and output data, controlling simulations and viewing results. The model can run on older computers and does not require more than the base 640K of memory, although usage is enhanced by a mouse for easy access and movement within and among windows within the user interface. Instructions are included with the installation package for running the program from Windows 3.1 with the included PIF and icon files or from Windows 95/98/NT.

Related resources

It is possible to completely re-parameterize the model for locations outside North America. The data required for adding species to the model can be found in basic dendrology texts. In North America we used the *Textbook of Dendrology* (Harlow *et al.*, 1979), the *Silvics of North America* volumes compiled by Burns and Honkala (1990a, b), and *Trees of the Southeastern United States* (Duncan and Duncan, 1988) as sources to generate parameters. For scientific and common names, we followed the nomenclature in *Checklist of United States Trees* (Little, 1979). *Oaks of North America* (Miller and Lamb, 1985) was a valuable resource for searches within specific families or genera. *Knowing Your Trees* (Collingwood and Brush, 1978) was used to assess tree-crown shape in assigning stocking classes, and *Important Forest Trees of the Eastern United States* (USDA Forest Service, 1991) was the source of information for each species. The 'How-To' publication (Colbert and Racin, 2001) provides a complete guide to building parameter files for the model.

Methodology

Forest conditions in the mid-Atlantic region of the USA are quite variable, with continental climate and annual rainfall from 50 to 180+ cm (generally in the range of

---

[3] The use of trade, firm or corporation names is for the information and convenience of the reader. Such use does not constitute an official endorsement or approval by the US Department of Agriculture or Forest Service of any product or service to the exclusion of others that may be suitable.

100–130 cm) in the areas where managed forests are most prevalent. Rainfall is generally well distributed throughout the year. The forests we studied are considered to be mixed mesophytic and consisted of oak-dominated, mixed oak–hickory, or oak–maple forest types. These forest stands are usually quite diverse in canopy composition and contain from five to 20 or more different tree species. Elevations range from the coastal plains of Delaware, just 2–10 m above sea level, to the forests of the more interior central Pennsylvania and West Virginia mountains, which range from 350 to 900 m above sea level. Here, our study sites range from 400 to 600 m elevation. In Ohio, stands were located on the Dorr Run management unit of the Athens district of the Wayne National Forest, which is 210–320 m elevation.

Analyses of four forest tree basal-area growth models were carried out using data from the states of Delaware (DE), Pennsylvania (PA), West Virginia (WV) and Ohio (OH) in the mid-Atlantic region of the USA. We fitted these with non-linear sigmoid growth models which are described in detail by Fekedulegn *et al.* (1999). The four-parameter models we chose to utilize in this study have been shown to be adequate for modelling sigmoid growth with sufficient flexibility and good statistical properties (Draper and Smith, 1981; Schnute, 1981; Myers, 1986; Vanclay, 1994). These models were fit to basal-area data derived from radial increment data and estimates of inner radius and age.

Source data

A total of 190 radial growth-increment samples were used. The data were obtained from southern red oak (*Quercus falcata*) and white oak (*Quercus alba*) from the coastal plain in central and southern DE; northern red oak (*Quercus rubra*) from the ridge-and-valley area of central PA; northern red oak on the Coopers Rock State Forest in north central WV; and northern red oak from OH. Table 29.1 provides the numbers of sample data and type. All samples were taken at breast height (137 cm). Increment cores were taken with 4.3 mm borers, and disc samples were taken from felled trees. All increment core samples were taken aiming through the tree centre, attempting to get as close to the pith as possible.

Increment cores were first dried and glued in place with water-soluble glue on top of wood mounts of approximately 18 mm high × 8 mm wide cross-section. The mount top is bevelled so that it contains a groove 4 mm across that runs the length of the mount. Samples were oriented vertically and sanded using fine (400–1200) grit to expose cell structure. Annual radial increments were measured to the nearest 0.001 mm on a measuring stage and the radius and age at the inner edge of the innermost ring were estimated using a 1-mm scaled circular ruler, taking into consideration the curvature of the earliest growth rings and the width of those same rings. In some instances, on increment core samples and on all disc samples we were

Table 29.1. Sample counts by location, type and tree species.

Location	Type	Species	Number
Delaware (DE)	IC[a]	*Quercus falcata*	24
	IC	*Quercus alba*	32
Pennsylvania (PA)	IC	*Quercus rubra*	26
West Virgina (WV)	Disc	*Quercus rubra*	21
West Virginia (WV)	IC	*Quercus rubra*	60
Ohio (OH)	IC	*Quercus rubra*	28

[a]IC = increment core sample.

able to provide data from the pith at age 0. Basal area (inside bark) series were then produced, assuming circular cross-sections at breast height.

Sigmoid models

The non-linear models fitted were the following:

$$\omega(t) = \beta_0/(1 + \beta_1 \exp(-\beta_2 t))^{1/\beta_3} + \varepsilon \qquad \text{(1) Richards}$$

$$\omega(t) = \beta_0 - \beta_1 \exp(-\beta_2 t^{\beta_3}) + \varepsilon \qquad \text{(2) Weibull}$$

$$\omega(t) = \beta_0(1 - \beta_1 \exp(-\beta_2 t))^{1/(1-\beta_3)} + \varepsilon \qquad \text{(3) Chapman–Richards}$$

$$\omega(t) = (\beta_0^{1-\beta_3} - \beta_1 \exp(-\beta_2 t))^{1/(1-\beta_3)} + \varepsilon \qquad \text{(4) von Bertalanffy}$$

proposed by (1) Richards (1959), (2) Ratkowsky (1983), (3) Turnbull (1963), Pienaar and Turnbull (1973) and (4) von Bertalanffy (1957). It should be noted that all parameters are assumed to be positive ($\beta_0, \beta_1, \beta_2, \beta_3 > 0$) for all models; in Equation 2, $\beta_1 < \beta_0$; in Equation 3, $\beta_1 < 1$; and in both Equations 3 and 4, $\beta_3 < 1$. For the tree species used, growing in North America, biological bounds were constructed. Maximum basal area for the red oaks is less than 2 m^2 and less than 5 m^2 for *Q. alba*. We set the upper bound for β_0 to be 8 m^2 and the lower bound to be 0.1 m^2 for all runs of each model. We used the NLIN procedure (SAS, 2000) with the Marquardt (1963) method to estimate the parameters for each model and basal-area series, supplying the partial derivatives (Fekedulegn *et al.*, 1999). We found convergence to be quite sensitive to the starting values used. We first used modal initial conditions, estimated from the data as described in Fekedulegn *et al.* (1999). Then we used the distribution of values resulting from the first application of the NLIN procedure to create a new initial search grid (Table 29.2) used for the final runs of the NLIN procedure.

To assess the predictive power of each model, we truncated each data set and refitted the models to each truncated data set. This allowed us to examine differences and produce tests of each model's ability to predict future basal area.

Table 29.2. Initial search grid (L, lower bound; U, upper bound; S, step size).

Model		β_0	β_1	β_2	β_3
Richards (R)	L	0.101	10^{-4}	10^{-3}	5×10^{-5}
	U	5.0	0.1	0.3	2.0
	S	0.6	10^{-2}	0.03	0.2
Weibull (W)	L	0.2	10^{-2}	10^{-4}	0.1
	U	5.0	4.9	10^{-3}	4.5
	S	0.6	0.59	2×10^{-4}	0.5
Chapman–Richards (C–R)	L	1.01	10^{-4}	10^{-3}	10^{-2}
	U	5.0	0.99998	0.1	0.99
	S	0.6	0.15	0.015	0.15
von Bertalanffy (VB)	L	0.2	0.36	0.002	0.36
	U	5.0	1.04	0.022	0.80
	S	0.4	0.05	0.003	0.03

Statistical summaries are provided for parameters for each model. We tested the differences among the models' ability to fit the data by examining the number of samples where the convergence criteria were met. Because the fit residuals are not normally distributed, we used a non-parametric test procedure to test for location and scale differences across a one-way classification (Wilcoxon). We calculated the mean square error (MSE) for each fit and tested difference among the models. We explored parameter interactions and consistency of fits among sites and among species. We also examined the interactions among parameters within models and the values of β_0 among models since this parameter is the asymptotic basal area as the tree reaches maturity.

Results

To judge the quality of models and fitting procedures, we looked at the number of samples where the convergence criteria were met. Of the 380 total series, the Chapman–Richards model consistently converged most often, followed by the Richards, Weibull and finally the von Bertalanffy model (Table 29.3).

For all models, the fitting procedure converged for the majority of these data and there was no significant difference among the numbers of series fitted for each model. It can also be seen that the length of the data series did not have a consistent effect on convergence. There was no data set that was significantly different in the number of series converging (Table 29.4).

Among data sets, the Chapman–Richards model consistently converged most often. It should be noted that we first attempted to use less dense initial grids and found that the procedure did not converge for a very large number of samples, across all models.

We examined MSE for each model, series length and source data set and found that there was no significant difference among these classifications. We found the choice of limits and mesh size to have a large effect on the number of converging series and the quality of the fit (MSE). We found good quality fits (MSE $< 10^{-4}$) in 87.5% of the non-converging samples. It should be noted that since the procedure used did not permit the inclusion of non-constant constraints (boundary conditions) among parameters, there was some data truncation for early growth for some samples when fitting the von Bertalanffy model. It is sufficient that $\beta_0^{1-\beta_3} > \beta_1$ for $\hat{\omega}$ to be positive for all t. When this condition is not met, $\hat{\omega}$ will remain undefined for small t, and the associated data are ignored during the fitting procedure.

Table 29.3. Adequacy of Marquardt fit to full and reduced series (counts in parentheses).

Model	Full	Reduced	Total
		%	
Richards (R)	83	72	77
	(158)	(136)	(294)
Weibull (W)	71	75	73
	(135)	(142)	(277)
Chapman–Richards (C–R)	96	94	95
	(182)	(178)	(360)
von Bertalanffy (VB)	63	70	67
	(120)	(133)	(253)

Table 29.4. Percentages of Marquardt fit among data sets and models.

Model	Q.a.	Q.f.	PA	IC	D	OH
			%			
Richards (R)	61	85	73	82	74	85
Weibull (W)	87	29	81	85	76	57
Chapman–Richards (C–R)	97	98	88	95	88	100
von Bertalanffy (VB)	62	81	62	64	45	85

Sites: *Q.a.* – DE, *Q. alba*; *Q.f.* – DE, *Q. falcata*; IC – West Virginia increment core samples; D – West Virginia disc samples.

Model fits: graphs and residual plots

While improvement can be made in obtaining convergence, results obtained on samples that did not converge often appear to be adequate and represent the data well throughout the range of those data. The errors about the non-converged fitted curves are often no worse or even better than another sample taken at another radius from the same tree where convergence criteria were met. The ability to find an adequate fit does not appear to be associated with either the length or starting point of the data series. Figure 29.1 shows the fitted summaries for such a sample. The truncated series did not converge. Convergence was obtained for both series from a second increment core sample *taken from the same tree*. It was found that problems with convergence did not appear to be associated consistently with either the full or truncated series. Influences such as individual tree release, weather or insect defoliation can cause fluctuations in the growth pattern that are not well represented by these models.

Predictions

We examined 5-, 10- and 15-year predictions. Figure 29.2 shows two trees where the truncated series demonstrates that a prediction from the truncated model can be used for predicting the trend of diameter growth. Table 29.5 shows the values for each of the predictions, along with actual data.

Table 29.5. Actual and predicted basal area for two trees; Y, measured basal area; $\hat{Y}$, predicted basal area, full data series; $\hat{Y}_t$, predicted basal area, truncated data series.

PI[a]	0	5	10	15
		E118234A		
Y	0.1133	0.1207	0.1283	0.1364
$\hat{Y}$	0.1120	0.1204	0.1284	0.1360
$\hat{Y}_t$	0.1121	0.1206	0.1287	0.1364
		T03B		
Y	0.233	0.2819	0.323	0.3764
$\hat{Y}$	0.2344	0.2789	0.3258	0.3747
$\hat{Y}_t$	0.234	0.2796	0.3295	0.3832

[a]Prediction interval length in years.

Model asymptotes (theoretical maximum values)

We examined the parameters for each model. β_0 was the only parameter that was comparable among models. We found that the truncated series ($P = 0.0104$) and the

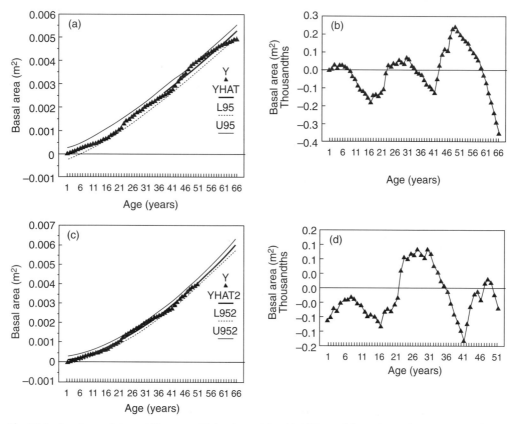

Fig. 29.1. Basal area data and Chapman–Richards model, with 95% confidence intervals: (a) converged to the full data set; (b) residuals about $\hat{Y}$ from (a); (c) non-convergence to the truncated data set; (d) residuals for (c). Note that the residuals are clustered closer to the curve in the truncated fit.

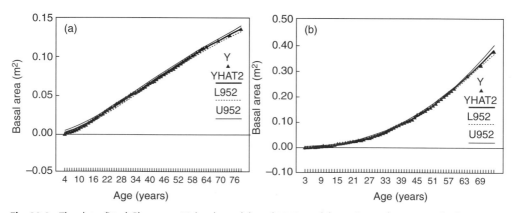

Fig. 29.2. The data, fitted Chapman–Richards model, and 95% confidence intervals (a) using the first 59 years of a Pennsylvania sample and (b) using the first 64 years of an Ohio sample (see Table 29.5).

von Bertalanffy model ($P < 0.001$) had greater predicted asymptotes. The asymptotes were the lowest and did not differ between the Richards and Weibull models while, for these two, β_0 did differ significantly from the other two models. There were a number of samples for which the predicted asymptote was well above the biological maximum for the tree species.

Other parameters

Parameters for each model were examined both graphically and statistically. Here we present what we think are the most relevant observations from that review. The parameter statistics are for the samples where the convergence criteria were met. Q1 and Q3 are the 25% and 75% quantiles.

Richards

	Q1	Median	Q3
β_0	0.100	0.143	0.270
β_1	0.11×10^{-3}	0.21×10^{-3}	1.61×10^{-3}
β_2	0.0196	0.0280	0.0428
β_3	0.02×10^{-3}	0.04×10^{-3}	0.22×10^{-3}

For this model, β_0 was the lowest across all fits; the 99th percentile estimate was 3.89 m²; β_1 showed the widest range of any parameter across all models; β_2 had the narrowest range when compared with the other models' t coefficients; β_3 had the second widest range across all models (over five orders of magnitude) but for most of the fits, estimates converged to values less than 10^{-3}.

Weibull

	Q1	Median	Q3
β_0	0.100	0.152	0.360
β_1	0.100	0.152	0.360
β_2	0.10×10^{-4}	0.61×10^{-4}	3.49×10^{-4}
β_3	1.66	1.98	2.33

As we improved the convergence across models and samples, we found just one striking relationship among parameters. An almost perfect linear relationship existed between β_0 and β_1 in this model; β_0 was above the estimates for the Richards model but it still tended to under-predict expected maximum diameter for these data. The time coefficient, β_2, showed the widest variation across models but, because t in this model form carries an exponential factor, β_3, the variation in these two parameters compensated for one another. These two parameters showed a strong log-linear relationship. Another interesting note is that while the convergence criteria were met, one sample showed no fit; $\hat{Y}$, the estimated or fitted value of the basal area growth function, was essentially constant.

Chapman–Richards

	Q1	Median	Q3
β_0	0.118	0.255	2.79
β_1	0.228	0.987	0.999
β_2	0.0055	0.0143	0.0245
β_3	0.509	0.689	0.966

Under the Chapman–Richards formulation, β_0 showed the most biologically reasonable range of values, but it did considerably overestimate on more that 10% of the samples. β_1 ranged over four orders of magnitude but did not show any noticeable interaction with other parameters. Except for the Weibull form, β_3 showed the narrowest range of variability, with the exception of one sample.

No consistent pattern emerged among species or among sites in the parameterization of these models except that β_3 in the Weibull model shows some slight location dependence. These differences are mitigated by the fact that the range in differences are similar to what is obtained within West Virginia data between those samples taken from whole-tree dissections and those taken from increment cores.

Discussion and Conclusions

Each of these models will provide reasonable fit to radial increment data and permit estimates of future basal area under nominal conditions. There is consistency between models in terms of MSE. Research to date suggests that MSE is a better screening criterion than meeting SAS's PROC NLIN convergence criteria. We attempted to use other fitting methods but found that the Marquardt method performed best. To strengthen our understanding of the power and consistency of these models to perform across this region, we plan to expand the data to include balance among species and to classify our analyses to account for canopy strata and site factor effects within species. We will fit data from older trees to ascertain how tree age affects parameters, particularly the asymptote. We will explore the ranges for parameters of these models that give rise to realistic trends for mature and over-mature trees. As mentioned earlier, convergence is highly dependent on starting values. We found that the use of a starting grid will usually provide good results. When boundary conditions are considered, convergence and the quality of final values can be further ensured. Some care must be taken to deal with non-linear boundary conditions, which may not be used under some procedures. We plan to compare these models with growth models used in forest management in this region and with the diameter growth model used as the basis for predicting diameter increment in forest gap simulators.

References

Botkin, D.B. (1993) *Forest Dynamics: an Ecological Model*. Oxford University Press, New York, 309 pp.

Burns, R.M. and Honkala, B.H. (technical coordinators) (1990a) *Silvics of North America. 1. Conifers*. Agriculture Handbook 654. US Department of Agriculture, Washington, DC, 675 pp.

Burns, R.M. and Honkala, B.H. (technical coordinators) (1990b) *Silvics of North America. 2. Hardwoods*. Agriculture Handbook 654. US Department of Agriculture, Washington, DC, 877 pp.

Colbert, J.J. and Racin, G. (1995) *User's Guide to the Stand-Damage Model: a Component of the Gypsy Moth Life System Model (Version 1.1)*. General Technical Report NE-207. US Department of Agriculture, Forest Service, Northeastern Forest Experiment Station, Radnor, Pennsylvania, 38 pp. [1 computer disk ($3\frac{1}{2}$ in.)].

Colbert, J.J. and Racin, G. (2001) *How to Use the Stand-Damage Model*, Version 2.0. General Technical Report NE-281. US Department of Agriculture, Forest Service, Northeastern Forest Experiment Station, Newtown, Pennsylvania, 79 pp. [1 computer disk ($3\frac{1}{2}$ in.)].

Colbert, J.J. and Sheehan, K.A. (1995) *Description of the Stand-Damage Model: Part of the Gypsy Moth Life System Model*. General Technical Report NE-208. US Department of Agriculture, Forest Service, Northeastern Forest Experiment Station, Radnor, Pennsylvania, 111 pp.

Colbert, J.J., Perry, P. and Onken, B. (1997) Preparing for the gypsy moth: design and analysis of stand management. Dorr Run, Wayne National Forest. In: *Communicating the Role of Silviculture in Managing the National Forests: Proceedings of the National Silviculture Workshop.* 19–22 May, Warren, Pennsylvania General Technical Report NE-238. US Department of Agriculture, Forest Service, Northeastern Forest Experiment Station, Radnor, Pennsylvania, pp. 76–84.

Collingwood, G.H. and Brush, W.D. (1978) *Knowing Your Trees.* American Forestry Association, Washington, DC, 389 pp.

Draper, N.R. and Smith, H. (1981) *Applied Regression Analysis,* 2nd edn. John Wiley & Sons, New York, 709 pp.

Duncan, W.H. and Duncan, M.B. (1988) *Trees of the Southeastern United States.* University of Georgia Press, Athens, Georgia, 322 pp.

Fekedulegn, D., Mac Siurtain, M.P. and Colbert, J.J. (1999) Parameter estimation of nonlinear growth models in forestry. *Silva Fennica* 33, 327–336.

Gottschalk, K.W., Colbert, J.J. and Guldin, J.M. (forthcoming). Modeling gypsy moth-related tree mortality under different outbreak and management scenarios in Ozark-Ouachita Interior Highlands forests. In: *Proceedings, Upland Oak Ecology, A Symposium on History, Current Conditions, and Sustainability:* 8–10 October 2002, Fayetteville, Arkansas. US Department of Agriculture, Forest Service, Southern Research Station, Asheville, North Carolina, GTR-SRS-xx.

Guldin, J.M., Thompson, F.R., Richards, L.L. and Harper, K.C. (forthcoming) Status and trends in vegetation. In: *Ozark-Ouachita Highlands Assessment: Terrestrial Vegetation and Wildlife.* US Department of Agriculture, Forest Service, Southern Research Station, Asheville, North Carolina.

Harlow, W.M., Harrar, E.S. and White, F.M. (1979) *Textbook of Dendrology: Covering the Important Forest Trees of the United States and Canada,* 6th edn. McGraw-Hill, New York, 510 pp.

Liebhold, A.M., Gottschalk, K.W., Muzkia, R.-M., Montgomery, M.E., Young, R., O'Day, K. and Kelley, B. (1995) *Suitability of North American Tree Species to Gypsy Moth: a Summary of Field and Laboratory Tests.* General Technical Report NE-211. US Department of Agriculture, Forest Service, Northeastern Research Station, Radnor, Pennsylvania, 34 pp.

Little, E.L. Jr (1979) *Checklist of United States Trees (Native and Naturalized).* Agriculture Handbook 541. US Department of Agriculture, Washington, DC, 375 pp.

Marquardt, D.W. (1963) An algorithm for least squares estimation of nonlinear parameters. *Journal of the Society of Industrial and Applied Mathematics* 11, 431–441.

Miller, H. and Lamb, S. (1985) *Oaks of North America.* Naturegraph Publishers, Happy Camp, California, 327 pp.

Myers, R.H. (1986) *Classical and Modern Regression with Applications.* Duxbury Press, Boston, Massachusetts, 359 pp.

Pienaar, L.V. and Turnbull, K.J. (1973) The Chapman-Richards generalization of von Bertalanffy's growth model for basal area growth and yield in even-aged stands. *Forest Science* 19(1), 2–22.

Racin, G. and Colbert, J.J. (1995) *Guide to the Stand-Damage Model Interface Management System.* General Technical Report NE-209. US Department of Agriculture, Forest Service, Northeastern Research Station, Radnor, Pennsylvania, 149 pp.

Ratkowsky, D.A. (1983) *Nonlinear Regression Modelling.* Marcel Dekker, New York, 276 pp.

Richards, F.J. (1959) A flexible growth function for empirical use. *Journal of Experimental Botany* 10, 290–300.

SAS (2000) *SAS Online Doc Version 8,* February 2000, SAS Institute, Cary, North Carolina.

Schnute, J. (1981) A versatile growth model with statistically stable parameters. *Canadian Journal of Fisheries and Aquatic Sciences* 38, 1128–1140.

Shugart, H.H. (1984) *A Theory of Forest Dynamics: the Ecological Implications of Forest Succession Models.* Springer-Verlag, New York, 278 pp.

Turnbull, K.J. (1963) Population dynamics in mixed forest stands: a system of mathematical models of mixed growth and structure. PhD dissertation, University of Washington.

United States Department of Agriculture, Forest Service (1991) *Important Forest Trees of the Eastern United States.* FS-466. US Department of Agriculture, Forest Service, Washington, DC, 111 pp.

Vanclay, J.K. (1994) *Modelling Forest Growth and Yield*. CAB International, Wallingford, UK, pp. 380.

von Bertalanffy, L. (1957) Quantitative laws in metabolism and growth. *Quarterly Review of Biology* 32, 218–231.

30 Landscape Visualization with Three Different Forest Growth Simulators

Falk-Juri Knauft[1]

Abstract

Individual-based tree models are well suited to represent growth processes at scales where trees interact through mutual shading and/or root competition. Management decisions in general are performed at larger scales even when selective thinning is considered. Unfortunately, no satisfactory scaling theory is available to upscale from the individual tree to the stand or even landscape level, but modern information technology provides several options to attempt 'brute-force-solutions' that circumvent the scaling problem, i.e. modelling a sufficient number of individual trees in a stand/landscape (depending on the application) to represent the spatial distributions of properties of simulated forests with meaningful precision. This chapter will present three different integrated systems/examples to approach this (scaling) task at various levels of detail and realism. In all three examples the simulator kernel is individual tree-based; it is linked to GIS functionality and a visualization interface. They differ in their purposes and hence in number/detail of processes included and the resolution of the visualization interface. The first example, VIWA, integrates the tree-growth model SILVA with the tree visualization package AMAP. Its main purpose is to serve as a communication interface between forest managers and the public, requiring an almost photo-realistic visualization. The second example, VIWA2, is aimed at forestry experts only and thus allows a more abstract presentation of trees. It combines the forestry-oriented simulator BWIN with a landscape visualization tool ENVISION. The third example is targeted at the communication between forest experts and ecosystem researchers. It combines TRAGIC++ with the AMAP package and has an interactive user interface JTRAGIC.

Introduction

Communication about such complex problems as the goal-oriented management of forest stands and landscapes is, especially in the case of discussion partners with different backgrounds, a very difficult task. Without appropriate visualization tools, the subjective perceptions will be strongly diverse. To clarify and (potentially) solve this problem we suggest focusing on an interactive simulation and visualization environment.

[1] Bayreuth Institute for Terrestrial Ecosystem Research, Germany
Correspondence to: falk-juri.knauft@bitoek.uni-bayreuth.de

A simulation is a software-based implementation of models and therefore a simplified representation of real world aspects. Common applications of computerized simulation in forestry can be found in the field of growth modelling. A number of different approaches can be identified. Practical forestry-related groups of model developers prefer empirical or holistic models, while groups oriented more towards ecosystem research prefer process models.

The visualization of simulated or empirical data is a traditional method for analysis but also for presentation (e.g. of predicted scenarios). Because of the superior human ability to identify/recognize complex patterns, it is an effective interface for the transfer of complex information. As the predominant traditional visualization technology in forestry, the spatial context of many pieces of information (data) led to a wide utilization of maps.

Interaction in the context of forest simulation has so far been realized inside the model as interaction among individual trees, and outside the model as interaction between the simulator and its user. In the context of this chapter we include the interaction between people, who might be, for instance, different users of the simulation tool.

The task of transferring information is supported by communication tools (media). Maps of forest units are one traditional example of documenting the context for silvicultural decisions. Table 30.1 shows a selection of traditional communication tools in forestry and related sciences. They can be differentiated by their partners, the type of possible interaction among them, and the corresponding temporal and spatial scales.

State of the Art

The recent development in information technology has had an enormous impact on this research-relevant technology. One of the biggest benefits can be found in the field of simulation technologies. Just a few years ago, detailed simulation of individual-based tree growth was a difficult and, in regard to computation capacity, expensive task. Today, a number of different simulation tools can run on nearly every modern PC. For the German region the most common are BWIN (Nagel, 1997) and SILVA (Kahn and Pretzsch, 1997; for a technical comparison see Windhager, 1999). Both were initially developed from empirical models, but later supplemented by

Table 30.1. Traditional communication tools between foresters (F), scientists (S) and the public (P) (t → point in time, t+dt → later point in time, × → position, x+dx → different position) (the entry 'no way' marks communication channels that face technical difficulties for the given task).

Communication tool	Partner	Space	Time	Direction
Forest maps	F(t) → F(t+dt) F(t) → P(t)	Stand (ha)	10–20 years	One way
Administrative records	F (t) → F (t+dt)	Stand (ha)	−160 years	One way
Joint visits to forest	F(x) → F (x+dx)	Group of trees	Hours–days	Interactive
Silvicultural books	F (t) → F (t+dt) F(x) → F (x+dx)	Variable	Personal memory	No way (?)
Publications	S(x) → S (x+dx)	Variable	Often < years	One way
Political programmes, visions, goals	P→ all	Variable	Variable	Interactive ?
Predictions	S(t) → all (t+dt)	Variable	Variable	One way, No way(?)

process models. The opposite development can be seen for the growth simulator TRAGIC++ (Hauhs *et al.*, 1993, 1995, 1999; BITOEK, 2002a).

Forestry-related individual-tree-based growth models, such as BWIN and SILVA, differentiate the individual tree into only a few compartments, such as root, stem and crown. Functional–structural growth models identify a much larger number of compartments. They model specific functionalities for the compartments, and the result of the simulation typically depends on the generated structure within the individual tree. One approach of this type uses so-called L-systems (Lindenmayer, 1968). The idea of a rule-based behaviour of tree growth has been implemented in the 'growth engines' of GROGRA (Kurth, 1994, 1999) and AMAP (de Reffye and Blaise, 1993).

Parallel to the development of sophisticated growth simulators, visualization technologies also advanced to new levels. Traditional media such as graphs and maps can now be generated with a few mouse clicks. The production of maps, especially, gained much greater efficiency with the introduction of geographical information systems (GIS). These systems even enable the user to create 3D maps with very little effort. One step further is the creation of 3D scenes based on technology standards such as VRML, OPENGL and DIRECTX (see Seifert, 1998). These are used at different levels of resolution and quality as visualization tools in all of the mentioned growth simulators. Most of the difference is caused by the high demand of computation capacities for these visualization technologies, and their varying availability.

The generation and processing of large quantities of data has become common through use of the described technologies along with effective tools for data storage and transfer. The spatial context of forestry-related data calls for GIS functionalities. To transfer data of complex and variable structure among independent modules of different standards, tools such as XML have been developed. The eXtensible Markup Language is a mechanism to identify structures (content such as words, pictures, etc. and some indication of what role that content plays) in a document (ASCII file, stream, etc.).

Besides the exchange of information among the software packages, one major task is the appropriate transfer to the user. For the realization of communication and interactivity between the user and the software, the graphical user interface (GUI) became the standard with the introduction of the Apple Macintosh. Today, convenient developer kits enable the engineering of adapted GUIs.

Methodology

The three modules concept

As the general structure of our forestry-related information system we concentrate on three main components: simulator, database and visualization tools (Knauft, 2000). This structure has been realized to various degrees of integration in professional forest information systems (ABIES, FOGIS–INTEND Kassel), as well as in academic systems (VIWA, TRAGIC, FORCITE).

The availability of efficient tools such as GIS, forest growth simulators and tree or landscape visualization tools, combined with limited financial resources, favours a modular concept. Table 30.2 presents an overview of the criteria which have influenced the selection in the presented applications.

Table 30.2. Case-specific criteria for software selection.

Partner	Case-specific selection criteria	Selected software
Forester/general public	All components already available, high claims on realistic visualization	SILVA – ARCINFO – AMAP
Forester/forester	BWIN and ARCINFO already available, visualization software as freeware	BWIN – ARCINFO – ENVISION
Forester/ scientific community	TRAGIC and ARCINFO already available, high claims on detailed visualization	TRAGIC – ARCINFO – AMAP

A first design concept – VIWA

Our first idea to develop an integrated forest simulation and visualization system grew from the demand for a tool to support forestry management decisions by illustrating their impact on the forest landscape over a time scale of decades (Knauft, 2000; Knauft and Sloboda, 2000). The task defines the participants in the communication problem: foresters and the general public. It demands a reliable simulation of forest growth and management up to a full rotation cycle, a flexible and powerful spatial database for the processing of stand and individual tree data and, finally, an almost photo-realistic visualization of the scenery including botanically correct three-dimensional forest tree models (in contrast to 'billboard-icons' as used in ENVISION (McGaughey, 2000)). The solution was found in an integration of the forest growth simulator SILVA, the GIS ARCINFO, and the tree architecture visualization toolkit AMAP.

The data processing starts in the GIS with sets of stand data and related geographical entities. A region can be selected using a map interface. The stand data (f.i. inventory data) of the selected region are transferred to the simulation tool, which generates samples of the initial individual tree structure of the stands. With these structure samples, the full stand structure is created inside the borders of the stand unit and stored in the GIS. Based on these samples, additional growth and management intervals can be simulated. These can be used in turn to create scenarios of forest landscape development. The final step is the transfer of the individual tree data and the digital elevation model to the visualization tool, where single images, time series or movements through the scene in image series or animations can be rendered and recorded. The implemented system VIWA (short for 'Virtueller Wald' - virtual forest) is costly and complicated, due to the use of commercial software (ARCINFO and AMAP) on three different computers with different operating systems.

The second example – VIWA 2

These resources were not affordable for a second application. The Forest Research Station of Lower Saxony asked for a system to support communication *within* the forestry community which should be efficient to use. Therefore, we chose, as well as the already licensed GIS ARCINFO, the simulator BWIN (developed at the research station) and the visualization tool ENVISION (McGaughey, 2000). For efficient use, all these products can be run on a single computer (operating system MS Windows-NT and upwards) and can be accessed over a GUI, which also controls the interfaces between the software components (see Fig. 30.1). The data handling of this system is generally the same as in the first system, but via the GUI it is much easier to handle and to control.

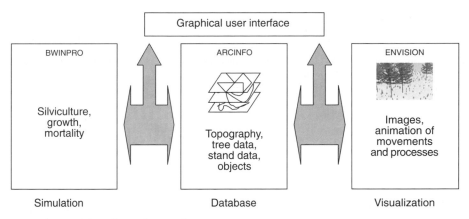

Fig. 30.1. Schema of the solution for intra-forestry communication, VIWA 2.

The TRAGIC concept

Finally, at the Bayreuth Institute for Terrestrial Ecosystem Research (BITÖK) we are developing a communication interface between ecosystem scientists and practical experts in silviculture. The interface for this specific case places much greater requirements on the system than the previous applications. Scientists hope to receive feedback about the quality of their models and the foresters look for a tool that enables them to document, communicate and plan their management activities more efficiently. Hence information is required to flow in both directions and the visualization tool becomes an input interface for the expert forester.

Waiving practical experience as an input and feedback to theoretical research would be a great loss. On the opposite side, theoretical research could be interrogated as to whether its results would have an impact on practical forestry (see Schanz, 1994). Therefore, we chose a highly realistic three-dimensional and interactive visualization allowing interactive movements in the virtual scene and interactive action such as the selection of trees for thinning. The system consists so far of the individual-based forest growth simulator TRAGIC++, the Internet-accessible visualization front-end JTRAGIC (Hauhs *et al.*, 1995, 1999; Scheerer, 2000; BITOEK, 2002b) and a database. The integration of a GIS (here also ARCINFO) will enable the visualization of landscapes and therefore allow the discussion about forest management at this scale. The generation and simulation of the data will be realized in the same way as in the first two applications.

The demand for interactive access to tree and stand data and for simulated movement through the stand does (under the constraint of the limited graphics capabilities of the sort of computers that are typically available in forestry offices) limit the possible detail level in the visualization. For special situations with higher expectations, the AMAP toolkit will be integrated at the server site to generate photorealistic still images and recorded animations. So far the transfer of all simulated structural information (TRAGIC++ even simulates branches and needles) into the virtual AMAP scene is not finished, but when completed it will enhance the discussion between scientists and foresters.

The architecture of the TRAGIC environment is illustrated in Fig. 30.2. While the data-server operates the spatial database (tree and stand data) as well as a knowledge repository (silvicultural activities), the simulation-server runs the TRAGIC server

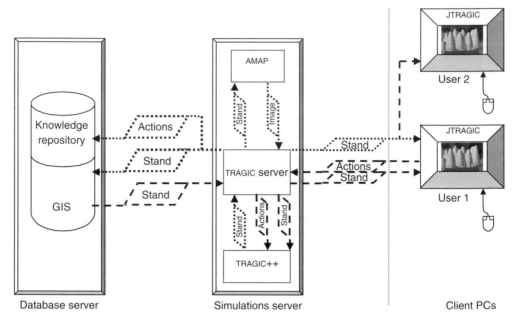

Fig. 30.2. Schema of the fully developed TRAGIC environment as server–client design with interactive user interface (dashed line, data before simulation; dotted line, data after simulation).

(central data interface), the simulator TRAGIC++ and the visualization tool AMAP. On the client side, the user can download an older stand, plan its activities, and send these back to the server, eventually with an additional call for an AMAP visualization. After the simulation, the stand data (as individual tree data) are stored in the database along with the related activities. The resulting stand can then be downloaded from the original user or any other user with appropriate permission.

Data sources

The following applications are based on data from three different regions in Lower Saxony.

For the case of communication with the public, we chose stand data from the exhibition ground 'ErlebnisWald' ('Experience Forest') near the town of Uslar. We included the stand parameters of species, basal area, upper height, stand age and regeneration type (planting or natural regeneration) in the generation of individual tree data samples by SILVA. Those were, as mentioned before, multiplied in the GIS for the full stand area and stored together with a digital elevation model. For the simulation of management and growth, the samples were again imported and processed within SILVA. Only tree position, individual height and species parameters were transferred to the visualization tool.

The application for the forestry internal discussion illustrates a plot from the Lueneburg Heathland. For this case we were supplied with a full tree data table. Besides position, tree height and species as well as the DBH and lower crown height were utilized for the visualization. The last example is based on a variety of stand and management information about the Lange Bramke catchments in the northern parts of the Harz mountains (see also Hauhs, Chapter 5, this volume).

Results

The VIWA example

The first scenario (Fig. 30.3) presents a possible discussion about the impact of changes in the forest management objectives from an intensively managed forest to a more extensive use with a 'reconstruction' of a potentially natural forest type, as is often demanded by environmental interest groups. To support the relatively abstract discussion, we generated a virtual scene of a specific forest landscape.

Especially for untrained persons, a photo-realistic visualization is suitable, as it corresponds best with their experience. Additionally, public expectations of virtual scenes are highly influenced by the quality of television and film productions. Therefore, highly detailed and botanically correct trees are demanded. Only very few, mostly professional, rendering systems (such as AMAP) are able to produce this quality. They are quite demanding in terms of computational resources. Important in the context of this chapter is the aggregation of stand structure information into a landscape view. Less weight is given to a high degree of detail and realism, which was not demanded for this situation (see also Knauft, 2000).

The VIWA 2 example

The second case focused on supporting discussions of real or potential situations of forest development within the community of foresters and people of similar background. This group is relatively homogeneous with respect to their knowledge about structure, growth and management of forests; they share a common expertise. On the other hand, the communication must be effective and precise. The main topics of their discussion are structure, growth rate and yield, but also stability and ecological value.

In this case, silvicultural research results are presented to a professional audience. To illustrate the method and for better understanding of the situation, the sampled areas are visualized using the more abstract tree models of ENVISION (see Fig. 30.4). Although the visualization seems to be more schematic, it contains

(a) (b)

Fig. 30.3. Comparison of visualizations of a forest landscape under recent stand-wise forest management (a) with the potential historical forest type (b). Both graphics generated with AMAP, reproduction from Knauft (2000).

Fig. 30.4. Visualization of a monitoring plot (graphic generated with ENVISION).

more verifiable information. While in the first application only the generated parameters of position, species and tree height were used for the visualization, in this second example, DBH and lower crown height were also taken into account for the visualization, and all data were actually measured for every individual tree.

The TRAGIC example

A combination of the visual realism as implied by the first VIWA system and the data representation in the second version can be realized within the TRAGIC environment. It consists of the simulator TRAGIC++, which includes a tabular interface and also a visualization tool (Fig. 30.5a) and the visual interface JTRAGIC. The latter does not include its own simulation module; it is only able to visualize a stand as transferred from a central database and record the silvicultural activities (movements or silvicultural decisions such as cutting or planting of trees) of the forester. These are sent back to the simulator and used there for the next simulation period. The example presented here is part of the restoration of historic stand development (for more

Fig. 30.5. Visualization of a historical plot showing the situation in 1752 as part of a reconstruction from the 18th century with TRAGIC (a) (see Hauhs, Chapter 5, this volume) and after its export to AMAP (b). The position and size of stems and crowns are the same in both scenes.

details please refer to Hauhs, Chapter 5, this volume). Some indication parameters are used to evaluate the simulated stands and their structure by the researcher, but also by foresters. A number of foresters who have been interviewed demanded an increase in the visual realism of the trees (Schanz, 1994). We included an interface to the AMAP toolkit (Fig. 30.5b). For a full stand impression, the plot can than be multiplied in the scene (Fig. 30.6).

Discussion and Conclusions

The characteristics of the communication partners

For the field of forest modelling we can identify three major interest groups.

1. The general public possesses little knowledge about the complex processes of forest ecosystems. The public is primarily interested in changes at the landscape scale, as they are interested in the aesthetics and recreational value of forest landscapes, rather than in individual plant dynamics. Their preferences become settled in politics, media and economy.
2. Experienced foresters have an inside view about processes which influence the dynamics of forests, but they are also more interested in the (economic) result on enterprise (or landscape) level rather than individual trees. Based on practical experience, they have the competence to achieve their goals for specific situations.
3. Less homogeneous are the ecosystem-related scientists, as they focus on individual organisms, populations or ecosystems. Their competence is based on theories, experiments and observations, often at the process level.

These three partners are already engaged in several communication processes, but most of these traditional channels are technically characterized as 'one-way' (see Table 30.1). If interactive communication occurs at all, then it is usually only within the individual groups. Here the question is whether this is a technical constraint (there being no proper interactive communication tool), a social constraint and/or whether it is due to the organization for the tasks at hand?

In the schematic summary of established communication tools in Table 30.1, the link between research and management is not covered (see also Schanz, 1994). It is argued in another chapter (Hauhs, Chapter 5, this volume) that the reason for this gap is that the relationship between empirical/heuristic models of management and

Fig. 30.6. Visualization of a full stand based on the historical plot reconstruction from the 18th century.

dynamic models in science resembles competition more than mutual support. It is clear that interfaces that remain useful between two (or even among all three) of these groups are highly non-trivial.

While growth is simulated in individual-based growth models at the level of individual plants, its evaluation is commonly done at the level of stands or even landscapes. Therefore, upscaling becomes mere aggregation. The individual-tree-based forest growth simulators TRAGIC++, SILVA and BWIN realize a solution for the step from the individual plant to the stand. The next level of aggregation requires different tools, such as forest information systems for management support, or the visualization system VIWA for communication support.

The general decision for a modular system and wide integration of available software components eased the transfer of the methodology from the first VIWA version up to the TRAGIC environment. While preserving the general concept, an exchange of a case-adapted simulation or visualization tool with another user required only small changes within the interface tools. Further, it enabled us to enhance the possibilities of the systems from case to case.

The major step from the first VIWA version to the second was the integration of the full system on a single computer under a common GUI, therefore greatly enhancing its efficiency. Simultaneously, it reduced the necessary technical knowledge for the user. This goes even further in the TRAGIC environment. Here, the interactive graphical interface providing the necessary technical knowledge for its application is similar to other desktop applications. This increases the group of potential users on a large scale. The server–client design opens the way for an interactive communication between the users over the graphical interface.

In science in general, and in ecosystem research in particular, a dynamic systems approach is the typical solution to this task, i.e. the behavioural (temporal changes) and structural complexities are derived from a realistic representation of stand dynamics. This model type brings the perceptional 'bias' of scientists close to the aesthetics of the public: more structural resolution (mapping of perceived complexity) is regarded as a path to more realistic dynamics, but the addition of complexity for its own sake, without an accompanying increase in predictive capacity or utility, almost never results in an improved model.

In forest management, another (speculative) type of grounding has been described (Hauhs and Lange, 2001), in which growth behaviour and response is the ultimate foundation of modelling (here empirical). The reduced complexity of forest growth as seen from the perspective of forestry occurs at a relatively coarse resolution: the rotation period in the temporal and the stand/group development in the spatial perspective. The standards of silvicultural management describe an interactive situation in which the current structure of a stand is used to assess and decide about the appropriate interference. It is typical that the goal of any given interference is a regulation of competition among trees that leaves the maximum number of choices for the next round of interference (rather than relying on a prediction far into the future of this particular stand). In contrast to many other modelling schemes, the visualization interface becomes the centrepiece through which other components are linked.

While the proposed three-component concept is not intended to solve all the problems of data transfer between practical forestry and modelling (it will be a future challenge to foresters and scientists as well), nor does it try to substitute other concepts of forestry information systems or silvicultural decision support systems, it has served us as a learning environment to study the various methods of interaction and information exchange around the topic of forest growth and management.

Acknowledgements

This work was supported by the German Ministry of Research (BMBF) (contract no. PT BEO 51-0339476 C), the Lower Saxony Forestry Research Station and the Faculty of Forestry and Forest Ecology at the University of Göttingen.

References

BITOEK (2002a) *Documentation for the Forest Growth Simulator TRAGIC++*. Bayreuth Institute for Terrestrial Ecosystem Research, Ecological Modelling. Available at: www.bitoek. uni-bayreuth.de/mod/html/tragic/documentation

BITOEK (2002b) *Homepage of the Interactive Forest Growth Simulator Interface JTRAGIC*. Bayreuth Institute for Terrestrial Ecosystem Research, Ecological Modelling. Available at: www.bitoek.uni-bayreuth.de/mod/html/webapps/jtragic

Hauhs, M. and Lange, H. (2001) Lessons from sustainable development. In: von Gadow, K., Pukkala, T. and Tomé, M. (eds) *Sustainable Forest Management*. Kluwer, Dordrecht, pp. 69–98.

Hauhs, M., Rost-Siebert, K., Kastner-Maresch, A.E. and Lange, H. (1993) *A New Model Relating Forest Growth to Input and Output Fluxes of Energy and Matter of the Corresponding Ecosystem: TRAGIC (Tree Response to Acidification of Groundwater in Catchments)*. Final Report of EC Project EV4V-0032-D(B), Institut für Bodenkunde und Waldernährung der Universität Göttingen.

Hauhs, M., Kastner-Maresch, A. and Rost-Siebert, K. (1995) A model relating forest growth to ecosystem-scale budgets of energy and nutrients. *Ecological Modelling* 83, 229–243.

Hauhs, M., Dorwald, W., Kastner-Maresch, A. and Lange, H. (1999) The role of visualization in forest growth modelling. In: Amaro, A. and Tomé, M. (eds) *Empirical and Process-based Models for Forest Tree and Stand Growth Simulation*. Edições Salamandra, Lisbon, pp. 403–418.

Kahn, M. and Pretzsch, H. (1997) Das Wuchsmodell SILVA: Parametrisierung der Version 2.1 für Rein- und Mischbestände aus Fichte und Buche. *Allgemeine Forst- und Jagdzeitung* 168 (6–7), 115–123.

Knauft, F.-J. (2000) Entwicklung von Methoden zur GIS-gestützten Visualisierung von Waldentwicklungsszenarien. Dissertation thesis, Fakultät für Forstwissenschaften und Waldökologie der Georg-August-Universität Göttingen. Available at: webdoc.sub.gwdg.de/diss/2000/knauft

Knauft, F.-J. and Sloboda, B. (2000) Visualisierung virtueller Waldlandschaften durch Integration individuenbasierter Modelle. In: *Tagungsband Herbstkolloquium 1999*. Deutscher Verband Forstlicher Forschungsanstalten und Internationale Biometrische Gesellschaft Deutsche Region, Ljubljana, pp. 293–308.

Kurth, W. (1994) *Growth Grammar Interpreter GROGRA 2.4: a Software-Tool for the 3-Dimensional Interpretation of Stochastic, Sensitive Growth Grammars in the Context of Plant Modelling: Introduction and Reference Manual*. Berichte des Forschungszentrum Waldökosysteme 38. Göttingen.

Kurth, W. (1999) *Die Simulation der Baumarchitektur mit Wachstumsgrammatiken*. Habilitationsschrift an der Georg-August-Universität Göttingen. Wissenschaftlicher Verlag, Berlin, 327 pp.

Lindenmayer, A. (1968) Mathematical models for cellular interaction in development. *Journal of Theoretical Biology* 54, 3–22.

McGaughey, R.J. (2000) *EnVision – Environmental Visualization System*. Homepage of the USDA Forest Service Pacific Coast Research Station. Available at: forsys.cfr.washington.edu/envision.html

Nagel, J. (1997) *BWIN: Programm zur Bestandesanalyse und Prognose*. Handbuch Niedersächsische Forstliche Versuchsanstalt Abteilung A, 41 pp.

de Reffye, P. and Blaise, F. (1993) Modélisation de l'architecture des Arbres. Applications Forestières et Paysagères. *Revue Forestière Française* 45, 128–136.

Schanz, H. (1994) *Forstliche Nachhaltigkeit aus der Sicht von Forstleuten in der Bundesrepublik Deutschland*. Working paper 1994, Institut für Forsteinrichtung und Forstliche Betriebslehre, Universität Freiburg.

Scheerer, B. (2000) Protokollierung von Durchforstungseingriffen in einem Waldwachstumssimulator. Diploma thesis, University of Bayreuth, 113 pp.

Seifert, S. (1998) Dreidimensionale Visualisierung des Waldwachstums. Diploma thesis, Fachbereich Informatik der Fachhochschule München, Munich.

Windhager, M. (1999) Evaluierung von vier verschiedenen Waldwachstumssimulatoren. Dissertation thesis, Universität für Bodenkultur Wien, Vienna, 217 pp.

31

How Good is Good Enough? Information Quality Needs for Management Decision Making

David D. Reed[1] and Elizabeth A. Jones[2]

Abstract

There are many aspects to information quality, and everything from measurement method documentation to data entry errors to metadata documentation falls under this topic. Even with good data quality assurance procedures in place, uncertainty exists because of random variation in the population and the sampling design utilized to collect the data. If data are used as the basis for model projections, further uncertainty is introduced due to model prediction errors. Adequacy of information quality can only be evaluated in the context for which it is used. The focus of this chapter deals with the uncertainty in information and its impact on decision making. The consequences of management decisions can only be assessed through consideration of both the cost of obtaining more precise information as well as the cost of making a poor decision.

Introduction

All resource management decisions are made in the face of uncertain and incomplete knowledge. Even if information regarding the current state of the system is known with high precision, future conditions are uncertain, to at least some degree, and this uncertainty typically increases with time. Managers must deal with these numerous sources of uncertainty when making decisions. Furthermore, many good managers have an intuitive understanding of their system of interest and the associated uncertainties, but they often cannot explicitly identify all of the sources of uncertainty that affect their information base. It is, therefore, impossible to develop perfect knowledge of current, much less future, conditions.

How, then, can managers utilize all of this imprecise information to improve their decision-making processes? How can managers act to minimize their total risk exposure? In many ways, resource management comes down to risk management. In trying to avoid an obvious source of risk due to uncertain information, a manager may inadvertently be exposed to a greater level of risk from an unidentified source.

[1] School of Forest Resources and Environmental Science, Michigan Technological University, USA
Correspondence to:ddreed@mtu.edu
[2] Department of Mathematical Sciences, Michigan Technological University, USA

The key to decision making is a full understanding of the risks encountered as a result of imprecise information. Managers must balance the cost of obtaining more precise information, where this is possible, with the cost of making a poor decision.

Decision Complexity

The simplest type of decision is a binary choice. Statisticians are most familiar with this through hypothesis testing. There is a factual statement (e.g. $\mu_0 = 0$) with the necessary decision being whether or not to reject the veracity of this statement. The usual types of uncertainty are known as Type I (α) and Type II (β) errors (Fig. 31.1). Type I error (the probability of rejecting the statement when it is true) is controlled by the analyst through statistical methodology. Type II error (the probability of not rejecting the statement when it is false) is understood to be inversely related to Type I error, but it is rarely explicitly considered in the decision-making process. The complexities of this simple decision-making scenario are immense and are specific to the context at hand – what is the appropriate statistical procedure? what are the implicit and explicit assumptions required for the procedure to be appropriate? Unfortunately, there are virtually no resource decisions that truly fit this relatively simple decision-making scenario.

A more complex decision-making scenario occurs when the choices are no longer binary, but there is a finite, identifiable set of choices. In this case, the decision does not consist of determining the veracity of a single statement, but in choosing among a finite group of identified alternatives. There is risk associated with each alternative, and the challenge is to identify the alternative that carries the minimum total risk to the manager. Again, the complexities are enormous and there are few resource decisions that fit this decision-making scenario.

The most complex scenario involves a choice among a very large number of alternatives, where most or all of the alternatives do not have explicitly defined measures of risk. This is, unfortunately, the most common situation in resource management. Managers often simplify the decision process by eliminating a large number of possible alternatives, resulting in a situation resembling scenario 2 (e.g. a decision between planting species A, B or C on a particular habitat type). While ignoring many possible alternatives, the process is then simplified. There is an implicit assumption that the alternatives considered in the reduced set carry less

Decision	Hypothesis	
	Statement True	Statement False
Statement True	Correct	Type II Error
Statement False	Type I Error	Correct

Fig. 31.1. Binary choice decision scenario, similar to simple statistical hypothesis testing.

total risk than do any of the eliminated alternatives. Decisions are often simplified even further into a situation resembling the binary scenario (e.g. should this stand be harvested during this management period or should it be left alone and evaluated again at some time in the future?).

As illustrated in the following examples, the available information may be adequate at a particular scale, such as a harvest scheduling application involving many stands within an ownership, but inadequate at a different scale, such as the choice of a management regime for a single stand. Since the cost of obtaining information is often substantial, it is imperative that the intended uses of the information and the management decisions that the information will support be identified prior to expending the resources to collect the information.

Inventory Example

As noted previously, it is not only important to consider the cost of improving the quality of information available to support a decision, but it is also important to consider the cost of making a poor decision. Cochran (1977) presents the use of the cost-plus-loss method in a sampling context, while Burkhart *et al.* (1978) present the use of this method in a multi-resource sampling setting. In this context, the total expected loss $(T(X))$ is a function of the cost of sampling to obtain a given precision $(C(X))$ plus the expected loss resulting from imprecise knowledge of the quantity of interest $(L(X))$:

$$T(X) = C(X) + L(X) \tag{1}$$

In a simple application, $C(X)$ is a linear function of the number of samples taken, $L(X)$ is a squared error loss function that accounts for both the sampling error and the value of the resource in question, and X is the decision variable, which, in this case, is the number of samples to install in the population of interest.

The question of sample size determination involves a balancing of the cost of installing n samples with the expected loss due to the uncertainty associated with a sample of size n. Burkhart *et al.* (1978) extend this to the situation of multiple inventories with multiple subsequent management decisions. The method can be extended to even more complex situations; Reed *et al.* (1984), for example, use the cost-plus-loss method to examine the question of whether or not to sample in order to adjust individual tree volume equations for application to a specific stand. Ståhl *et al.* (1994) used the cost-plus-loss framework to examine the timing and precision of inventory activities, explicitly recognizing that the loss due to imprecision increases with the time since the inventory, and that the cost of the inventory is strongly related to the resulting precision of the estimates.

The important point here is that decision makers must function as risk managers, not just risk avoiders. The explicit consideration of the loss function forces managers to consider not just the costs of implementing a particular option, but also both the risks of poor performance associated with an option and the opportunity costs associated with foregoing the same option. Good managers implicitly incorporate these considerations into their decision-making process, but forcing their explicit consideration can lead to significant insights even for experienced managers.

Harvest Scheduling Example

The simplest cases of the prior example involve a decision regarding the inventory method or timing for a single stand. A second example is a harvest scheduling application that combines inventory results from many stands. The goal is to develop a schedule of management activities, such as harvests, for a number of different stands over several management periods. In this particular example, an inventory of 200 quaking aspen (*Populus tremuloides*) stands forms the resource base. Inventory information for these stands was generated from a sample of the US Forest Service Forest Inventory and Analysis data from the state of Minnesota as retrieved from the Eastwide database (Hanson *et al.*, 1992). A stand size is assigned to each stand from an independent inventory of stands. In other words, the data used in this example are a composite from several sources in the north-central region of the USA and are not derived from any particular area or ownership. From the inventory data, annual volumes were estimated using the FVS,LS (Bush and Brand, 1995) stand simulator. In this example, each management period is 1 year in length, only clearcut harvesting is considered, and the goal is to obtain a schedule of stands to harvest each year over a 30-year planning horizon.

The HABPLAN (Van Deusen, 1999) harvest scheduling program was used. In contrast to harvest scheduling optimization programs that use linear programming techniques to obtain a single, optimum schedule that maximizes a given objective function (total volume production in this example), HABPLAN uses the Metropolis algorithm to iteratively evaluate schedules and identify those that produce the largest harvested volumes. HABPLAN was used to iteratively evaluate harvest timing for each stand in an attempt to identify the combination of stand activities that resulted in the largest harvested volume over the 30-year planning horizon. Schedules were evaluated over 10,000 iterations.

Figure 31.2 shows the 200 (of 10,000) schedules produced by HABPLAN for the example data that have the largest total volumes over the planning horizon. The total volume harvested over the 30-year period is expressed as a proportion of the maximum produced over the 10,000 iterations. Over the 200 iterations shown here, there was a difference of 0.02% between the volume production of the maximum and minimum schedules. There are, in essence, a very large number of schedules that produce essentially equivalent volume production over the planning horizon. While there are certainly schedules that produce far less volume than those identified here, there are many alternative schedules that produce essentially equivalent results.

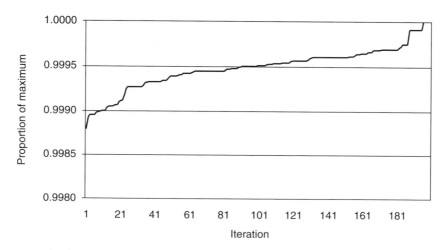

Fig. 31.2. Total volume production over the 30-year management horizon for 200 HABPLAN iterative solutions, expressed as a proportion of the largest value.

What happens if the original inventory data contain sampling errors? To investigate this, three scenarios were examined. Using Monte Carlo techniques, the initial inventory for each stand was randomly varied by ±5%, ±10% and ±20% (assuming a uniform error distribution). Each stand was assigned a random error factor and this was applied to all projected volumes for the stand as well as to the initial starting volume. The resulting harvest schedules did not differ ($P < 0.05$) in total volume production from those obtained from the original data. In other words, random symmetrical errors applied to individual stands had virtually no impact on the estimated total volume harvested from the harvest scheduling exercise.

Discussion

Resource managers are concerned with very complex systems, where they are presented with an essentially infinite variety of choices. It is important that managers do not just consider the cost associated with improving the quality of information available to support decision making, but that they also understand and consider the risk associated with making a poor decision. Avoiding the relatively easy-to-quantify costs of obtaining higher quality information may result in greater total risk due to the loss of possible gains from management options which are either not explicitly considered or are rejected on the basis of the poorer quality information which is available.

At the stand level, this is manifest in the cost of obtaining more precise volume estimates as opposed to the potential gains that might be received if better quality information led to the consideration or implementation of different management regimes. In a financial analogy, there is relatively little risk of loss to investments in fixed-interest-bearing instruments, but there is considerable, and often poorly quantified, risk that such 'safe' investments may really lead to an erosion of value in an inflationary environment. The additional information necessary to evaluate this requires some cost to obtain, but it may lead to reduced total risk by allowing more adequate consideration of both the cost and the potential loss associated with a particular investment decision.

In the harvest scheduling example, the examination of stand-level errors in volume estimates provides some important insights. If the individual stand estimates are unbiased (i.e. symmetrical about the estimate with an expected value of the error equal to zero), then over the planning horizon, there is little impact of precision in the individual stand estimates on the final estimated total volume production. The presence of bias in the estimates (reflected through non-symmetrical errors about the estimate with an expected value of the error not equal to zero) could impact the final estimated total volume production over the planning horizon. While error in the stand-level volume estimates is important at the scale of selecting a management regime for the individual stand and can result in suboptimal harvest schedules for individual stands, it has little impact on the larger scale question of selecting a harvest schedule for an area consisting of many stands. Hierarchical considerations may be important in understanding the impact of imprecise information in decision making, analogous to their impact on physiological processes, where variables and timesteps affecting individual cell or organ processes are quite different from those affecting populations or ecosystems.

The distribution of the uncertainties can also affect their impact on decision making. It is logical to consider sampling errors as being symmetrical about the mean. Model projection errors, however, are often not symmetrical since many resource models are logarithmic. If the usual regression error assumption is met using the logarithmically transformed developmental data ($\varepsilon \sim N(0, \sigma^2)$), then conversion of the predicted values back to arithmetic units results in a skewed, non-symmetrical distribution of errors in the final predicted values, with the modal arithmetic predicted value being less than the mean value (Finney, 1941; Baskerville, 1972). In this case, it is logical to expect a greater impact on the total harvested volume estimated from the harvest scheduling exercise. A harvest scheduling algorithm may tend to preferentially select stands and treatment schedules for which the predicted volumes are overestimates. While symmetrical individual stand errors did not impact the final total volume production over the management period, asymmetrical errors could have an impact on the estimated total volume production over the 30-year period.

In most forest resource situations, inventory precision goals are usually expressed either as percentages or as units per unit area. In a total volume or value context, an error of ±5% in a large-area stand has a much greater impact than an error of ±5% in a smaller stand. When individual stands are combined, as they are in the harvest scheduling example discussed above, estimates for large-area stands may have a proportionally greater impact on a quantity such as total volume production over a 30-year period than estimates of similar precision for a small-area stand, even though both are unbiased. In the harvest scheduling example, it might make sense to have greater precision requirements for large-area stands than small-area stands. This would obviously need to be balanced with cost, but many simulation studies have assumed equal stand sizes, and many managers may not fully appreciate the advantages that might be received in a harvest scheduling situation by relaxing precision requirements for small stands while maintaining or increasing precision goals in larger area stands. In other words, the appropriate quantity to use to estimate the needed sample size may not be the precision expressed as a percentage of the mean/ha, but in many situations might be the precision of total volume or value expressed in absolute numerical units of total volume or value at the stand level.

Summary

Natural resource managers are concerned with very complex decision-making scenarios; there are often an essentially infinite variety of options from which to select, and precision of available information concerning the risk associated with each option is difficult to assess. It is important for managers to consider and balance both the cost of obtaining more precise information and the cost of making a poor decision. These costs vary by situation, and often precision needs differ for the same piece of information when it is used to support different decisions. In a harvest scheduling context, the sampling errors associated with estimates of stand volumes have little impact on the final projected harvest volume provided that the estimates are unbiased, while error associated with stand-level inventories can have large impacts on the consequences of individual stand management decisions. Managers, therefore, must also consider the context in which information will be used to support decision making in order to judge the adequacy of the available information for supporting the decision at hand.

References

Baskerville, G.L. (1972) Use of logarithmic regression in the estimation of plant biomass. *Canadian Journal of Forest Research* 2, 49–53.

Burkhart, H.E., Stuck, R.D., Leuschner, W.A. and Reynolds, M.R. (1978) Allocating inventory resources for multiple-use planning. *Canadian Journal of Forest Research* 8, 100–110.

Bush, R.R. and Brand, G.J. (1995) The Lake States TWIGS variant of the Forest Vegetation Simulator. US Department of Agriculture, Forest Service, North Central Forest Experiment Station, St Paul, Minnesota.

Cochran, W.G. (1977) *Sampling Techniques*, 3rd edn. John Wiley & Sons, New York, 330 pp.

Finney, D.I. (1941) On the distribution of a variate whose logarithm is normally distributed. *Journal of the Royal Statistical Society, Series B* 7, 155–161.

Hansen, M.H., Frieswyk, T., Glover, J.F. and Kelly, J.F. (1992) *The Eastwide Forest Inventory Data Base: Users Manual*. General Technical Report NC-151, US Department of Agriculture, Forest Service, North Central Forest Experiment Station, St Paul, Minnesota, 48 pp.

Reed, D.D., Jones, E.A. and Green, E.J. (1984) Sampling to adjust individual tree volume equations for application to a specific stand. In: La Bau, V.J. (ed.) *Proceedings of the International Conference on Inventorying Forest and Other Vegetation of the High Latitude and High Altitude Regions*, pp. 289–292.

Ståhl, G., Carlsson, D. and Bordensson, L. (1994) A method to determine optimal stand data acquisition policies. *Forest Science* 40, 630–649.

Van Deusen, P.C. (1999) Multiple solution harvest scheduling. *Silva Fennica* 33, 207–216.

Part 5

Model Archives and Metadata

Forest models are a valuable summary of the forest data from which they are derived. Such models encapsulate the information and knowledge within the source data and, as such, are a primary means of understanding forest structure and processes.

There are no standards for the description or presentation of modelling metadata. As a consequence, metadata are presented in varying ways; in many cases, there is little common information available for models developed for similar systems. This makes it virtually impossible for third-party users to adequately evaluate available models and make decisions on model suitability for their situation.

Having considered the need for, and implementation of, a *forest model archive* to conserve and enhance the sharing of forest models of the past, it is natural to consider the environment for the development, publication and sharing of future forest models. Workshop participants reviewed the current state of the Forest Model Archive, which was initiated following an earlier IUFRO meeting in Greenwich, UK, in 2001.

The editors thank Keith Rennolls (UK) for his coordination of the contributions on this topic.

32 Forest Modelling: Conserving the Past and Building the Future

Keith Rennolls[1]

Abstract

Forest models are a valuable summary of the forest data from which they are derived. Such models encapsulate the information and knowledge within the source data and as such are a primary means of understanding forest structure and processes. At a time when there is much discussion and activity concerned with building distributed data warehouses of forest data, it is incumbent on forest modellers to consider the conservation of forest models that have been developed to-date. Accordingly, the establishment of a forest model archive (FMA) has been proposed by Rennolls *et al.* at an IUFRO 4.11 meeting in Greenwich, June 2001, and such an initiative has been adopted by IUFRO 4.11, and subsequently also by IUFRO 4.01, at the IUFRO 4.01 Vancouver conference, 2001. The first part of this chapter reports on the progress in this initiative since then. In particular, some of the features of a 'small' prototype FMA (www.forestmodelarchive.info) are described and discussed.

Having considered the need for, and implementation of, a forest model archive to conserve and enhance the sharing of forest models of the past, it is natural to consider the environment for the development, publication and sharing of future forest models. This has led to the formation of a new peer-reviewed e-journal, *Forest Biometry, Modelling, and Information Sciences*, (FBMIS), which was launched in the summer of 2002 (www.FBMIS.info). The aims of FBMIS are presented and discussed.

Introduction

Forest modellers, whether forest biostatisticians, mathematical or computer modellers, or information scientists, generally content themselves with dealing with the forest data that they have available to them, the process of storing or managing it, building models based upon it, and making such models and their interpretations available to forest managers. This work is crucial to the scientific process of research in the forest sciences, because the models we produce are a valuable concise descriptive summary of the patterns, information and knowledge that are embedded in the forest data that we work with. Further than this concise capture of information and

[1] CMS, University of Greenwich, UK
Correspondence to: k.rennolls@gre.ac.uk

patterns in calibrated descriptive models, the forest modelling community has developed a variety of modelling approaches that give insight into, and understanding of, forest structure and process, and hence allow the possibility of predicting forest behaviour in contexts that go beyond the range of available forest data. Stand-level models have been developed as appropriately parameterized non-linear models of forest processes such that the fitted parameters relate to observable physical features of the forest, and the forest environment. Spatial individual-tree models of forests allow the processes of tree competition, growth and mortality to be modelled with a framework that can be easily used to simulate forest behaviour under novel management regimes. Process-based models allow the flow of forest growth resources through trees and the stand to be modelled, and simulations from such models allow the impact of environmental pollution on the growth and production of the forest to be evaluated. Clearly, forest models are of paramount importance in the representation and understanding of forests, not only for research purposes, but also for forest management purposes. There are also substantial modelling contributions relating to populations in forests, and populations dependent on forests.

Now, at the start of the 21st century, we live in a vast electronic community of the World Wide Web. Large international efforts are under way to set up large distributed data warehouses of forest data (e.g. the IUFRO GFIS taskforce). There is much talk of data-mining such large data warehouses in order to increase our knowledge and the understanding of the world's forests, both for forest management and for conservation purposes. However, forest modellers have been doing such 'data-mining' of forest data resources for several generations, and the models that have been obtained are the 'discovered information and knowledge'. It is particularly apposite, therefore, that the forest modelling community take stock of what it has achieved, and consider whether the valuable heritage that exists in forest modelling is being satisfactorily conserved, so that it may be shared, both on a worldwide basis currently, and with subsequent generations of forest modellers. Such considerations led Rennolls *et al.* (2001) to suggest the need for a forest model archive (FMA). Some initial considerations for an FMA, and a report on the establishment of a simple prototype FMA are described in the next section.

However, modellers are always keen to move on to the next, and better, model. Hence, while establishment of an FMA is an important consideration for the conservation and sharing of forest modelling heritage, we should also consider the environment of the future in which models may be developed, published and shared. Clearly, the FMA could develop in time and grow with future modelling achievements, and in so doing would provide a valuable means of sharing best modelling practice on a worldwide scale. However, publications about forest modelling studies are scattered across a wide variety of theoretical and applications journals; there is no focused publishing outlet for the forest modelling disciplines. While such publication outlets have played, and will continue to play, a major role in the publication and communication of forest modelling studies, these publication outlets are not generally available in the developing world, because of the high cost of subscriptions to such journals. IUFRO, with its conferences organized throughout the world by various IUFRO groups and working parties, has been a primary forum for communication between forest modellers. However, such IUFRO conferences are generally only attended by scientists from the developed world, because of the high costs of attendance at such meetings, and consequently the proceedings of such conferences usually do not reach potentially needy recipients in the developing world, because no worldwide dissemination framework exists.

It is therefore concluded that a new and more focused publication outlet in forest modelling is required which is freely accessible by forest researchers, and in particular by forest researchers and managers in the developing world. Accordingly, a new peer-reviewed e-journal, *Forest Biometry, Modelling, and Information Sciences* (FBMIS), was launched in September 2002. FBMIS is intended to provide free access to anyone with access to the Internet and it is hoped thereby that the journal will enhance communication worlwide in the forest modelling sciences. Details about FBMIS are presented here, together with a discussion of links between FBMIS and the FMA.

A Forest Model Archive

Initial considerations

Rennolls *et al.* (2001), in a session on 'Information Sharing' at a IUFRO 4.11 conference (Greenwich, UK), suggested the need for a forest model archive (FMA) to conserve the heritage of forest models that had been developed and calibrated over several decades, and to enhance sharing on a worldwide basis. The main points of justification have been outlined in the Introduction of this chapter, and are covered more fully in Rennolls *et. al.* (2001).

Rennolls *et. al.* (2001) considered three alternative approaches to development of such a FMA, while indicating that an FMA of the future might have features of all of them.

The first approach was based on state-of-the-art information systems techniques, and was described as:

> ...a system that is a multi-tiered client/server system with clients running Web browsers linked to Web servers and, as required, to application servers and/or the data and metadata servers. The application software architecture adopts the software engineering principle of separation of concerns and is thus multi-layered. Each layer is designed to perform a coherent function or a set of closely related functions.
>
> (Berners-Lee *et al.*, 1999, 2000, 2001)

Such an architecture is consistent with professional IS standards within which (well-resourced) forest information systems are currently being developed. However, such an approach requires major inputs from a widely skilled team of information scientists and programmers, experienced in distributed database technology (e.g. CORBA, SOAP) and a variety of programming languages (Java, C++, XML). Such development would undoubtedly need substantial funding for development to lead to a usable system

The second approach considered by Rennolls *et al.* (2001) was based on recent developments in the computer communications industry. Partly based upon the success of the MP3 standard as a basis for sharing music across the Internet, it was considered that *Gnutella*, a newly introduced distributed communications system, might provide the basis for a distributed network of model archive nodes, with each node being the home location of a subscribing forest modeller.

However, as in the first approach, development of such a system would require high-level skills in state-of-the-art systems and programming languages. Such skills are not easy to come by, and are usually in high demand and therefore very costly. In fact, it now seems that *Gnutella* has gone out of fashion in the communications industry and that the .net environment is now proffered as the new hope for a worldwide operating system that would allow the aims of this approach to be achieved.

The third approach considered by Rennolls *et al.* (2001) was a simple website using ASP and Microsoft as standard:

The FMA would be a simple web site with an ASP/HTML interface that is accessed via a standard browser, making use of a simple ACCESS database for the storage of various registers and model metadata, with the community contributing to a single archive, which might possibly be mirrored. The archive would be maintained by an 'archivist' (group). The model contributors could merely email the archivist the relevant files which would then be made available on the FMA site.

(Paraphrased slightly from the original.)

Simplicity is, in fact, an important design criterion. The recent history of software developments is littered with expensive and over-ambitious projects that never delivered the project's aims. Simple prototypes usually do deliver, do so cheaply, and naturally lead to realistic extensions in aims, design and implementation. Also, choice of a software environment that has proved itself effective, even if rather outdated in comparison with the most recent and fashionable software offerings available, is a reasonable prototyping approach in the rapidly changing world of software fashion.

The reader will undoubtedly see that it was almost inevitable that the only progress possible in the year since the Greenwich Initiative, in a totally unfunded context, was the third approach listed above. This has indeed been the case, and a description of the resulting prototype FMA is presented. However, the humble prototype FMA presented here does not preclude future developments moving towards the first and second approaches described above, particularly if supporting funding were to be available.

The Greenwich Initiative

Following the presentation of Rennolls *et al.* (2001) at the Greenwich meeting, an impromptu discussion of the 'Trafalgar Group' (consisting of Risto Päivinen, Tim Richards, Keith Rennolls, Keith Reynolds, Moh Ibrahim, Alex Fedoric, Peter Smith, Robert Muetzelfeldt and John Palmer) considered various factors and other approaches to the development of an FMA. The result was the Greenwich Initiative on FMA development. The following were the main factors and considerations thought to be of particular importance in this initiative.

1. The initiative would be named Forest Model Archive (FMA).
2. That consideration would be given to logic frameworks relevant to the FMA. This discussion would focus on use of logic models as higher-order abstractions for specifying logical relationships among potentially diverse models.
3. That consideration be given to the delivery framework of the FMA.
4. That metadata and user-orientation be given particular consideration.
5. That model specifications and implementations be considered.
6. That the Greenwich Initiative should be linked to the IUFRO GFIS project.
7. That the initiative could possibly be extended to include the co-activities of other IUFRO groups, in particular IUFRO 4.01, Mensuration, Growth and Yield.

In order to focus the actions of the initiative, a 1-year time horizon was set, with the following two main actions determined:

A. That a Yahoo chat thread on the various aspects of the initiative be set up.
B. That a workshop be conducted in Summer 2002 to discuss any progress.

At the IUFRO 4.01 meeting in Vancouver in August 2001, an invitation was extended to IUFRO 4.01 to join the Greenwich Initiative. This was accepted and point 7 fulfilled. Action B is fulfilled to some extent by this chapter. Action A was implemented by K. Reynolds, but unfortunately the discussion never really took off.

A simple prototype FMA

A simple FMA prototype has been developed by the author along the lines of the third approach described in the previous section. It is available online at www. forestmodelarchcive.info. It is an open-access website which allows model submission in two forms.

First, there is a Register of Forest Models (RFM) that has been modelled on the Register of Ecological Models (REM) based at the University of Kassel, in an attempt to provide consistency of metadata standards between the Registers. See www.ibilio.org/london/agriculture/links/1/msg00128.html for the original announcement of this REM service or go to eco.wiz.uni-kassel.de/ecobas.html to see the REM.

Secondly, a full submission of forest models is possible (subject to peer review), and this leaves the copyright and IPR of contributed models with the model authors. User registration, which assumes an acceptance of these conditions, is required before access is granted to these full models. The ideal requirements of such a full model submission are stated in the FMA site to be the following:

The names given in parentheses should be either the section headings in a single submitted file, or the file names if each component is included in separate files.

1. A description of the theory of the model (THEORY)
   This may be either a Word file, a pdf file, or a reference to a published article in a freely accessible e-journal, such as FBMIS. Full mathematical details should be given.

2. A listing of the code for the model algorithm (MODEL)
   This should be a txt file, or a Word rtf file with the code in Courier font. You may include an executable form of the code if it may be run on a PC server platform. However, if your model is very large, or has other system requirements, you may possibly wish to reference your own web site from which the model may be demonstrated.

3. The fitting criterion used, if applicable (CRITERION)
   This may not be submitted if the THEORY section contains this information. However, if the model fitting process has been complex, possibly maximum likelihood or a non-standard weighting procedure, then the contributor may wish to put this in a separate section.

4. The fitting algorithm used, if applicable (FITTING)
   That is, the optimization algorithm by which the Criterion is maximized or minimized. If a statistical package is used then this may not be known. However, if the fitting has been achieved by use of known algorithms, such as from Numerical Recipes, or the NAG library, then reference to the routine used would be useful. Details of the starting parameter values used for fitting the DATA to give the RESULTS, and how they were determined, should be given.

5. Test data, with metadata (DATA)
   This may be real data, if the owner of the data has given permission for the data to be made public. Full details of the nature of the data should be given: for example, species, location, time span, environmental factors and conditions, the way the data were collected. Graphical displays of the data would also enhance the accessibility and comprehensibility of the data. Full metadata should be given including the variable names and what they represent, the units used, and any approximations or rounding used. If the data have been selected from a larger data set then the method of selection should be documented.
   If real data cannot be supplied, possibly for confidentiality or ownership reasons, then one of the following options may be followed:
   (i) Anonymization of the data by withholding some of the metadata mentioned above, e.g. location, specific dates, species.

    (ii) Fuzzing of the data, by the addition of random values to the actual data values.

    (iii) Provision of simulated data. The method of simulation of the data should be described. For example it would be nice to know if the data are simulated from the model fitted to to an earlier, raw, data set, since model mis-specification errors would not be expected in such a case.

6. The results of fitting the model to the test data using the fitting algorithm (RESULTS) The fitted parameters of the MODEL on the DATA using FITTING on CRITERION, together with any standard errors, etc., if available. A display of the fitted model superimposed on the DATA would be desirable.

Not all of these will be available for all models submitted to the FMA, particularly 'legacy' models, or 'historical' models. Submission is still welcomed if it is considered they make a contribution to the overall forest modelling heritage.

The current structure of the FMA website is shown in Fig. 32.1.

It may be seen that there are a variety of user facilities, including an ability to search for models, either by author, or by keyword. Some facilities are still under development. The models currently populating the FMA are only provided as examples. The first listed in Fig. 32.1 is concerned with a model of the relationship between the number of species and the sample plot area for a forest site in Sumatra, a simple functional measure of species diversity. An accumulation of such data and models in the FMA might assist a meta-analysis of such diversity relationships on a global scale. The second example listed could include an indexed family of transformed polymorphic Chapman–Richards growth functions in which the rate of movement from one extreme curve to another varies with age (Fig. 32.2), a model of some potential use in height–age studies. Other entries (Fig. 32.1) are an integrated forest process model (IFPM), also a part of this author's forest modelling legacy, and the well-known and much-used FVS model (information extracted from REM, and the author's own site).

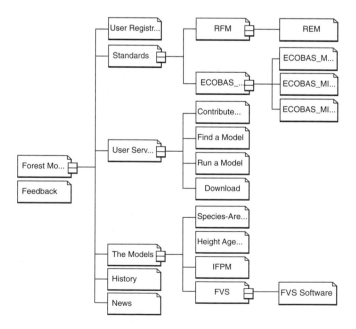

Fig. 32.1. Structure of the prototype FMA.

FMA Futures

First, all forest modellers are invited to examine the FMA site, to make suggestions and comments, and hopefully contribute their models.

Beyond the original aims of the FMA to provide a means of conserving and sharing forest models, we ought to consider some further possibilities that are enabled by the existence of an FMA. The following is an incomplete list.

1. The FMA could become a test-bed of models. Benchmark data sets could be included on which various models might be compared. Feedback from model users potentially could be included in the FMA, indicating areas where particular models have been found to be suitable, and the limits of applicability of the models.
2. The FMA could become a basis for collaboration between modellers working on similar, related or even contrasting models. Theme-related discussion threads on particular issues could easily be set up.
3. The FMA could become a means of presenting situations and scenarios where various types of model development are needed. The need for more extensive efforts to model tropical forests, and their impact on local communities, is an important example.
4. The FMA could be a starting point for international collaborative efforts to integrate forest models of different forest components, possibly at differing scales.

These possibilities may be ambitious, but it seems that all of them would be very worthwhile. Such endeavours are normally only approached by fairly small teams, often those that happen to be successful in competition for funding. Such exclusive and focused team approaches might be best for the development of models with well-focused aims. However, for the rather wide-ranging aims listed above, it might be that a more 'open-club' approach would be more appropriate, and more effective, and that the FMA would provide an underpinning framework. Let's try!

A New e-Journal: FBMIS

The case for a new free-access e-journal in the forest modelling sciences has been made briefly in the Introduction. The FMA as described above is in itself a new form of publication and communication suited to the nature of the modelling product, which makes use of the unique opportunities provided by the World Wide Web. A

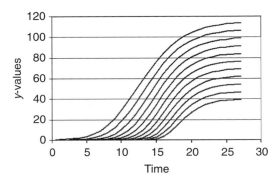

Fig. 32.2. Indexed family of Chapman–Richards growth curves with age-variable rates on convergence.

close link between the new e-journal and the FMA would provide a unique publishing forum in the forest modelling sciences that has not existed previously.

This new e-journal, *Forest Biometry, Modelling and Information Sciences* (FBMIS), was launched in September 2002 on www.fbmis.info. FBMIS is intended to be a freely-accessible, peer-reviewed journal, with hardcopy deposited in archival libraries, and to have Volumes and sequential page numbers as in conventional journals. The journal includes facilities for searching by author and by keyword, and an optional free journal subscription. Many features of FBMIS have been modelled on the successful e-journal *Conservation Ecology*. Construction of the FBMIS website has followed the same principles of simplicity that have been described above in relation to the construction of the FMA website. Modern fashions in web design and flashy dynamic features have been shunned in order to make the site and its published articles easily accessible, particularly in the developing world, where the more modern web browsers might not be available. Articles are available in Adobe pdf format, with links to the authors' own home sites for the display of more advanced features. Authors of articles in the FBMIS are encouraged to submit their models to the FMA where such features may also be made available.

The following is an extract from the FBMIS coverage page.

> Articles are expected to be explicitly concerned with issues relating to forests or trees, processes in forests or trees, populations associated with forest or trees, the forest environment, and the social and economic impact of forests on human communities. This wide range of potential application areas is referred to below, briefly, as 'forests'.
>
> There are several types of article that are welcomed. The main types are the following:
>
> - Articles which report original research and development of new mathematical, statistical or computing models or techniques which arise from forestry problems or contexts.
> - Articles discussing general methodological issues that are related to use of biometric, modelling or information systems in forestry.
> - Articles which present a meta-analysis of the models or results of previous studies.
>
> The main methodological themes of FBMIS, as indicated in the title of the journal are:
>
> Forest Biometry, which includes
>
> (i) Data collection methods:
>
> - Measurement, mensuration and remote sensing: theory relating to the use of instruments or the analysis and interpretation of data resulting from the use of them.
> - Experiments: novel planned investigations of the influence of controlled management treatments and environmental variables and factors.
> - Sampling and inventory: for the collection of tree or forest data, or data relating to processes and populations that occur within forests. New methods of design or analysis, or novel applications of standard techniques.
>
> (ii) Biometrics: A vast disciplinary area, overlapping all other methodological themes of the FBMIS (and some would argue, including them), but summarized here as the use of statistical methods to analyse, summarize and interpret forest data.
>
> Forest Modelling,
>
> ... which includes use of mathematical, statistical, stochastic and computer software models to represent the structure and processes occurring in the forest, and the use of statistical methods for fitting such models to forest data. Models need not necessarily be statistical in nature, and consequently data-fitting of the model is not required, though of course desirable. Forest growth and yield models are a major application area, but models of animal populations in forests, of the spread of

disease or fire through a forest, or any other forest process are welcomed. Models for the investigation of the impact of forests on the economic and social welfare of human communities dependent on forests, particularly in the developing world are sought. Similarly articles with models relating to forest biodiversity, and its conservation and management, are invited.

Forest Information Sciences;
> … which include techniques for the storage, warehousing and archiving of data, metadata and information, and its management for the purposes of analysis, modelling, knowledge extraction and the building of Forest Management Information and Decision Support Systems.

The current Editorial Board of FBMIS includes (in no particular order): Keith Rennolls (Editor-in-chief), University of Greenwich, London, UK; Biing Guam, National Taiwan University, Taiwan 10764, China – Taipei; George Gertner, University of Illinois, USA; Dave Reed, Michigan Technological University, USA; Oscar Garcia, University of Northern British Columbia, Canada; Steen Magnussen, Natural Resources Canada, Canada; Margarida Tomé, Technical University of Lisbon, Portugal; Christoph Kleinn, University of Göttingen, Germany; Jerry Vanclay, Southern Cross University, Lismore, NSW, Australia; Jens Peter Skovsgaard, Danish Forest and Landscape Research Institute, Hoersholm, Denmark; Ronald E. McRoberts, North Central Research Station, USDA Forest Service, St Paul, Minnesota, USA; Harold E. Burkhart, Virginia Polytechnic Institute and State University, Blacksburg, Virginia, USA; H. Gyde Lund, Forest Information Services, Virginia, USA; Chris J. Cieszewski, The University of Georgia, Athens, Georgia, USA; Norma Beatriz Esper, Secretaría de Ambiente y Desarrollo Sustentable, Buenos Aires, Argentina; Daniel Mandallaz, ETH Zentrum, Zurich, Switzerland; Keith Reynolds, PNW Research Station, USDA Forest Service, Corvallis, Oregon, USA; Shouzheng Tang, Chinese Academy of Forestry, Beijing, China; Hans Schreuder, Rocky Mountain Forest and Range Experiment Station, USDA FS, Colorado, USA; Ana Amaro, Instituto Superior de Gestao, Lisbon, Portugal; Joachim Saborowski, University of Göttingen, Germany; Joe Landsberg, Australian National University; Frits Mohren, Wageningen University, The Netherlands; Annikki Mäkelä, University of Helsinki, Finland; Klaus von Gadow, University of Göttingen, Germany; Robert Monserud, USDA FS; Bo Ranneby, SLU, Sweden.

This Editorial Board is expected to expand, and it is hoped that in time more Editorial Board members will be drawn from the continents of South America, Africa and Asia. Nominations would be welcomed.

Forest modellers of the world are invited to submit articles to FBMIS!

Acknowledgement of Discussions

Most of this chapter has consisted of discussion material, or the reports of discussions. Most of the issues raised and discussed have arisen out of the author's long history and association with colleagues in IUFRO 4.11, 4.01 and 4.02. Also, from: (i) experiences and contacts made while working as a forest biometrician for various periods over the last 5 years in Indonesia, on the EU-funded FIMP project; (ii) exposure to various forestry and developmental projects in the developing world, through involvement with the work of the Natural Resources Institute of the University of Greenwich, over the last 5 years; and most recently (iii) contacts and discussions with forest modellers in the IUFRO 4.11 2001 Greenwich meeting, and IUFRO 4.01 2001 Vancouver meeting. I have not assigned credit for various ideas that have been suggested to me because the various discussants have been so large in number and on so many issues that I do not recall the precedence of the idea contributors. Any credits are widely shared, but I do accept that the limitations in what has been presented and achieved so far are my responsibility.

References

Berners-Lee, T., Connolly, D. and Swick, R.R. (1999) *W3C W3C Web Architecture: Describing and Exchanging Data*. W3C Note, 7 June 1999. Available at: www.w3.org/1999/04/WebData

Berners-Lee, T. (2000) *Semantic Web – XML2000*. Available at: www.w3.org/2000/Talks/1206-xml2k-tbl

Berners-Lee, T., Hendler, J. and Lassila, O. (2001) *The Semantic Web*. Available at: www.scientificamerican.com/2001/0501issue/0501berners-lee.html

Rennolls, K., Ibrahim, M. and Smith, P. (2001) A forest model archive? In: Rennolls, K. (ed.) *Forest Biometry, Modelling and Information Science*. IUFRO 4.11, Greenwich June 2001. Available at: cms1.gre.ac.uk/conferences/iufro/proceedings

33

A Logic Approach to Design Specifications for Integrated Application of Diverse Models in Forest Ecosystem Analysis

Keith M. Reynolds[1]

Abstract

The Forest Model Archive (FMA) was proposed as a new IUFRO section 4.11 initiative at the Greenwich, UK section 4.11 conference in June 2001. A discussion group, sponsored by IUFRO section 4.11, initiated the design process for the FMA in Autumn 2001. Key features of the FMA Initiative include a directory of forest model applications, and an associated meta database describing key attributes of forest models which facilitates selection and appropriate use of models for specific applications.

As an extension to the FMA concept, this chapter discusses the use of logic models as metadata specifications for the integrated application of possibly numerous forest models needed to address the broader, more complex and abstract problems posed by modern ecosystem management and ecosystem research issues. A hypothetical example is presented to demonstrate the use of logic models as metadata specifications for multi-model applications (meta models), and the potential for hypermedia linkage from meta-model topics to FMA metadata for specific forest models.

Introduction

The Forest Model Archive (FMA) was proposed as a new IUFRO section 4.11 initiative at the Greenwich, UK section 4.11 conference in June 2001 (see Chapter 32, this volume). A discussion group, sponsored by IUFRO section 4.11, initiated the design process for the FMA in autumn 2001. Key features of the FMA Initiative include a directory of forest model applications, and an associated metadatabase describing key attributes of forest models which facilitates selection and appropriate use of models for specific applications.

As an extension to the FMA concept, this chapter discusses the use of logic models as metadata specifications for the integrated application of possibly numerous forest models needed to address the broader, more complex and abstract problems posed by modern ecosystem management and ecosystem research issues. To provide a concrete example, this chapter illustrates the concept with what could be

[1] USDA Forest Service, Pacific Northwest Research Station, USA
Correspondence to: kreynolds@fs.fed.us

described as a wood balance sheet for a hypothetical geographical region of the world in which the logic considers the production capacity of forest land, value and volume of wood products produced, wood consumption and forest carbon cycling. The logic model represents a modest subset of topics covered in a much more comprehensive specification for evaluation of sustainable forest management (Reynolds *et al.*, 2003).

Logic and Knowledge Representation

All modelling is ultimately concerned with knowledge representation, but most readers are likely to come from a background in simulation modelling, so I briefly discuss the topic of knowledge representation in order to place logic modelling in context.

From the early 1960s through to at least the late 1970s, simulation as a scientific endeavour was primarily concerned with the representation of process as a way of understanding systems. Given such a procedural perspective, the venerable flow diagram has become one of the most familiar means by which science encapsulates pertinent knowledge about systems.

Collins and Loftus (1975) introduced semantic networks as an alternative form of representation that emphasized objects and their relationships. Various modelling methods evolved subsequently that have their origins in semantic networks. Perhaps the most well-known modern manifestation is the universal modelling language (UML; Boggs and Boggs, 1999) for supporting object-oriented analysis and design (Booch, 1994). Other later approaches include Bayesian belief networks that represent interdependencies among conditional probabilities of events (Pearl, 1988), and propositional networks that represent interdependencies among conclusions and premises (Halpern, 1989).

Logic Models as Integrating Frameworks

Logic models, often referred to as knowledge bases (Walters and Nielsen, 1988), provide a formal specification for organizing and interpreting information, and are a form of metadatabase. In our design of a specification for evaluating sustainable forest management (SFM), my colleagues and I have been using the NetWeaver Developer system (Rules of Thumb, North East, PA)[2] which represents a problem in terms of propositions about topics of interest and their logical inter-relationships. In the design of a NetWeaver model, a topic for analysis is translated into a testable proposition. For example, if the topic is forest sustainability, the associated proposition might be as simple as 'the forest ecosystem is sustainable'. The statement of the proposition by itself is inherently ambiguous because sustainability is an abstract concept. However, the full formal logic specification underlying a proposition makes the semantic content of the proposition clear and precise. The biophysical, socioeconomic and framework topics are logical premises of forest sustainability. The proposition about forest ecosystem sustainability evaluates as *true to the degree that* integrity of the biophysical environment, suitable socioeconomic conditions, and a suitable framework exists.

---

[2] The use of firm or trade names in this publication is for reader information and does not imply endorsement by the US Department of Agriculture of any product or service.

The phrase, *true to the degree that*, in the previous sentence is intended to emphasize that strength of support for propositions in NetWeaver models is evaluated by what might be termed 'evidence-based reasoning'. More specifically, this form of reasoning is implemented in NetWeaver with fuzzy maths (Reynolds, 1999), a branch of applied mathematics that implements qualitative reasoning as a method for modelling lexical uncertainty (FuzzyTech, 1999):

> Stochastic uncertainty deals with the uncertainty of whether a [particular] event will take place and probability theory lets you model this. In contrast, lexical uncertainty deals with the uncertainty of the definition of the event itself. Probability theory cannot be used to model this [because] the combination of subjective categories in human decision processes does not follow its axioms. …Even though most concepts used are not precisely defined, humans can use them for quite complex evaluations and decisions that are based on many different factors.

The logical discourse on forest ecosystem sustainability is extended by providing a logic specification for each premise. Each iteration of discourse extends the logic structure another level deeper by defining a logic specification for each topic in the level above. The pattern of discourse generally proceeds from abstract to concrete propositions, with a tendency for premises of a particular proposition to be less abstract than that proposition. Eventually, each logic pathway terminates in a premise, or set of premises, each of which can be evaluated by reference to data. Logic pathways in a knowledge base can thus be construed as a cognitive map of the problem that provides a formal data specification. The specification not only describes what data are to be evaluated, but how the data are to be interpreted to arrive at conclusions.

According to Prabhu *et al.* (2001), 'Possibly one of the biggest challenges facing researchers currently is the identification and quantification of … thresholds' for SFM indicators or more specific measurement endpoints. In the context of endpoint evaluation, thresholds refer to critical values that distinguish between, for example, fully acceptable, fully unacceptable or partially acceptable values. 'Of course, it often happens in science that things are too complicated to be calculated exactly, so that one has to be content with a rough, qualitative understanding' (Dyson, 1979: 48). In other words, perhaps more often than not, definitions of thresholds for endpoints used in evaluation of SFM will require scientifically-based judgements in lieu of more precise calculations. Fuzzy membership functions provide an effective approach to representing such qualitative or semi-quantitative relationships (Zadeh, 1976).

Knowledge-base Design

Although the initial motivation behind design of a knowledge base for landscape evaluation commonly is to provide an interpretation of existing data, knowledge-based applications can equally well be used prospectively; that is, to interpret data pertaining to a future, predicted state of a system. This and the following sections present a small, but concrete, example of using a knowledge base as a logic framework for interpreting data from model outputs. The primary goal of the evaluation is to assess the overall balance (at, say, 50 years in the future) among wood production capacity, actual wood production and carbon budgets on a regional or national scale. The topics of interest in this problem are outlined on the left-hand side of Fig. 33.1. Data relevant to the topic outline are listed on the right-hand side of the figure. A variety of biometric, demographic and econometric models might be used to arrive at predictions for the indicated data elements. In the context of this chapter, the specific models employed to arrive at the predictions are not particularly of interest. Instead, the focus here is, given a set of model predictions, design of a logic specification that provides a synthesis of model outputs.

For compactness, the topic structure of the knowledge base is shown in outline form, omitting details of the logic specification that involve calculations (Fig. 33.1). The outline on the left-hand side of the figure illustrates the basic hierarchical organization of primary knowledge base topics and the logic operator by which each topic synthesizes information about its premises. For example, the overall evaluation of wood balance depends on the three premises, production capacity, production and carbon cycle, and the contributions of the premises are evaluated by the AND operator in the specification of wood balance (shown parenthetically after the wood balance topic). Notice that production capacity, for example, is both a premise of wood balance and a topic with its own proposition and set of premises (timber, plantation and removals).

The AND and UNION operators evaluate their set of underlying premises as limiting factors and incremental contributions, respectively. The AND operator returns a measure of the strength of evidence that is heavily weighted toward the premise with least support. In contrast, premises underlying the UNION operator incrementally contribute evidence in support of their proposition, and can therefore compensate for one another to some degree. CALC indicates a specialized form of logic network that performs a set of standard mathematical operations (details omitted).

The list of items on the right-hand side (Fig. 33.1) represents both data and elementary topics that evaluate a single datum. Items connected to a UNION operator indicate primitive logic networks that evaluate strength of evidence provided by an individual datum. Items connected to a CALC operator indicate data used directly in a calculation. Notice that several items on the right-hand side (Fig. 33.1) represent both primitive networks and data. For example, timber area is evaluated as a primitive network in the context of the timber topic, but is also used in the three sets of calculations related to the carbon cycle.

Results

Application of the knowledge base (Fig. 33.1) to forest modelling predictions is demonstrated under the following hypothetical situation:

1. Forest research scientists in a hypothetical region are interested in evaluating overall wood balance for the region 50 years in the future.

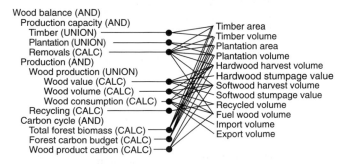

Fig. 33.1. Topic outline of knowledge base for the wood balance sheet.

2. The scientists have run a series of models to predict ecosystem states for the data elements defined for the knowledge base (Fig. 33.1).

3. Suitable subsets of the models have been run to test possible consequences of implementing various policies.

4. The policies are codified as a set of five scenarios, including a base case under which current policies remain in effect (Table 33.1).

Results of evaluating the five scenarios are presented in Fig. 33.2. Although the region and its data are hypothetical, some care was taken to develop a reasonably realistic data set, representative of a developed country with significant forest resource potential and implementing reasonably plausible policy changes.

A detailed accounting of system response to policy changes as represented in evaluation of the scenarios would be appropriate were the analysis based on real data. Because the data in this example are hypothetical, such an exercise is not particularly edifying for the purposes of this chapter. On the other hand, it is instructive to reflect back on the relatively complex web of inter-relationships depicted in the knowledge-base specification (Fig. 33.1). The knowledge-base specification is a

Table 33.1. Wood balance scenarios for a hypothetical region.

Scenario number	Name	Description
1	Base	Base case, default projection after 50 years
2	More recycling	Scenario 1 with improved recycling of wood
3	More timber land	Scenario 2 with additional timber land
4	More plantation growth	Scenario 3 with improved plantation growth rate
5	More plantation land	Scenario 4 with additional land in plantations

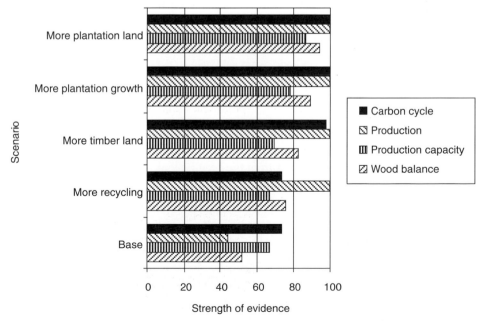

Fig. 33.2. Evaluation of the balance among wood production capacity, actual wood production and carbon cycling for a hypothetical region. The five scenarios are described in Table 33.1.

synthesis of data interpretations that encapsulates our understanding of the web of relationships, some synergistic, some antagonistic. Reasoning 'in one's head' about the implications of the 12 predicted data items with respect to wood balance would be problematic at best, and the knowledge base presented here as an example is relatively small and simple compared with typical applications.

Discussion and Conclusions

As a practical matter, the example of evaluating the balance among wood production capacity, actual wood production and carbon cycling was deliberately kept rather small. Certainly, it would not be difficult to imagine a simulation model, for example, that could have performed a similar type of analysis in this situation. However, logic models also can easily accommodate much larger, more abstract problems. For example, the small example presented in this chapter was extracted from a knowledge-base specification that provides an evaluation of SFM based on the complete set of criteria and indicators outlined in the Montreal Process (WGCIC-SMTBF, 1995).

Forest research organizations have been accumulating an arsenal of sophisticated computational tools such as simulators, linear programs, expert systems, spatial analysis tools and so forth since at least the mid-1960s. Although there are numerous, scientifically rigorous models to address numerous questions about the state of forest ecosystems, it is probably fair to say that most such models only address very narrow questions concerning at most several ecosystem properties. Meanwhile, over the same time period, forest research and forest management have been steadily shifting toward ecosystem-based perspectives on science and management in which the important questions have become much broader, and involve simultaneous analyses of potentially numerous, interacting facets.

Perhaps more often than not, there has been a tendency among modellers in forest research, since the introduction of expert systems for forest applications in the mid-1970s, to regard logic models as yet another type of alternative analytical tool. The origins of this point of view are understandable, considering that early expert systems, based on Boolean logic, were similarly constrained to narrow, well-defined problems. However, the advent of general fuzzy systems such as NetWeaver opens up an entirely new realm of logic-based applications as frameworks for addressing the broad, often abstract, questions posed by ecosystem management problems. Used as analytical frameworks, logic systems do not simply represent yet another analytical approach to problem solving, but have potential to provide real, additional value by organizing and synthesizing results from many other types of models.

Finally, this chapter started from the premise that the concept of a forest model archive might be usefully extended in some fashion by means of logic models. The FMA *per se*, as originally proposed, is a metadatabase for forest modelling applications. Considering the large numbers of models, problems and geographically specific adaptations of models for problems, the potential value of the FMA seems compelling. In much the same way, specifications for the integrated application of diverse forest models through knowledge-base specifications may be equally compelling. Knowledge bases used in this manner can be thought of not only as metadata specifications for interpretation of data, but also as meta-models. Well-designed knowledge bases are also fully self-documenting. In the context of FMA, there is the intriguing possibility of designing meta-models as knowledge bases that can actively hyperlink to FMA model metadata for specific data requirements of the meta-model.

References

Boggs, W. and Boggs, M. (1999) *Mastering UML with Rational Rose*. Sybex, San Francisco, California.

Booch, G. (1994) *Object-oriented Analysis and Design*. Benjamin Cummings, New York.

Collins, A. and Loftus, E.F. (1975) A spreading activation theory of semantic processing. *Psychological Review* 82, 407–428.

Dyson, F. (1979) *Disturbing the Universe*. Basic Books, New York.

FuzzyTech (1999) *FuzzyTech User Guide*. FuzzyTech, Berlin.

Halpern, D.E. (1989) *Thought and Knowledge: an Introduction to Critical Thinking*. Lawrence Erlbaum Associates, Hillsdale, New Jersey.

Pearl, J. (1988) *Probabilistic Reasoning in Intelligent Systems: Networks of Plausible Inference*. Morgan Kaufmann, San Mateo, California.

Prabhu, R., Ruitenbeek, H.J., Boyle, T.J.B. and Colfer, C.J.P. (2001) Between voodoo science and adaptive management: the role and research needs for indicators of sustainable forest management. In: Raison, R.J., Brown, A.G. and Flinn, D.W. (eds) *Criteria and Indicators for Sustainable Forest Management*. IUFRO 7 Research Series, CAB International, Wallingford, UK, pp. 39–66.

Reynolds, K.M. (1999) *NetWeaver for EMDS Version 2.0 User Guide: a Knowledge Base Development System*. General Technical Report PNW-471. US Department of Agriculture, Forest Service, Pacific Northwest Research Station, Portland, Oregon.

Reynolds, K.M., Johnson, K.N. and Gordon, S.N. (2003) The science/policy interface in logic-based evaluation of forest ecosystem sustainability. *Forest Policy and Economics* (in review).

Walters, J.R. and Nielsen, N.R. (1988) *Crafting Knowledge-based Systems: Expert Systems Made Realistic*. John Wiley & Sons, New York.

WGCICSMTBF (Working Group on Criteria and Indicators for the Conservation and Sustainable Management of Temperate and Boreal Forests) (1995) *Criteria and Indicators for the Conservation and Sustainable Management of Temperate and Boreal Forests: the Montreal Process*. Canadian Forest Service, Hull, Quebec.

Zadeh, L.A. (1976) The concept of a linguistic variable and its application to approximate reasoning. III. *Information Science* 9, 43–80.

Part 6

Conclusions

———————————

34 Emerging Trends and Future Directions: a Workshop Synthesis

David D. Reed,[1] Ana Amaro,[2] Ralph Amateis,[3] Shongming Huang[4] and Margarida Tomé[5]

Abstract

Workshop participants divided into four working groups to examine emerging trends and future research directions of forest ecosystem modelling from different perspectives. Each group reported their individual discussions in detail in a common session. Here, a synthesis of common items resulting from the four discussions is presented. While not being all-inclusive, seven general issues were identified in three general research areas: (i) the design or development of forest models or development of modelling methodology; (ii) the use of forest models; and (iii) the evaluation and documentation of forest models. Taken together, this synthesis provides a workshop consensus on the future of forest growth modelling.

Introduction

Following presentation of the technical papers, workshop participants divided into four working groups to examine the emerging trends and future research directions of forest ecosystem modelling from different perspectives. The four working groups were organized around the following topics:

- forest reality and modelling strategies;
- mathematical properties and reasoning;
- estimation processes; and
- validation and uncertainty.

Here, the goal is to synthesize the four discussions and identify common items that emerged from the respective discussions.

[1] School of Forest Resources and Environmental Science, Michigan Technological University, USA Correspondence to:ddreed@mtu.edu
[2] Department of Mathematics, Instituto Superior de Gestão, Portugal
[3] Department of Forestry, Virginia Tech, USA
[4] Forest Management Branch, Land and Forest Division, Alberta Sustainable Resource Development, Canada
[5] Department of Forestry, Instituto Superior de Agronomia, Portugal

© CAB International 2003. *Modelling Forest Systems* (eds A. Amaro, D. Reed and P. Soares)

Emerging Trends and Future Research Directions

Seven general issues emerged from the discussions. Some of these were mentioned by all groups, some by fewer, but all surfaced in at least two of the discussions. These seven issues loosely fall in the following three areas:

- the design or development of forest models or development of modelling methodology;
- the use of forest models; or
- the evaluation and documentation of forest models.

The individual issues are discussed below. The discussions were wide-ranging and these summaries cannot possibly capture all of the points that arose. They are intended to capture the general essence of the discussions and serve as a summary of the consensus of the workshop participants regarding emerging trends and future research directions in the field of forest modelling.

Forest model development and modelling methodology

Methods used in developing models of forest growth and development are constantly evolving. Workshop participants identified at least two general areas that require further development to address current issues and needs, and suggested activities that may increase the efficiency of developing methodological advances in forest modelling.

Scale

Forest modelling really began with methods to estimate individual tree volume and stand level yield beginning in the 18th century. Today's information needs require the ability to develop models that can function across temporal, spatial and hierarchical (organizational) scales. Temporally, questions of gaseous pollutant impact may require modelling of physiological processes at the timescale of seconds to minutes, while questions of sustainability and succession require timescales of decades to centuries. Spatially, many contemporary questions must be addressed at landscape levels (10^3–10^4 ha) as opposed to focusing on individual trees or single stands (10^0–10^1 ha). This requires understanding and modelling of physiological processes at the cell or organ level to understand pollutant impacts on forest health and productivity, but assessment of long-term sustainability requires extrapolating those results to the organismal, community and population levels. Currently available modelling methodology can only begin to address issues crossing several orders of magnitude temporally or spatially, and integration across hierarchical levels is still in its early stages of development, despite efforts over the last couple of decades.

Complex systems

Forest ecosystems are complex, non-linear systems with many feedback mechanisms and interactions with external factors such as weather, as well as human intervention. All models are simplifications that may or may not consider various aspects of system structure and behaviour, depending on the purpose of the modelling exercise. Factors such as climate have often been assumed to be constant in forest modelling exercises, but advances in understanding of climate emphasize its variability and non-static nature. Questions requiring the use of models for predictive purposes may require that models have the ability to incorporate future climate regimes for which there are no data on ecosystem response. Utilization of modelling methodology with the flexibility to represent system behaviour under such changing conditions is becoming increasingly important. It is interesting to note the analogy to automobiles; those manufactured in 2002 are much more complex than those manufactured in 1913, but this added complexity increases functionality, reliability and comfort for the occupants. Technologies that do not improve performance have been added, but they usually end up disappearing after some time. It is important that models are not made more complex simply for the sake of adding complexity, but that model complexity be added to improve functionality or reliability.

Interaction with other disciplines

One needs to look no further than R.A. Fisher to know that applied problems in agriculture and natural resources have spurred the development of modelling and analytical methods applicable to an extremely wide range of disciplines. Similarly, forest modelling practitioners have always borrowed liberally from other fields such as econometrics to develop and utilize new modelling technologies. The extreme efficiency of adapting modelling technologies to new applications, as opposed to developing similar technologies or methodologies from scratch, should drive all forest modelling practitioners to actively pursue interactions with theoretical and applied scientists from other disciplines as an efficient route to identifying new approaches and methodologies for particular problems. While this may seem obvious, it is easy to become insulated and focus interactions within a narrow group interested in the same problem. It is critical that effort be continually expended to maintain and expand contacts with investigators in other disciplines, whether formally through membership in broad professional organizations, or informally through day-to-day interactions within an investigator's institution.

Integration with other resources

To date, models of forest growth and production, hydrological models, wildlife population models and models of other associated resources have been developed independently of each other, even when the intent was to apply them to the same geographical area. Resource management is requiring knowledge of multiple resources and their interactions, yet models combining resources are only available in very rudimentary forms. To provide information required by managers, and ultimately by society, forest modellers must increase their interactions with specialists from other resource areas. This implies not only an increase in the level of communication, but the development of truly integrated models that will of necessity require the cooperation of specialists from many disciplines.

Use of forest models

Many of the barriers to adoption of forest models are due to models not satisfying the perceived information needs of the user. This may occur for several reasons. Models may not provide information in a form that is of utility to users, or they may not provide critical information about a process that managers need to manage the system. There are also cases where model results are difficult to interpret or are not suitable to the cultural background of the ultimate user of the information.

User involvement

Model utility can be maximized if users are involved in the design of model output, regarding both the content of the information presented and the form of presentation. User information needs must be considered in model development. There are many cases, of course, where models are applied for uses the developer did not intend or consider. In this case, documentation of the limitations, as well as the capabilities, of a model is critical. Model developers may also have opportunities to interact with people interested in using the model in situations that were not anticipated during model development; it is important to take advantage of such opportunities.

The value of allowing model users to interact with models during execution, providing feedback during interactive use of the model and direction for further execution, is being increasingly realized. In many ways, this leads to a modeller thinking of a user as analogous to a black-box component or subroutine – information is sent and processed, and new information is returned for use in further model execution by other model components. This is very different from considering a user to be a passive recipient of model results, or little more than a provider of directions for model execution. This viewpoint leads to dramatically different approaches to the entering of information or display of model results that we are only just beginning to explore. It also expands the possibilities of model utility far beyond currently perceived boundaries. The development of new methods and approaches to integrate user interaction into model execution will lead to great changes in both the development and use of forest models.

Communication of results

The presentation of model results has evolved from simple tables, to charts, to complex maps, and more recently to real-time, web-based dissemination. As presented by several authors in this volume, it is now possible to provide realistic visualization of stands and landscapes. It is critical to remember the intended audience and their capacity and skills in interpreting such information. Scientists in a research institute have very different capacities for interpreting different types of displays, and associated expectations of information content, than do rural farmers. It is important to remember that such scientists are equally unable to interpret information as typically perceived and processed by the farmers. Cultural differences in such things as the perceptions of shapes and colours, or methods of describing location, can make maps perceived as excellent by some almost indecipherable by others. As model output inevitably moves to becoming more graphical, model developers are going to be forced to develop and utilize dramatically different methods of communicating results to model users.

Forest model evaluation and documentation

The question of when to trust model results is not new. Model users usually have to either accept the claims of model developers or undergo extensive (and expensive) model evaluation efforts themselves, which is often neither feasible nor possible.

While not eliminating the need for users to implement their own evaluation processes, the establishment of commonly accepted guidelines and standards of model evaluation, and establishment of widely accepted descriptions of model performance, would greatly simplify the process of model evaluation.

Benchmark data, standards and guidelines

Model evaluation and documentation has almost always been in the hands of the developers. Validation data are often unavailable, and in the rare cases when they are available, they are often only a subset of the development data. There are no commonly accepted guidelines for model evaluation, and no commonly accepted performance standards for particular model uses. It is obvious that the purposes of developing a model drive the choice of appropriate benchmarks and performance standards. Unfortunately, this variety has contributed to a form of paralysis, where it is impossible to assemble consistent comparisons, even for a particular, well-defined use. The development and acceptance of benchmark data, standards of model performance, and guidelines of model evaluation for a variety of purposes are needed to provide potential users with unbiased assessment of model performance. Workshop participants identified several important actions in this regard. One is the establishment of model documentation standards, such as those incorporated in the proposed Forest Model Archive. The genomics disciplines have established the submission of raw data to public archives as a broadly accepted practice that is even considered mandatory by publishers under certain circumstances. A similar understanding could possibly be reached with publishers of forest modelling research and documentation. A third area concerns the establishment of benchmark data sets. Obviously, it would not be reasonable to establish benchmark data sets of every possible species and management situation, and it would not be reasonable to evaluate a *Eucalyptus* model using benchmark *Picea* data. It might be possible, though, to develop comprehensive, empirical descriptions of forest system behaviour that could be compared with model results in an evaluation process. All these efforts will require considerable coordination among members of the forest modelling community, but establishment of infrastructure such as the Forest Model Archive is enabling these developments in ways that have not been possible in the past.

Summary

Workshop participants identified emerging trends and future directions of forest growth modelling. There were seven specific trends that were identified in three general research areas.

- The design or development of forest models or development of modelling methodology:
  - *issues of scale*
  - *complex systems*
  - *interaction with other disciplines*
  - *integration with other resources*
- The use of forest models:
  - *user involvement*
  - *communication of results*
- The evaluation and documentation of forest models:
  - *benchmark data, standards and guidelines*

Aspects of each of these issues are presented and discussed in the context of their impact on the development and utilization of forest models. These discussions synthesize the perspectives and thoughts of workshop participants and, as such, can serve as guidance for those interested in the development of this field in the near future.

Index

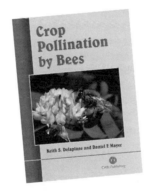
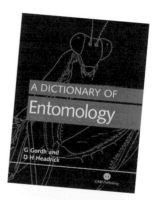
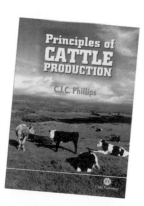